SELECTED PHILOSOPHICAL AND

SCIENTIFIC WRITINGS

THE
OTHER VOICE
IN
EARLY MODERN
EUROPE

A Series Edited by Margaret L. King and Albert Rabil Jr.

RECENT BOOKS IN THE SERIES

Emilie Du Châtelet

SELECTED PHILOSOPHICAL AND SCIENTIFIC WRITINGS

ᴈ

Edited and with an Introduction by
Judith P. Zinsser
and Translated by
Isabelle Bour and Judith P. Zinsser

THE UNIVERSITY OF CHICAGO PRESS
Chicago & London

Emilie Du Châtelet, 1706–1749

Judith P Zinsser is professor emerita of history at Miami University (Oxford, OH).

Isabelle Bour is professor of eighteenth-century British studies at the University of the Sorbonne Nouvelle.

The University of Chicago Press, Chicago 60637
The University of Chicago Press, Ltd., London
Introductions, bibliography, and notes © 2009 by Judith P. Zinsser
English translation © 2009 by Isabelle Bour and Judith P. Zinsser
© 2009 by The University of Chicago
All rights reserved. Published 2009
Printed in the United States of America

18 17 16 15 14 13 12 11 10 09   1 2 3 4 5

ISBN-13: 978-0-226-16806-7 (cloth)
ISBN-13: 978-0-226-16807-4 (paper)
ISBN-10: 0-226-16806-9 (cloth)
ISBN-10: 0-226-16807-7 (paper)

The University of Chicago Press gratefully acknowledges the generous support of James E. Rabil, in memory of Scottie W. Rabil, toward the publication of this book.

Library of Congress Cataloging-in-Publication Data
Du Châtelet, Gabrielle Emilie Le Tonnelier de Breteuil, marquise, 1706–1749.
Selected philosophical and scientific writings / Emilie du Châtelet ; edited and with an introduction by Judith P. Zinsser and translated by Isabelle Bour and Judith P. Zinsser.
    p.   cm. — (The other voice in early modern Europe)
Includes bibliographical references and index.
ISBN-13: 978-0-226-16806-7 (cloth : alk. paper)
ISBN-13: 978-0-226-16807-4 (pbk. : alk. paper)
ISBN-10: 0-226-16806-9 (cloth : alk. paper)
ISBN-10: 0-226-16807-7 (pbk. : alk. paper) 1. Mandeville, Bernard, 1670–1733.
Fable of the bees. 2. Fire—Early works to 1800. 3. Physics—Early works
to 1800. 4. Newton, Isaac, Sir, 1642–1727. Principia. 5. Bible—Criticism,
interpretation, etc. I. Zinsser, Judith P. II. Title. III. Series: Other voice in early
modern Europe.
PQ1981.D55A2 2009
848'.509—dc22
2008055106

CONTENTS

ACKNOWLEDGMENTS

A scholarly work is rarely possible without institutional support. Miami University provided Judith P. Zinsser with a research leave in the fall of 2007. The project had its genesis at an NEH Summer Institute on the Enlightenment at Stanford University and subsequently evolved while she held a Camargo Foundation Fellowship. Albert J. Rabil, Jr., the coeditor of the Other Voice in Early Modern Europe series, won an NEH Collaborative Grant that enabled the translators to meet in Paris and in Oxford (Ohio) and helped to defray the costs of publication. In addition, Professor Rabil, a true gentleman-scholar, has been the most thoughtful and careful of editors, respectful of the texts and of the translators' sensitivities.

A translation project requires many favors of colleagues and friends. Scholars generously shared their work and their expertise. Bertram Eugene Schwarzbach, Ulla Kölving, and Anne Soprani allowed us to use writings of Du Châtelet's that they have edited for publication but that have not yet appeared. Many helped with details that defeated the Internet and the best reference librarians: Edwin Yamauchi, Alan Dainard and Penny Arthur of the Graffigny Project, and Paul Veitch Moriarty. Others read chapters and saved the text from egregious errors: Adam J. Apt, Antoinette Emch-Dériaz and Gérard Emch, and Anna Klosowska. Ronald K. Smelzer, in addition to reading chapters and identifying some of the most mysterious of Du Châtelet's references, provided the illustrations and continual encouragement. Roger J. Millar helped in untold ways. The cover portrait is used by the kind permission of Henri-François de Breteuil and his wife, Séverine Decazes de Glückbierg, who have, in so many ways, gained Du Châtelet the historical recognition she deserves.

Randy Petilos at the University of Chicago Press shepherded the manuscript through the publishing process and calmly solved perceived crises. Dawn Hall made copyediting a pleasure, and Maia Rigas dealt with last-minute changes with equanimity and imagination.

THE OTHER VOICE IN
EARLY MODERN EUROPE:
INTRODUCTION TO THE SERIES

Margaret L. King and Albert Rabil Jr.

THE OLD VOICE AND THE OTHER VOICE

In western Europe and the United States, women are nearing equality in the professions, in business, and in politics. Most enjoy access to education, reproductive rights, and autonomy in financial affairs. Issues vital to women are on the public agenda: equal pay, child care, domestic abuse, breast cancer research, and curricular revision with an eye to the inclusion of women.

These recent achievements have their origins in things women (and some male supporters) said for the first time about six hundred years ago. Theirs is the "other voice," in contradistinction to the "first voice," the voice of the educated men who created Western culture. Coincident with a general reshaping of European culture in the period 1300–1700 (called the Renaissance or early modern period), questions of female equality and opportunity were raised that still resound and are still unresolved.

The other voice emerged against the backdrop of a three-thousand-year history of the derogation of women rooted in the civilizations related to Western culture: Hebrew, Greek, Roman, and Christian. Negative attitudes toward women inherited from these traditions pervaded the intellectual, medical, legal, religious, and social systems that developed during the European Middle Ages.

The following pages describe the traditional, overwhelmingly male views of women's nature inherited by early modern Europeans and the new tradition that the "other voice" called into being to begin to challenge reigning assumptions. This review should serve as a framework for understanding the texts published in the series The Other Voice in Early Modern Europe. Introductions specific to each text and author follow this essay in all the volumes of the series.

TRADITIONAL VIEWS OF WOMEN, 500 B.C.E.–1500 C.E.

Embedded in the philosophical and medical theories of the ancient Greeks were perceptions of the female as inferior to the male in both mind and body. Similarly, the structure of civil legislation inherited from the ancient Romans was biased against women, and the views on women developed by Christian thinkers out of the Hebrew Bible and the Christian New Testament were negative and disabling. Literary works composed in the vernacular of ordinary people, and widely recited or read, conveyed these negative assumptions. The social networks within which most women lived—those of the family and the institutions of the Roman Catholic Church—were shaped by this negative tradition and sharply limited the areas in which women might act in and upon the world.

GREEK PHILOSOPHY AND FEMALE NATURE. Greek biology assumed that women were inferior to men and defined them as merely childbearers and housekeepers. This view was authoritatively expressed in the works of the philosopher Aristotle.

Aristotle thought in dualities. He considered action superior to inaction, form (the inner design or structure of any object) superior to matter, completion to incompletion, possession to deprivation. In each of these dualities, he associated the male principle with the superior quality and the female with the inferior. "The male principle in nature," he argued, "is associated with active, formative and perfected characteristics, while the female is passive, material and deprived, desiring the male in order to become complete."[1] Men are always identified with virile qualities, such as judgment, courage, and stamina, and women with their opposites—irrationality, cowardice, and weakness.

The masculine principle was considered superior even in the womb. The man's semen, Aristotle believed, created the form of a new human creature, while the female body contributed only matter. (The existence of the ovum, and with it the other facts of human embryology, was not established until the seventeenth century.) Although the later Greek physician Galen believed there was a female component in generation, contributed by "female semen," the followers of both Aristotle and Galen saw the male role in human generation as more active and more important.

---

1. Aristotle, *Physics* 1.9.192a20–24, in *The Complete Works of Aristotle*, ed. Jonathan Barnes, rev. Oxford trans., 2 vols. (Princeton, 1984), 1:328.

In the Aristotelian view, the male principle sought always to reproduce itself. The creation of a female was always a mistake, therefore, resulting from an imperfect act of generation. Every female born was considered a "defective" or "mutilated" male (as Aristotle's terminology has variously been translated), a "monstrosity" of nature.[2]

For Greek theorists, the biology of males and females was the key to their psychology. The female was softer and more docile, more apt to be despondent, querulous, and deceitful. Being incomplete, moreover, she craved sexual fulfillment in intercourse with a male. The male was intellectual, active, and in control of his passions.

These psychological polarities derived from the theory that the universe consisted of four elements (earth, fire, air, and water), expressed in human bodies as four "humors" (black bile, yellow bile, blood, and phlegm) considered, respectively, dry, hot, damp, and cold and corresponding to mental states ("melancholic," "choleric," "sanguine," "phlegmatic"). In this scheme the male, sharing the principles of earth and fire, was dry and hot; the female, sharing the principles of air and water, was cold and damp.

Female psychology was further affected by her dominant organ, the uterus (womb), *hystera* in Greek. The passions generated by the womb made women lustful, deceitful, talkative, irrational, indeed—when these affects were in excess—"hysterical."

Aristotle's biology also had social and political consequences. If the male principle was superior and the female inferior, then in the household, as in the state, men should rule and women must be subordinate. That hierarchy did not rule out the companionship of husband and wife, whose cooperation was necessary for the welfare of children and the preservation of property. Such mutuality supported male preeminence.

Aristotle's teacher Plato suggested a different possibility: that men and women might possess the same virtues. The setting for this proposal is the imaginary and ideal Republic that Plato sketches in a dialogue of that name. Here, for a privileged elite capable of leading wisely, all distinctions of class and wealth dissolve, as, consequently, do those of gender. Without households or property, as Plato constructs his ideal society, there is no need for the subordination of women. Women may therefore be educated to the same level as men to assume leadership. Plato's Republic remained imaginary, however. In real societies, the subordination of women remained the norm and the prescription.

2. Aristotle, *Generation of Animals* 2.3.737a27–28, in *The Complete Works*, 1: 1144.

The views of women inherited from the Greek philosophical tradition became the basis for medieval thought. In the thirteenth century, the supreme Scholastic philosopher Thomas Aquinas, among others, still echoed Aristotle's views of human reproduction, of male and female personalities, and of the preeminent male role in the social hierarchy.

ROMAN LAW AND THE FEMALE CONDITION. Roman law, like Greek philosophy, underlay medieval thought and shaped medieval society. The ancient belief that adult property-owning men should administer households and make decisions affecting the community at large is the very fulcrum of Roman law.

About 450 B.C.E., during Rome's republican era, the community's customary law was recorded (legendarily) on twelve tablets erected in the city's central forum. It was later elaborated by professional jurists whose activity increased in the imperial era, when much new legislation was passed, especially on issues affecting family and inheritance. This growing, changing body of laws was eventually codified in the *Corpus of Civil Law* under the direction of the emperor Justinian, generations after the empire ceased to be ruled from Rome. That *Corpus*, read and commented on by medieval scholars from the eleventh century on, inspired the legal systems of most of the cities and kingdoms of Europe.

Laws regarding dowries, divorce, and inheritance pertain primarily to women. Since those laws aimed to maintain and preserve property, the women concerned were those from the property-owning minority. Their subordination to male family members points to the even greater subordination of lower-class and slave women, about whom the laws speak little.

In the early republic, the *paterfamilias*, or "father of the family," possessed *patria potestas*, "paternal power." The term *pater*, "father," in both these cases does not necessarily mean biological father but denotes the head of a household. The father was the person who owned the household's property and, indeed, its human members. The *paterfamilias* had absolute power—including the power, rarely exercised, of life or death—over his wife, his children, and his slaves, as much as his cattle.

Male children could be "emancipated," an act that granted legal autonomy and the right to own property. Those over fourteen could be emancipated by a special grant from the father or automatically by their father's death. But females could never be emancipated; instead, they passed from the authority of their father to that of a husband or, if widowed or orphaned while still unmarried, to a guardian or tutor.

Marriage in its traditional form placed the woman under her husband's authority, or *manus.* He could divorce her on grounds of adultery, drinking wine, or stealing from the household, but she could not divorce him. She could neither possess property in her own right nor bequeath any to her children upon her death. When her husband died, the household property passed not to her but to his male heirs. And when her father died, she had no claim to any family inheritance, which was directed to her brothers or more remote male relatives. The effect of these laws was to exclude women from civil society, itself based on property ownership.

In the later republican and imperial periods, these rules were significantly modified. Women rarely married according to the traditional form. The practice of "free" marriage allowed a woman to remain under her father's authority, to possess property given her by her father (most frequently the "dowry," recoverable from the husband's household on his death), and to inherit from her father. She could also bequeath property to her own children and divorce her husband, just as he could divorce her.

Despite this greater freedom, women still suffered enormous disability under Roman law. Heirs could belong only to the father's side, never the mother's. Moreover, although she could bequeath her property to her children, she could not establish a line of succession in doing so. A woman was "the beginning and end of her own family," said the jurist Ulpian. Moreover, women could play no public role. They could not hold public office, represent anyone in a legal case, or even witness a will. Women had only a private existence and no public personality.

The dowry system, the guardian, women's limited ability to transmit wealth, and total political disability are all features of Roman law adopted by the medieval communities of western Europe, although modified according to local customary laws.

CHRISTIAN DOCTRINE AND WOMEN'S PLACE. The Hebrew Bible and the Christian New Testament authorized later writers to limit women to the realm of the family and to burden them with the guilt of original sin. The passages most fruitful for this purpose were the creation narratives in Genesis and sentences from the Epistles defining women's role within the Christian family and community.

Each of the first two chapters of Genesis contains a creation narrative. In the first "God created man in his own image, in the image of God he created him; male and female he created them" (Gn 1:27). In the second, God created Eve from Adam's rib (2:21–23). Christian theologians relied principally on Genesis 2 for their understanding of the relation between man and woman, interpreting the creation of Eve from Adam as proof of her subordination to him.

The creation story in Genesis 2 leads to that of the temptations in Genesis 3: of Eve by the wily serpent and of Adam by Eve. As read by Christian theologians from Tertullian to Thomas Aquinas, the narrative made Eve responsible for the Fall and its consequences. She instigated the act; she deceived her husband; she suffered the greater punishment. Her disobedience made it necessary for Jesus to be incarnated and to die on the cross. From the pulpit, moralists and preachers for centuries conveyed to women the guilt that they bore for original sin.

The Epistles offered advice to early Christians on building communities of the faithful. Among the matters to be regulated was the place of women. Paul offered views favorable to women in Galatians 3:28: "There is neither Jew nor Greek; there is neither slave nor free, there is neither male nor female; for you are all one in Christ Jesus." Paul also referred to women as his coworkers and placed them on a par with himself and his male coworkers (Phlm 4:2–3; Rom 16:1–3; 1 Cor 16:19). Elsewhere, Paul limited women's possibilities: "But I want you to understand that the head of every man is Christ, the head of a woman is her husband, and the head of Christ is God" (1 Cor 11:3).

Biblical passages by later writers (although attributed to Paul) enjoined women to forgo jewels, expensive clothes, and elaborate coiffures; and they forbade women to "teach or have authority over men," telling them to "learn in silence with all submissiveness" as is proper for one responsible for sin, consoling them, however, with the thought that they will be saved through childbearing (1 Tm 2:9–15). Other texts among the later Epistles defined women as the weaker sex and emphasized their subordination to their husbands (1 Pt 3:7; Col 3:18; Eph 5:22–23).

These passages from the New Testament became the arsenal employed by theologians of the early church to transmit negative attitudes toward women to medieval Christian culture—above all, Tertullian (*On the Apparel of Women*), Jerome (*Against Jovinian*), and Augustine (*The Literal Meaning of Genesis*).

THE IMAGE OF WOMEN IN MEDIEVAL LITERATURE. The philosophical, legal, and religious traditions born in antiquity formed the basis of the medieval intellectual synthesis wrought by trained thinkers, mostly clerics, writing in Latin and based largely in universities. The vernacular literary tradition that developed alongside the learned tradition also spoke about female nature and women's roles. Medieval stories, poems, and epics also portrayed women negatively—as lustful and deceitful—while praising good house-keepers and loyal wives as replicas of the Virgin Mary or the female saints and martyrs.

There is an exception in the movement of "courtly love" that evolved in southern France from the twelfth century. Courtly love was the erotic love between a nobleman and noblewoman, the latter usually superior in social rank. It was always adulterous. From the conventions of courtly love derive modern Western notions of romantic love. The tradition has had an impact disproportionate to its size, for it affected only a tiny elite, and very few women. The exaltation of the female lover probably does not reflect a higher evaluation of women or a step toward their sexual liberation. More likely it gives expression to the social and sexual tensions besetting the knightly class at a specific historical juncture.

The literary fashion of courtly love was on the wane by the thirteenth century, when the widely read *Romance of the Rose* was composed in French by two authors of significantly different dispositions. Guillaume de Lorris composed the initial four thousand verses about 1235, and Jean de Meun added about seventeen thousand verses—more than four times the original—about 1265.

The fragment composed by Guillaume de Lorris stands squarely in the tradition of courtly love. Here the poet, in a dream, is admitted into a walled garden where he finds a magic fountain in which a rosebush is reflected. He longs to pick one rose, but the thorns prevent his doing so, even as he is wounded by arrows from the god of love, whose commands he agrees to obey. The rest of this part of the poem recounts the poet's unsuccessful efforts to pluck the rose.

The longer part of the *Romance* by Jean de Meun also describes a dream. But here allegorical characters give long didactic speeches, providing a social satire on a variety of themes, some pertaining to women. Love is an anxious and tormented state, the poem explains: women are greedy and manipulative, marriage is miserable, beautiful women are lustful, ugly ones cease to please, and a chaste woman is as rare as a black swan.

Shortly after Jean de Meun completed *The Romance of the Rose*, Mathéolus penned his *Lamentations*, a long Latin diatribe against marriage translated into French about a century later. The *Lamentations* sum up medieval attitudes toward women and provoked the important response by Christine de Pizan in her *Book of the City of Ladies*.

In 1355, Giovanni Boccaccio wrote *Il Corbaccio*, another antifeminist manifesto, although ironically by an author whose other works pioneered new directions in Renaissance thought. The former husband of his lover appears to Boccaccio, condemning his unmoderated lust and detailing the defects of women. Boccaccio concedes at the end "how much men naturally surpass women in nobility" and is cured of his desires.[3]

WOMEN'S ROLES: THE FAMILY. The negative perceptions of women expressed in the intellectual tradition are also implicit in the actual roles that women played in European society. Assigned to subordinate positions in the household and the church, they were barred from significant participation in public life.

Medieval European households, like those in antiquity and in non-Western civilizations, were headed by males. It was the male serf (or peasant), feudal lord, town merchant, or citizen who was polled or taxed or succeeded to an inheritance or had any acknowledged public role, although his wife or widow could stand as a temporary surrogate. From about 1100, the position of property-holding males was further enhanced: inheritance was confined to the male, or agnate, line—with depressing consequences for women.

A wife never fully belonged to her husband's family, nor was she a daughter to her father's family. She left her father's house young to marry whomever her parents chose. Her dowry was managed by her husband, and at her death it normally passed to her children by him.

A married woman's life was occupied nearly constantly with cycles of pregnancy, childbearing, and lactation. Women bore children through all the years of their fertility, and many died in childbirth. They were also responsible for raising young children up to six or seven. In the propertied classes that responsibility was shared, since it was common for a wet nurse to take over breast-feeding and for servants to perform other chores.

Women trained their daughters in the household duties appropriate to their status, nearly always tasks associated with textiles: spinning, weaving, sewing, embroidering. Their sons were sent out of the house as apprentices or students, or their training was assumed by fathers in later childhood and adolescence. On the death of her husband, a woman's children became the responsibility of his family. She generally did not take "his" children with

---

3. Giovanni Boccaccio, *The Corbaccio, or The Labyrinth of Love*, trans. and ed. Anthony K. Cassell, rev. ed. (Binghamton, N.Y., 1993), 71.

her to a new marriage or back to her father's house, except sometimes in the artisan classes.

Women also worked. Rural peasants performed farm chores, merchant wives often practiced their husbands' trades, the unmarried daughters of the urban poor worked as servants or prostitutes. All wives produced or embellished textiles and did the housekeeping, while wealthy ones managed servants. These labors were unpaid or poorly paid but often contributed substantially to family wealth.

WOMEN'S ROLES: THE CHURCH. Membership in a household, whether a father's or a husband's, meant for women a lifelong subordination to others. In western Europe, the Roman Catholic Church offered an alternative to the career of wife and mother. A woman could enter a convent, parallel in function to the monasteries for men that evolved in the early Christian centuries.

In the convent, a woman pledged herself to a celibate life, lived according to strict community rules, and worshiped daily. Often the convent offered training in Latin, allowing some women to become considerable scholars and authors as well as scribes, artists, and musicians. For women who chose the conventual life, the benefits could be enormous, but for numerous others placed in convents by paternal choice, the life could be restrictive and burdensome.

The conventual life declined as an alternative for women as the modern age approached. Reformed monastic institutions resisted responsibility for related female orders. The church increasingly restricted female institutional life by insisting on closer male supervision.

Women often sought other options. Some joined the communities of laywomen that sprang up spontaneously in the thirteenth century in the urban zones of western Europe, especially in Flanders and Italy. Some joined the heretical movements that flourished in late medieval Christendom, whose anticlerical and often antifamily positions particularly appealed to women. In these communities, some women were acclaimed as "holy women" or "saints," whereas others often were condemned as frauds or heretics.

In all, although the options offered to women by the church were sometimes less than satisfactory, they were sometimes richly rewarding. After 1520, the convent remained an option only in Roman Catholic territories. Protestantism engendered an ideal of marriage as a heroic endeavor and appeared to place husband and wife on a more equal footing. Sermons and treatises, however, still called for female subordination and obedience.

THE OTHER VOICE, 1300–1700

When the modern era opened, European culture was so firmly structured by a framework of negative attitudes toward women that to dismantle it was a monumental labor. The process began as part of a larger cultural movement that entailed the critical reexamination of ideas inherited from the ancient and medieval past. The humanists launched that critical reexamination.

THE HUMANIST FOUNDATION. Originating in Italy in the fourteenth century, humanism quickly became the dominant intellectual movement in Europe. Spreading in the sixteenth century from Italy to the rest of Europe, it fueled the literary, scientific, and philosophical movements of the era and laid the basis for the eighteenth-century Enlightenment.

Humanists regarded the Scholastic philosophy of medieval universities as out of touch with the realities of urban life. They found in the rhetorical discourse of classical Rome a language adapted to civic life and public speech. They learned to read, speak, and write classical Latin and, eventually, classical Greek. They founded schools to teach others to do so, establishing the pattern for elementary and secondary education for the next three hundred years.

In the service of complex government bureaucracies, humanists employed their skills to write eloquent letters, deliver public orations, and formulate public policy. They developed new scripts for copying manuscripts and used the new printing press to disseminate texts, for which they created methods of critical editing.

Humanism was a movement led by males who accepted the evaluation of women in ancient texts and generally shared the misogynist perceptions of their culture. (Female humanists, as we will see, did not.) Yet humanism also opened the door to a reevaluation of the nature and capacity of women. By calling authors, texts, and ideas into question, it made possible the fundamental rereading of the whole intellectual tradition that was required in order to free women from cultural prejudice and social subordination.

A DIFFERENT CITY. The other voice first appeared when, after so many centuries, the accumulation of misogynist concepts evoked a response from a capable female defender: Christine de Pizan (1365–1431). Introducing her *Book of the City of Ladies* (1405), she described how she was affected by reading Mathéolus's *Lamentations:* "Just the sight of this book . . . made me wonder how it happened that so many different men . . . are so inclined to express both in speaking and in their treatises and writings so many wicked insults about women and their behavior."[4] These statements impelled her to detest herself "and the entire feminine sex, as though we were monstrosities in nature."[5]

The rest of *The Book of the City of Ladies* presents a justification of the female sex and a vision of an ideal community of women. A pioneer, she has received the message of female inferiority and rejected it. From the fourteenth to the seventeenth century, a huge body of literature accumulated that responded to the dominant tradition.

The result was a literary explosion consisting of works by both men and women, in Latin and in the vernaculars: works enumerating the achievements of notable women; works rebutting the main accusations made against women; works arguing for the equal education of men and women; works defining and redefining women's proper role in the family, at court, in public; works describing women's lives and experiences. Recent monographs and articles have begun to hint at the great range of this movement, involving probably several thousand titles. The protofeminism of these "other voices" constitutes a significant fraction of the literary product of the early modern era.

THE CATALOGS. About 1365, the same Boccaccio whose *Corbaccio* rehearses the usual charges against female nature wrote another work, *Concerning Famous Women*. A humanist treatise drawing on classical texts, it praised 106 notable women: ninety-eight of them from pagan Greek and Roman antiquity, one (Eve) from the Bible, and seven from the medieval religious and cultural tradition; his book helped make all readers aware of a sex normally condemned or forgotten. Boccaccio's outlook nevertheless was unfriendly to women, for it singled out for praise those women who possessed the traditional virtues of chastity, silence, and obedience. Women who were active in the public realm—for example, rulers and warriors—were depicted as usually being lascivious and as suffering terrible punishments for entering the masculine sphere. Women were his subject, but Boccaccio's standard remained male.

Christine de Pizan's *Book of the City of Ladies* contains a second catalog, one responding specifically to Boccaccio's. Whereas Boccaccio portrays female virtue as exceptional, she depicts it as universal. Many women in his-

---

4. Christine de Pizan, *The Book of the City of Ladies*, trans. Earl Jeffrey Richards, foreword by Marina Warner (New York, 1982), 1.1.1, pp. 3–4.

5. Ibid., 1.1.1–2, p. 5.

tory were leaders, or remained chaste despite the lascivious approaches of men, or were visionaries and brave martyrs.

The work of Boccaccio inspired a series of catalogs of illustrious women of the biblical, classical, Christian, and local pasts, among them Filippo da Bergamo's *Of Illustrious Women*, Pierre de Brantôme's *Lives of Illustrious Women*, Pierre Le Moyne's *Gallerie of Heroic Women*, and Pietro Paolo de Ribera's *Immortal Triumphs and Heroic Enterprises of 845 Women*. Whatever their embedded prejudices, these works drove home to the public the possibility of female excellence.

THE DEBATE. At the same time, many questions remained: Could a woman be virtuous? Could she perform noteworthy deeds? Was she even, strictly speaking, of the same human species as men? These questions were debated over four centuries, in French, German, Italian, Spanish, and English, by authors male and female, among Catholics, Protestants, and Jews, in ponderous volumes and breezy pamphlets. The whole literary genre has been called the *querelle des femmes*, the "woman question."

The opening volley of this battle occurred in the first years of the fifteenth century, in a literary debate sparked by Christine de Pizan. She exchanged letters critical of Jean de Meun's contribution to *The Romance of the Rose* with two French royal secretaries, Jean de Montreuil and Gontier Col. When the matter became public, Jean Gerson, one of Europe's leading theologians, supported de Pizan's arguments against de Meun, for the moment silencing the opposition.

The debate resurfaced repeatedly over the next two hundred years. *The Triumph of Women* (1438) by Juan Rodríguez de la Camara (or Juan Rodríguez del Padron) struck a new note by presenting arguments for the superiority of women to men. *The Champion of Women* (1440–42) by Martin Le Franc addresses once again the negative views of women presented in *The Romance of the Rose* and offers counterevidence of female virtue and achievement.

A cameo of the debate on women is included in *The Courtier*, one of the most widely read books of the era, published by the Italian Baldassare Castiglione in 1528 and immediately translated into other European vernaculars. *The Courtier* depicts a series of evenings at the court of the duke of Urbino in which many men and some women of the highest social stratum amuse themselves by discussing a range of literary and social issues. The "woman question" is a pervasive theme throughout, and the third of its four books is devoted entirely to that issue.

In a verbal duel, Gasparo Pallavicino and Giuliano de' Medici present the main claims of the two traditions. Gasparo argues the innate inferiority of women and their inclination to vice. Only in bearing children do they profit the world. Giuliano counters that women share the same spiritual and mental capacities as men and may excel in wisdom and action. Men and women are of the same essence: just as no stone can be more perfectly a stone than another, so no human being can be more perfectly human than others, whether male or female. It was an astonishing assertion, boldly made to an audience as large as all Europe.

THE TREATISES. Humanism provided the materials for a positive coun-terconcept to the misogyny embedded in Scholastic philosophy and law and inherited from the Greek, Roman, and Christian pasts. A series of humanist treatises on marriage and family, on education and deportment, and on the nature of women helped construct these new perspectives.

The works by Francesco Barbaro and Leon Battista Alberti—*On Mar-riage* (1415) and *On the Family* (1434–37)—far from defending female equal-ity, reasserted women's responsibility for rearing children and managing the housekeeping while being obedient, chaste, and silent. Nevertheless, they served the cause of reexamining the issue of women's nature by placing do-mestic issues at the center of scholarly concern and reopening the pertinent classical texts. In addition, Barbaro emphasized the companionate nature of marriage and the importance of a wife's spiritual and mental qualities for the well-being of the family.

These themes reappear in later humanist works on marriage and the education of women by Juan Luis Vives and Erasmus. Both were moderately sympathetic to the condition of women without reaching beyond the usual masculine prescriptions for female behavior.

An outlook more favorable to women characterizes the nearly unknown work *In Praise of Women* (ca. 1487) by the Italian humanist Bartolommeo Gog-gio. In addition to providing a catalog of illustrious women, Goggio argued that male and female are the same in essence, but that women (reworking the Adam and Eve narrative from quite a new angle) are actually superior. In the same vein, the Italian humanist Mario Equicola asserted the spiritual equality of men and women in *On Women* (1501). In 1525, Galeazzo Flavio Capra (or Capella) published his work *On the Excellence and Dignity of Women*. This humanist tradition of treatises defending the worthiness of women cul-minates in the work of Henricus Cornelius Agrippa *On the Nobility and Pre-eminence of the Female Sex*. No work by a male humanist more succinctly or ex-plicitly presents the case for female dignity.

THE WITCH BOOKS. While humanists grappled with the issues pertaining to women and family, other learned men turned their attention to what they perceived as a very great problem: witches. Witch-hunting manuals, explorations of the witch phenomenon, and even defenses of witches are not at first glance pertinent to the tradition of the other voice. But they do relate in this way: most accused witches were women. The hostility aroused by supposed witch activity is comparable to the hostility aroused by women. The evil deeds the victims of the hunt were charged with were exaggerations of the vices to which, many believed, all women were prone.

The connection between the witch accusation and the hatred of women is explicit in the notorious witch-hunting manual *The Hammer of Witches* (1486) by two Dominican inquisitors, Heinrich Krämer and Jacob Sprenger. Here the inconstancy, deceitfulness, and lustfulness traditionally associated with women are depicted in exaggerated form as the core features of witch behavior. These traits inclined women to make a bargain with the devil—sealed by sexual intercourse—by which they acquired unholy powers. Such bizarre claims, far from being rejected by rational men, were broadcast by intellectuals. The German Ulrich Molitur, the Frenchman Nicolas Rémy, and the Italian Stefano Guazzo all coolly informed the public of sinister orgies and midnight pacts with the devil. The celebrated French jurist, historian, and political philosopher Jean Bodin argued that because women were especially prone to diabolism, regular legal procedures could properly be suspended in order to try those accused of this "exceptional crime."

A few experts such as the physician Johann Weyer, a student of Agrippa's, raised their voices in protest. In 1563, he explained the witch phenomenon thus, without discarding belief in diabolism: the devil deluded foolish old women afflicted by melancholia, causing them to believe they had magical powers. Weyer's rational skepticism, which had good credibility in the community of the learned, worked to revise the conventional views of women and witchcraft.

WOMEN'S WORKS. To the many categories of works produced on the question of women's worth must be added nearly all works written by women. A woman writing was in herself a statement of women's claim to dignity.

Only a few women wrote anything before the dawn of the modern era, for three reasons. First, they rarely received the education that would enable them to write. Second, they were not admitted to the public roles—as administrator, bureaucrat, lawyer or notary, or university professor—in which they might gain knowledge of the kinds of things the literate public thought worth writing about. Third, the culture imposed silence on women, considering speaking out a form of unchastity. Given these conditions, it is remarkable that any women wrote. Those who did before the fourteenth century were almost always nuns or religious women whose isolation made their pronouncements more acceptable.

From the fourteenth century on, the volume of women's writings rose. Women continued to write devotional literature, although not always as cloistered nuns. They also wrote diaries, often intended as keepsakes for their children; books of advice to their sons and daughters; letters to family members and friends; and family memoirs, in a few cases elaborate enough to be considered histories.

A few women wrote works directly concerning the "woman question," and some of these, such as the humanists Isotta Nogarola, Cassandra Fedele, Laura Cereta, and Olympia Morata, were highly trained. A few were professional writers, living by the income of their pens; the very first among them was Christine de Pizan, noteworthy in this context as in so many others. In addition to *The Book of the City of Ladies* and her critiques of *The Romance of the Rose*, she wrote *The Treasure of the City of Ladies* (a guide to social decorum for women), an advice book for her son, much courtly verse, and a full-scale history of the reign of King Charles V of France.

WOMEN PATRONS. Women who did not themselves write but encouraged others to do so boosted the development of an alternative tradition. Highly placed women patrons supported authors, artists, musicians, poets, and learned men. Such patrons, drawn mostly from the Italian elites and the courts of northern Europe, figure disproportionately as the dedicatees of the important works of early feminism.

For a start, it might be noted that the catalogs of Boccaccio and Alvaro de Luna were dedicated to the Florentine noblewoman Andrea Acciaiuoli and to Doña María, first wife of King Juan II of Castile, while the French translation of Boccaccio's work was commissioned by Anne of Brittany, wife of King Charles VIII of France. The humanist treatises of Goggio, Equicola, Vives, and Agrippa were dedicated, respectively, to Eleanora of Aragon, wife of Ercole I d'Este, duke of Ferrara; to Margherita Cantelma of Mantua; to Catherine of Aragon, wife of King Henry VIII of England; and to Margaret, Duchess of Austria and regent of the Netherlands. As late as 1696, Mary Astell's *Serious Proposal to the Ladies, for the Advancement of Their True and Greatest Interest* was dedicated to Princess Anne of Denmark.

These authors presumed that their efforts would be welcome to female patrons, or they may have written at the bidding of those patrons. Silent themselves, perhaps even unresponsive, these loftily placed women helped shape the tradition of the other voice.

THE ISSUES. The literary forms and patterns in which the tradition of the other voice presented itself have now been sketched. It remains to highlight the major issues around which this tradition crystallizes. In brief, there are four problems to which our authors return again and again, in plays and catalogs, in verse and letters, in treatises and dialogues, in every language: the problem of chastity, the problem of power, the problem of speech, and the problem of knowledge. Of these the greatest, preconditioning the others, is the problem of chastity.

THE PROBLEM OF CHASTITY. In traditional European culture, as in those of antiquity and others around the globe, chastity was perceived as woman's quintessential virtue—in contrast to courage, or generosity, or leadership, or rationality, seen as virtues characteristic of men. Opponents of women charged them with insatiable lust. Women themselves and their defenders—without disputing the validity of the standard—responded that women were capable of chastity.

The requirement of chastity kept women at home, silenced them, isolated them, left them in ignorance. It was the source of all other impediments. Why was it so important to the society of men, of whom chastity was not required, and who more often than not considered it their right to violate the chastity of any woman they encountered?

Female chastity ensured the continuity of the male-headed household. If a man's wife was not chaste, he could not be sure of the legitimacy of his offspring. If they were not his and they acquired his property, it was not his household, but some other man's, that had endured. If his daughter was not chaste, she could not be transferred to another man's household as his wife, and he was dishonored.

The whole system of the integrity of the household and the transmission of property was bound up in female chastity. Such a requirement pertained only to property-owning classes, of course. Poor women could not expect to maintain their chastity, least of all if they were in contact with high-status men to whom all women but those of their own household were prey.

In Catholic Europe, the requirement of chastity was further buttressed by moral and religious imperatives. Original sin was inextricably linked with the sexual act. Virginity was seen as heroic virtue, far more impressive than, say, the avoidance of idleness or greed. Monasticism, the cultural institution that dominated medieval Europe for centuries, was grounded in the renunciation of the flesh. The Catholic reform of the eleventh century imposed a similar standard on all the clergy and a heightened awareness of sexual requirements on all the laity. Although men were asked to be chaste, female unchastity was much worse: it led to the devil, as Eve had led mankind to sin.

To such requirements, women and their defenders protested their innocence. Furthermore, following the example of holy women who had escaped the requirements of family and sought the religious life, some women began to conceive of female communities as alternatives both to family and to the cloister. Christine de Pizan's city of ladies was such a community. Moderata Fonte and Mary Astell envisioned others. The luxurious salons of the French *précieuses* of the seventeenth century, or the comfortable English drawing rooms of the next, may have been born of the same impulse. Here women not only might escape, if briefly, the subordinate position that life in the family entailed but might also make claims to power, exercise their capacity for speech, and display their knowledge.

THE PROBLEM OF POWER. Women were excluded from power: the whole cultural tradition insisted on it. Only men were citizens, only men bore arms, only men could be chiefs or lords or kings. There were exceptions that did not disprove the rule, when wives or widows or mothers took the place of men, awaiting their return or the maturation of a male heir. A woman who attempted to rule in her own right was perceived as an anomaly, a monster, at once a deformed woman and an insufficient male, sexually confused and consequently unsafe.

The association of such images with women who held or sought power explains some otherwise odd features of early modern culture. Queen Elizabeth I of England, one of the few women to hold full regal authority in European history, played with such male/female images—positive ones, of course—in representing herself to her subjects. She was a prince, and manly, even though she was female. She was also (she claimed) virginal, a condition absolutely essential if she was to avoid the attacks of her opponents. Catherine de' Medici, who ruled France as widow and regent for her sons, also adopted such imagery in defining her position. She chose as one symbol the figure of Artemisia, an androgynous ancient warrior-heroine who combined a female persona with masculine powers.

Power in a woman, without such sexual imagery, seems to have been indigestible by the culture. A rare note was struck by the Englishman Sir Thomas Elyot in his *Defence of Good Women* (1540), justifying both women's participation in civic life and their prowess in arms. The old tune was sung by the Scots reformer John Knox in his *First Blast of the Trumpet against the Mon-*

strous Regiment of Women (1558); for him rule by women, defects in nature, was a hideous contradiction in terms.

The confused sexuality of the imagery of female potency was not reserved for rulers. Any woman who excelled was likely to be called an Amazon, recalling the self-mutilated warrior women of antiquity who repudiated all men, gave up their sons, and raised only their daughters. She was often said to have "exceeded her sex" or to have possessed "masculine virtue"—as the very fact of conspicuous excellence conferred masculinity even on the female subject. The catalogs of notable women often showed those female heroes dressed in armor, armed to the teeth, like men. Amazonian heroines romp through the epics of the age—Ariosto's *Orlando Furioso* (1532) and Spenser's *Faerie Queene* (1590–1609). Excellence in a woman was perceived as a claim for power, and power was reserved for the masculine realm. A woman who possessed either one was masculinized and lost title to her own female identity.

THE PROBLEM OF SPEECH. Just as power had a sexual dimension when it was claimed by women, so did speech. A good woman spoke little. Excessive speech was an indication of unchastity. By speech, women seduced men. Eve had lured Adam into sin by her speech. Accused witches were commonly accused of having spoken abusively, or irrationally, or simply too much. As enlightened a figure as Francesco Barbaro insisted on silence in a woman, which he linked to her perfect unanimity with her husband's will and her unblemished virtue (her chastity). Another Italian humanist, Leonardo Bruni, in advising a noblewoman on her studies, barred her not from speech but from public speaking. That was reserved for men.

Related to the problem of speech was that of costume—another, if silent, form of self-expression. Assigned the task of pleasing men as their primary occupation, elite women often tended toward elaborate costume, hairdressing, and the use of cosmetics. Clergy and secular moralists alike condemned these practices. The appropriate function of costume and adornment was to announce the status of a woman's husband or father. Any further indulgence in adornment was akin to unchastity.

THE PROBLEM OF KNOWLEDGE. When the Italian noblewoman Isotta Nogarola had begun to attain a reputation as a humanist, she was accused of incest—a telling instance of the association of learning in women with unchastity. That chilling association inclined any woman who was educated to deny that she was or to make exaggerated claims of heroic chastity.

If educated women were pursued with suspicions of sexual misconduct, women seeking an education faced an even more daunting obstacle: the assumption that women were by nature incapable of learning, that reasoning was a particularly masculine ability. Just as they proclaimed their chastity, women and their defenders insisted on their capacity for learning. The major work by a male writer on female education—that by Juan Luis Vives, *On the Education of a Christian Woman* (1523)—granted female capacity for intellection but still argued that a woman's whole education was to be shaped around the requirement of chastity and a future within the household. Female writers of the following generations—Marie de Gournay in France, Anna Maria van Schurman in Holland, and Mary Astell in England—began to envision other possibilities.

The pioneers of female education were the Italian women humanists who managed to attain a literacy in Latin and a knowledge of classical and Christian literature equivalent to that of prominent men. Their works implicitly and explicitly raise questions about women's social roles, defining problems that beset women attempting to break out of the cultural limits that had bound them. Like Christine de Pizan, who achieved an advanced education through her father's tutoring and her own devices, their bold questioning makes clear the importance of training. Only when women were educated to the same standard as male leaders would they be able to raise that other voice and insist on their dignity as human beings morally, intellectually, and legally equal to men.

THE OTHER VOICE. The other voice, a voice of protest, was mostly female, but it was also male. It spoke in the vernaculars and in Latin, in treatises and dialogues, in plays and poetry, in letters and diaries, and in pamphlets. It battered at the wall of prejudice that encircled women and raised a banner announcing its claims. The female was equal (or even superior) to the male in essential nature—moral, spiritual, and intellectual. Women were capable of higher education, of holding positions of power and influence in the public realm, and of speaking and writing persuasively. The last bastion of masculine supremacy, centered on the notions of a woman's primary domestic responsibility and the requirement of female chastity, was not as yet assaulted—although visions of productive female communities as alternatives to the family indicated an awareness of the problem.

During the period 1300–1700, the other voice remained only a voice, and one only dimly heard. It did not result—yet—in an alteration of social patterns. Indeed, to this day they have not entirely been altered. Yet the call for justice issued as long as six centuries ago by those writing in the tradition of the other voice must be recognized as the source and origin of the mature feminist tradition and of the realignment of social institutions accomplished in the modern age.

We thank the volume editors in this series, who responded with many suggestions to an earlier draft of this introduction, making it a collaborative enterprise. Many of their suggestions and criticisms have resulted in revisions of this introduction, although we remain responsible for the final product.

VOLUME EDITOR'S
INTRODUCTION

THE OTHER VOICE IN THE EARLY EIGHTEENTH CENTURY

Gabrielle Emilie le Tonnelier de Breteuil, Marquise Du Châtelet (1706–1749), accepted the important conventions of her day and rank. Born to a prominent courtier's family, she enjoyed every privilege owed a nobleman's daughter. At the age of nineteen she married the eldest son of one of the most ancient lineages of Lorraine.[1] Within the first years of her marriage Du Châtelet produced the requisite son and daughter. After that, however, she made a series of unorthodox choices that place her in the ranks of "the Other Voice of Early Modern Europe." She studied widely in both Classical and contemporary texts in Latin, French, and English; she wrote on philosophy, physics, and religion; she published books and pamphlets on science, including the *Foundations of Physics* [Institutions de physique] (1740), her inclusive study of the natural world that began with Leibnizian metaphysics and ended with Newtonian mechanics.[2] She mastered the new mathematics so successfully that she not only translated Isaac Newton's *Principia* [Mathematical Principles of Natural Philosophy], but also corrected and expanded upon it in her accompanying commentary (1759). Other works on religion and human behavior circulated in manuscript, so daring that they would have offended the government censors, the royal guardians of morality, the Catholic religion, and the king's authority. There could have been no other official response to the over seven hundred pages of critique of the Old and New Testament that Du Châtelet composed to discredit the ancient por-

---

1. During Du Châtelet's time, Lorraine was a semi-independent duchy in eastern France, and strictly speaking not part of the French kingdom.

2. Du Châtelet's *Institutions de physique* has usually been translated as "Institutions of Physics." However, as Patricia Fara, a historian of science, has pointed out, "Foundations" is truer to the French meaning.

trayal of God and the miracles and divinity of Jesus. Thus, she challenged all expectations about a woman's place and function in European elite society in the first half of the eighteenth century.

Du Châtelet, like the other women in this series, demonstrates the persistence of the practical and intellectual constraints upon women's accomplishments and their equal persistence in contesting and overcoming them. As in the fifteenth-century Italy of the Humanists, the first decades of eighteenth-century France's Enlightenment offered an era when the gifted amateur scholar could read and learn outside university lecture halls, when even recently constituted institutions such as Paris's Royal Academy of Sciences had not reestablished exclusive control of the new knowledge of the natural world. Like other young enthusiasts, men as well as the few women, Du Châtelet sought out mathematicians who could teach her analytic geometry and calculus, subjects not yet part of university curricula. She borrowed texts from the Royal Library and ordered them from booksellers. With her reading, writing, and publications she became one of the still rarefied group of intellectuals who styled themselves members of the Republic of Letters, critics of the old knowledge and purveyors of the new.[3]

The selections of writings included in this collection are intended to show the range of the Marquise Du Châtelet's interests and abilities as an author and thinker and to suggest their evolution. Together they demonstrate why her male contemporaries, including Voltaire, her lover and companion of fifteen years, called her a "genius," and why she merits the Enlightenment's most revered title, that of "philosophe."[4]

A LIFE IN FOUR WORLDS

Du Châtelet offers an interesting paradox for those who seek feminist forebears. There is no evidence that she had read or even heard of any of the female precursors to her daring writings and activities. In an early draft of her preface for her translation of Bernard Mandeville's *The Fable of the Bees*, she made brief mention of Anne Dacier, a seventeenth-century French translator of Greek and Roman texts, and of Dacier's contemporary, Antoinette Duligier de la Garde Deshoulières, called by admirers of her poetry and prose France's "Tenth Muse." She may have heard of these learned women from her

---

3. On the origins of the Republic of Letters, see Anne Goldgar, *Impolite Learning: Conduct and Community in the Republic of Letters, 1680–1750* (New Haven, CT: Yale University Press, 1995).

4. "Philosophe" has the literal meaning of "philosopher" in French. In the eighteenth century, however, it meant a thinker with diverse interests from literature to science. Voltaire, Diderot, and Rousseau are among the most well known.

father, Louis Nicholas, Baron de Breteuil, who enjoyed reading novels and probably attended the weekly salons of the *précieuses* who authored them.⁵ The *querelle des femmes*, the traditional argument about women's nature and capabilities, might have been a topic of conversation at his Thursday evening gatherings. Voltaire could have recommended Dacier's version of the *Iliad* or the *Odyssey*. But Du Châtelet dropped the references in the final version. Even after she embarked on her program of self-education and authorship at the age of twenty-eight, she did not reject the expectations of her class and society. She merely found ways to satisfy the myriad demands made upon her and those decreed by her ambitions. In fact, the marquise, in the course of her brief life, was an active participant in four worlds: that of the family, of the royal courts at Versailles and Lunéville, of "le monde [the high society]" of Paris, and of the Republic of Letters.⁶

She admired and respected her military husband, Florent-Claude, Marquis Du Châtelet-Lomont. From their marriage in 1725, when she was eighteen and he appropriately twelve years her senior, she worked assiduously to advance his career. By 1745, he had risen to become one of the eleven commanders of Louis XV's army. Her society assumed, and she actively concurred, that their marriage represented a partnership of families to advance their lineages. As he was often away on campaign in the numerous wars of the middle decades of Louis XV's reign, this sent her to Brussels to pursue his contested inheritance, to the château of Cirey in Champagne to see to the revenues from the forges and estates, and to extended service in Lunéville at the court of the Duke of Lorraine, King Stanislas of Poland, her husband's feudal lord. She accepted that the marquis must approve the choice of her son's tutor, and that as a noblewoman she could not visit England without him, a great tragedy for this admirer of John Locke and Sir Isaac Newton. She wrote to Francesco Algarotti, an Italian expert on Newton's optics and an early guest at Cirey, "I desire my trip to England with more passion than ever," but this visit to "the country of people who think," was never to be.⁷

Although their first child, a daughter, Gabrielle-Pauline, born in June

---

5. The *précieuses* held salons in Paris where participants came for conversation and refined entertainment. On their gatherings and their links to the salons of the eighteenth century, see, for example, Benedetta Craveri, *The Age of Conversation*, trans. Teresa Waugh (New York: New York Review Books, 2005).

6. This brief version of Du Châtelet's life is based on Judith P. Zinsser, *Émilie Du Châtelet: Daring Genius of the Enlightenment* (New York: Penguin, 2006) (originally published as *La Dame d'Esprit: A Biography of the Marquise Du Châtelet*).

7. Du Châtelet to Algarotti, 15 June 1736, No. 66, *Lettres de la marquise Du Châtelet*, ed. Theodore Besterman (Geneva: Institutions et Musée Voltaire, 1958) 1: 117; Du Châtelet to Algarotti, 11 January 1737, No. 88, *Lettres de la marquise Du Châtelet*, 1: 159.

1726, learned Latin and showed exceptional intelligence, there is no evidence that her mother imagined anything but a conventional future for the little girl. Instead, Du Châtelet lavished intellectual tutelage on her son, Florent-Louis, born in November 1727, and assumed that her daughter would fulfill a traditional function, creating a marriage alliance that enhanced the family's prestige and position. Gabrielle-Pauline found satisfaction in this life, marrying at sixteen into the Neapolitan Papal nobility and rising at the Italian court to become a principal lady-in-waiting to the Queen of Naples.[8] Perhaps the daughter's remarkable success was the result of a different kind of education that she had received from her mother. For Du Châtelet worked the networks of politics and patronage at Versailles and Fontainebleau, and then in Lunéville at the court of Lorraine, with patience and skill. She cultivated family connections and used them to win military advancement for her husband and her son, and the advantageous marriage for her daughter. She and the marquis arranged protection for Voltaire from royal prosecution when his writings offended the royal censors, won him a court position, and ultimately, with the tacit consent of Paris's reigning *salonnières*, Mme Du Deffand and Mme de Tencin, the honor he coveted, election to the French Royal Academy.[9]

Du Châtelet assumed differences of birth and the respect owed to her *état*, or rank. She dismissed without self-reproach servants who behaved disloyally, such as a *femme de chambre* who wrote to a friend about her mistress. The Swiss mathematician who became her calculus tutor chafed at being a paid member of her household. When he talked of leaving her service, she described him as a "lackey and a badly raised one."[10] She looked down upon a distant relative, Françoise d'Issembourg d'Happoncourt, Mme de Graffigny, and expected the unlucky widow to be grateful for the favors shown her by the marquis. These attitudes came to Du Châtelet naturally and would not have surprised her peers. Her father had been an intimate of Louis XIV, his master of protocol, and one of the privileged courtiers allowed to view the king's last days in August 1714. Her mother, Anne de Froullay, came from an even more distinguished family, who also had served the royal household. In the late 1730s and 1740s when Du Châtelet sought favors at Louis XV's Versailles and Fontainebleau, she could appeal to cousins who had risen to

---

8. Du Châtelet bore two other children, Victor Esprit in 1733, Stanislas Adelaïde in 1749. Both died before their second birthdays.

9. Their full names were: Marie de Vichy-Champrond Du Deffand and Claudine Alexandrine Guérin de Tencin.

10. Du Châtelet to Johann Bernoulli II, 28 December 1739, No. 229, *Lettres de la marquise Du Châtelet*, 1: 386.

royal cabinet posts or other court appointments, and to relatives by marriage, such as Louis François Armand Du Plessis, Duke de Richelieu, a commanding general of the royal army and one of the four gentlemen of the bedchamber who served at the king's side from the moment he woke in the morning until he retired at night.

It was to the duke that she wrote of her worries that her husband would not approve of her liaison with Voltaire.[11] Adultery caused no concern in this world where love and lust were assumed to occur outside of the companionship and business of marriage. Rather, the problem was one of rank. In 1735, though Voltaire was considered France's greatest poet—for his verse epic of the French Wars of Religion, *The Henriade*—and its most successful playwright, rival to the great Racine and Corneille, he was still the son of a notary, a bourgeois. In addition, his disrespectful verses and ill-considered remarks to his social betters had twice landed him in the Bastille, and once sent him to formal exile in England. Richelieu's reassurances must have helped reconcile Du Châtelet's husband, but the marquis also could view this as adding the renowned poet to his household. Voltaire, in turn, was rich and generous, and contributed to expenses both at Cirey and in Paris. In fact, the marquis remained a loyal patron from 1734 when he first intervened with royal officials on the poet's behalf.[12] Although neither Du Châtelet nor her husband was ever able to negotiate cessation of the order to arrest the poet philosopher, the family château at Cirey in Champagne remained an accepted refuge.

Du Châtelet and Voltaire had probably met when she was a child; he visited a relative of the baron's at his nearby country château and attended the baron's Thursday evening salon. As a young married woman, Du Châtelet probably exchanged pleasantries with him at the various events in Paris she attended. At this point in her life, in her early twenties, Du Châtelet indulged in what she later called "les choses frivoles," frivolous things such as how she dressed her hair, the correct number of satin bows needed to decorate her skirt, an invitation to a new play from her older friend the Duchess de Saint-Pierre, or the prospect of gambling at cards at a late night supper at the Villars-Brancas, one of the wealthiest and most elite families of "le

11. See, for example, Du Châtelet to Duke de Richelieu, 22 [May 1735], No. 36, *Lettres de la marquise Du Châtelet*, 1: 67.

12. In addition to its controversial content, Voltaire's *Philosophical Letters* had appeared without the royal approbation required of all publications in France. In 1734 the marquis was granted the "official authorization to sequester" Voltaire, a situation similar to modern-day house arrest. In particular, the condemned author could not return to Paris without the government's permission.

monde," Parisian society. These were activities Voltaire enjoyed as well. It was, however, in May of 1733 when Du Châtelet returned from Semur, the provincial garrison town where her husband held the military governorship, that the fifty-five-year-old Voltaire first marveled at her genius and delighted in her company. The offer of asylum in the spring of 1734 at the Du Châtelet country château of Cirey did not signal an improper liaison to anyone of their society. Even her two-month visit that fall troubled no one. It was her decision in June of 1735 to leave Paris altogether, to retire to the countryside to be with Voltaire, that angered her cousin and the titular head of her lineage, François-Victor le Tonnelier de Breteuil, Marquis de Fontenay-Trésigny, the well-connected former member of the queen's household, soon to be Louis XV's secretary for war. Had her husband not accepted the arrangement, Du Châtelet would have been the object of ridicule and disdain.

The extended period of time from June of 1735 until 1739 that Du Châtelet spent at Cirey with Voltaire, making only occasional visits to Paris or to the court, brought her from interested and admiring novice to full-scale participant in the Republic of Letters. Her turn from "frivolous things," her realization, as she described it in her preface to Mandeville's *The Fable of the Bees,* that she "was a thinking creature," meant the discovery of all the learning that had been denied her as a female. She must have studied Latin as a child, and Euclid's geometry, but it was during these months and years that she acquired much of the classical education offered at a *collège* (an elite secondary school).[13] She read Latin authors in the original and in translation, including Petronius, Virgil, Cicero, Ovid, Horace, Lucretius. Her knowledge of Italian came from reading Ariosto's *Orlando Furioso* (1516) aloud with Voltaire. He introduced her to English writers such as John Wilmot, Lord Rochester, Alexander Pope, and Joseph Addison. She followed the journals, such as the *Mercure de France,* the *Gazette d'Amsterdam,* and the *Journal des savants* that gave the latest news of English authors and titles, and of Continental works of philosophy, science, and mathematics. It was in this period that her interests began to focus on the study of "natural philosophy," as the English called it, knowledge of the cosmos, of Cartesian and Leibnizian physics,

---

13. In Du Châtelet's era, a young French noblewoman would have learned reading, writing, and basic counting at home. She might also be taught appropriate female skills like embroidery, dancing, and singing, both at home and in the convent where she might be sent for a few years to learn deportment and to read appropriate religious works. In contrast, a boy would have had a tutor in languages and other academic subjects, and then been sent to a *collège* for further study in areas such as mathematics, philosophy, history, and literature.

metaphysics, and mathematics, and of Newton's optics and his concept of universal attraction.

Voltaire boasted that Du Châtelet learned English in fifteen days. It was probably at his suggestion that she embarked on her translation of the "Enquiry into the Origin of Moral Virtue" and the "Remarks" that Bernard Mandeville wrote to accompany his verse "The Fable of the Bees." For much of 1735 she and Voltaire seem to have been arguing over the rules of human behavior. Du Châtelet believed Mandeville, a Dutch émigré Huguenot, to be the English Montaigne, and her free translation of his thoughts represented a contribution to the dialogue.[14] As part of these same conversations, she and Voltaire began a study of the Old and New Testament. Du Châtelet made summaries of the seventeenth-century English deist Thomas Woolston's pamphlets on Jesus's miracles to aid in this project. The notes from their reading and discussions became her *Examinations of the Bible* [Examens de la Bible], probably completed sometime in 1744 or early in 1745.

At Cirey she never gave up the course in mathematics first pursued in Paris under the tutelage of Pierre Louis Moreau de Maupertuis, an academician, frequenter of the salons, and flirtatious tutor to women of the nobility and to men like Voltaire. Du Châtelet began her lessons sometime in 1733 and must have surprised and later annoyed her mentor with her enthusiasm. From René Descartes' analytic geometry, she went on to reading: Descartes' and Newton's works of natural philosophy, and those in astronomy by Galileo, the Cassinis, and Edmund Halley; the public lectures in physics such as those by the Dutchman Petrus van Musschenbroek, the Frenchmen Joseph Privat de Molières and Etienne-Simon Gamaches, and the British physicist John Keill. She had probably discovered John Locke's *Essay concerning Human Understanding* (1690) at Semur, but certainly studied him with renewed interest given Voltaire's admiration for the English philosopher. Beginning in 1736, Frederick of Prussia sent them French translations of the writings of Christian Wolff, a student and follower of Gottfried Wilhelm Leibniz. She also studied Leibniz's ideas through his articles for the prestigious German

---

14. Voltaire's thoughts formed the basis of his *Treatise of Metaphysics* [Traité de métaphysique], which, by tradition, he composed for Du Châtelet when he was courting her in 1734–35. The version of the *Treatise* in the Collection of Occidental Manuscripts at the State Library in St. Petersburg has her writing in two or three sections. Also, as part of these conversations, Du Châtelet composed a short treatise "On Liberty" [liberty or free will], which she intended to include in her *Foundations of Physics*. She never published the essay, but it is likely that the manuscript was sent to Frederick of Prussia in 1737 as if Voltaire had written it. For the first discussion of this ruse and the reasons for it, see Linda Gardiner Janik, "Searching for the Metaphysics of Science: The Structure and Composition of Madame Du Châtelet's *Institutions de physique*, 1737–40," *SVEC* 201 (1982): 85–113.

Acta Eruditorum and his written interchange with Samuel Clarke, a follower of Newton.[15]

In the summer of 1737, in the midst of this ambitious reading program, Du Châtelet aided Voltaire in his experiments with fire, in preparation for his entry to the Academy of Sciences prize competition of 1738. When she disagreed with his conclusions, and what to her was a blind adherence to Newton's ideas of attraction, she decided to write her own essay: *Dissertation on the Nature and Propagation of Fire* [Dissertation sur la nature et la propagation du feu]. Neither won, but through Voltaire's agitation, and because of the novelty of a poet and a "young woman of quality" writing in such a context, the Academy published both of their essays as well as those of the three winners. Hers appeared twice under the Academy's aegis and then, in a revised form in 1744 by her own arrangement.[16]

During this period Du Châtelet also collaborated with Voltaire on his *Elements of the Philosophy of Newton* [Eléments de la philosophie de Newton]. Modeled after Count Algarotti's *Newtonianism for the Ladies* [Il Newtonianismo per le donne], this was an account of Newton's optics and of his system of universal attraction for a French audience. When it appeared in 1738 with an extravagant introduction praising her contributions, Du Châtelet had already embarked on the *Foundations of Physics*, her own description of the cosmos. Not content with simple description, as Voltaire and so many other Newtonians had been, she insisted on also examining the causes of the phenomena that English and Continental astronomers, physicists, and mathematicians were presenting in their learned and popular books and treatises. In addition, she hoped that, in formulating a metaphysical basis for the cosmos, she could answer the critics of Newton's universe in France and elsewhere on the Continent, who remained loyal to Descartes because his system began with philosophy and theology and only then turned to the natural world. Thus her *Foundations* begins with rules of reasoning, how knowledge is constructed, and continues to a discussion of the Creator's role, with chapters on the nature of matter, space, and time. It is after presenting this metaphysical and philosophical "foundation" that she offers her descriptive synthesis of Newtonian and Leibnizian mechanics.

With the appearance of the *Foundations* in 1740 and in a revised edition

15. Du Châtelet also consulted the French translation of Locke's *Essay* by Pierre Coste (1700). The interchange between Leibniz and Clarke was commissioned by Caroline of Ansbach, the Princess of Wales, and published in French and in English in 1717.

16. The descriptive phrase appeared as part of the frontmatter for the Royal Academy publication of her *Dissertation*. Neither she nor Voltaire was identified by name, only by the number given to their submission, and phrases such as this.

in 1742, Du Châtelet had time to return to her study of the Old and New
Testament. She probably completed her *Examinations of the Bible* in 1744 and
then at the beginning of 1745 returned to science. Initially, she thought
of translating a commentary by John Keill on Newton's *Principia*, but encouraged by one of Europe's most eminent mathematicians and Newtonians, Father Jacquier, professor of experimental physics at the College of
Rome, she embarked instead on the mammoth task of translating the work
itself.[17] Wishing to bring it up to date with the latest confirmations of Newton's hypotheses about the effects of attraction on the earth and in the universe, she decided to write an accompanying commentary. As the months
and years passed, these explanations and mathematical studies expanded in
scope until they came to fill almost two-thirds of the second volume when it
was finally ready for the printer by the late summer of 1749.

Du Châtelet had completed her *Discourse on Happiness* [Discours sur le
bonheur] earlier in 1748. A treatise on happiness was a popular subject with
the men of the Republic of Letters. Voltaire wrote his version as poetry, his
Discourse in Verse on Man [Discours en vers sur l'homme]. One of their visi-
tors to Cirey, Claude Adrien Helvétius, also composed one. Du Châtelet's
initial pages probably formed part of discussions among the three. However, in the last section of her *Discourse*, she wrote in a completely different
voice and revealed more personal and immediate concerns than was customary in such essays. There she explained how she came to accept that
Jean-François de Saint-Lambert was to be trusted as the great love of her
life. He and Du Châtelet must have met at the court of Lunéville when she
came with her husband to fulfill the marquis's feudal duties. However, it was
not until the spring of 1748 that Du Châtelet's and Saint-Lambert's feelings passed quickly from flirtation to passion and love. Upon learning of
the liaison Voltaire was angry and petulant, even though Du Châtelet had
been forgiving when he had begun an affair with his favorite niece three
years before.

Du Châtelet's letters for this period, 1748–49, show the complexity of
her worlds, her efforts to fulfill her duties to her family, to her old friend
and companion, and to her new lover; and her obligations to preserve her
"honor" as a contributing member of the Republic of Letters. Throughout
these months, she continued with her *Commentary* on the *Principia*, her consul-
tations with Alexis-Claude Clairaut, her second mentor in the complicated

17. Fathers François Jacquier and Thomas Le Seur published an edition of Newton's *Principia*
with an extensive continuous commentary in their annotation. Known as the "Geneva edition"
of the *Principia*, it appeared in two volumes, 1739–42.

mathematics of his own and others' treatises.[18] She quieted Voltaire's chagrin over her affair and persevered with her activities to insure her husband's
position both at Versailles and at Lunéville, and her son's advancement. Although she became the subject of ridicule in the salons of Paris and the
corridors of Versailles when contemporaries learned of her pregnancy—a
forty-one-year-old marquise did not become pregnant and certainly not by
a man other than her husband—she had the proofs of the translation and
the *Commentary* ready before her lying-in on 4 September 1749. Her husband
had already accepted paternity, and when she died suddenly of complications occasioned by the birth, the marquis and Saint-Lambert mourned her
with an intensity that surpassed Voltaire's more public display.[19]

One of Du Châtelet's acts in the last days of her life was to send what
remained of her manuscript for her *Commentary* on the *Principia* to the royal
librarian, Abbé Claude Sallier, perhaps in anticipation that her authorship
might be contested in the future. On its publication in 1759, the reviewer for
the *Bibliothèque universelle* did not question her accomplishment and applauded
her many-layered translation of Newton's work for a French audience: the
literal translation, and the commentary with its mathematical and textual
explanations. Ironically, it was subsequent generations of biographers and
Newton scholars, such as its most famous modern commentator, editor, and
translator, I. Bernard Cohen, who robbed her of the fame so willingly given
by her contemporaries.

AN ENLIGHTENMENT *PHILOSOPHE*

The eighteenth century, the era of the Enlightenment, calls to mind the
names of many learned men, especially in France: Voltaire, Diderot, and
Rousseau. Historians have written of women's contributions, but primarily
in their role as the gifted hostesses of the salons that were an integral part of
the world of the Republic of Letters. Neither these women nor their guests
questioned the traditional views of women's nature and capabilities. If anything, by the end of the eighteenth century reason had added more layers to
the denigrating characterizations of women's minds and, as with Rousseau,

18. On the men who could be considered Du Châtelet's mentors, both intentional and unin-
tentional, see Judith P. Zinsser, "Mentors, the Marquise Du Châtelet, and Historical Memory,"
Notes and Records of the Royal Society 61 (2007): 89–108.

19. Du Châtelet died of a pulmonary embolism, a not uncommon consequence of childbirth
even in the twenty-first century, though it is not usually fatal today.

glorified the female body and its reproductive functions.[20] Europe's universities continued to exclude women with the one exception of Bologna, which granted a doctorate and the right to teach to its illustrious Newtonian expert, Laura Bassi. Again, with the exception of the Bologna Academy, the major learned institutions elected only men as members. Even informal gatherings such as those at the Café Gradot, where Parisian mathematicians and physicists met to discuss their ideas and writings, barred women.

Du Châtelet's principal anger about the constraints of an elite woman's life concerned the lack of education equal to that offered to a young man. In her "Translator's Preface" for *The Fable of the Bees*, she asked, "Can someone tell me the reason, if one can, why these creatures whose understanding appears in everything to be the same as men's always seem to be stopped by an invincible force on this side of a barrier?" In an early draft she described this as "the last injustice practiced in the good name of decency." She promised, "If I were king, I would establish *collèges* for women."[21] Voltaire and Du Châtelet's younger brother both had this opportunity to attend the equivalent of a modern elite secondary school, guaranteeing their entry into university, should they choose to attend one.[22] Although she omitted the last two more extreme comments in the final version, with her successes in science she remedied the lack she did note: that no great work of literature, history, art, or science had been written by a woman.

20. On the evolution of the role of *salonnière* and the ways in which women used it to exercise authority, see Joan DeJean, *Tender Geographies: Women and the Origins of the Novel in France* (New York: Columbia University Press, 1991); Daniel Gordon, *Citizens without Sovereignty: Equality and Sociability in French Thought, 1670–1780* (Princeton, NJ: Princeton University Press, 1994); Dena Goodman, *The Republic of Letters: A Cultural History of the French Enlightenment* (Ithaca, NY: Cornell University Press, 1994); and Antoine Lilti, *Le monde des salons: Sociabilité et mondanité à Paris au XVIIIe siècle* (Paris: Fayard, 2005). See on the renewal of negative attitudes toward women: Londa Schiebinger, *The Mind Has No Sex? Women in the Origins of Modern Science* (Cambridge, MA: Harvard University Press, 1989).

21. Emilie Du Châtelet, "Preface du traducteur," in *Studies on Voltaire with Some Unpublished Papers of Mme Du Châtelet*, ed. Ira O. Wade (Princeton, NJ: Princeton University Press, 1947), 135–36. There are four drafts of Du Châtelet's "Translator's Preface" in the Voltaire Collection of the St. Petersburg National Library. Only one, probably the last one, has been published. These quotations come from what I have previously designated as B226 of the early draft. On these drafts and Du Châtelet's translation, see Judith P. Zinsser, "Entrepreneur of the 'Republic of Letters': Émilie de Breteuil, Marquise Du Châtelet, and Bernard Mandeville's *The Fable of the Bees*," *French Historical Studies* 25, no. 4 (2002): 595–624.

22. A little girl of Du Châtelet's rank would have been educated by her mother and a governess in reading, writing, basic counting, and the etiquette she would need in her life as a courtier. Perhaps sent to a nunnery for the years surrounding her confirmation, there she would have been schooled in religious doctrine. Her family would also have seen to lessons in music and dancing. For speculations about Du Châtelet's education, see Zinsser, *Émilie Du Châtelet*, 22–31.

Unlike most French women writers of the seventeenth and eighteenth centuries, Du Châtelet did not take literature as her means to fame, and to her reputation as "une vraie femme savante [a truly learned woman]" as she was called in the *Journal universel* after her election to the Bologna Academy.[23] She never expressed any interest in writing poetry or a novel. When she translated, it was not the Roman classics, or a devotional text, but first, Mandeville's outrageous secular commentary on human behavior, and second, one of the most difficult works of physics, a tome that most of her learned male contemporaries did not even attempt to read.[24] In her published works she did not identify herself as a woman. Partly, this was, as with other women, to insure that her writings would not be dismissed simply because of the sex of the author, but also because she believed that the "truths" she was presenting were more important than any statement she might be making about women's capabilities.

Simply, she did not see herself as like other women. She accepted Galen's characterization of women's physiology as cold and moist, but believed she had more of the masculine humors, heat and dryness. As she explained in her *Discourse on Happiness*, because "I have an abundance of fire in my nature, I spend my morning drenching myself with liquids."[25] The judges of the entries for the Academy's essay on fire only discovered that hers was one of the fourteen submissions after they had made their choice. Du Châtelet addressed the *Foundations of Physics* to her son, but from the preface readers would assume that the author was the father, not the mother. Her more controversial *Examinations of the Bible*, except for a few careless feminine endings, gives no indication that the methodical critic of the Old and New Testaments might be a woman.[26] Similarly, the first third of her *Discourse on Happiness* suggests a male rather than a female *philosophe*. Ironically, after her death no such precautions were needed. The title page of her final work, the translation of and commentary on Newton's *Principia*, makes no mention of

23. See the *Journal universel* 10 (1746). Note the denigrating connotations of the phrase: "a truly learned woman."

24. As Voltaire noted in his 1752 "eulogy [*éloge*]" about her, which he published first in the *Bibliothèque impartiale*, and then as the "historical preface" to her translation of and commentary on the *Principia* when they were finally published in 1759.

25. Emilie Du Châtelet, *Réflexions sur le bonheur*, ed. Robert Mauzi (Paris: Société d'éditions Les Belles Lettres, 1961), 10. A subsequent edition of the *Discourse* was printed with a preface by Elisabeth Badinter (Paris: Rivages poche, 1997).

26. Only a scholar such as Bertram Eugene Schwarzbach, with his training and attentiveness to detail would have found these slips of the pen. See his introduction to his critical edition of the *Examens de la Bible* (Paris: Honoré Champion Editeur, forthcoming 2009). Note that an earlier draft was consulted for this translation project.

the Englishman: "Principes mathématiques de la philosophie naturelle, par feue Madame la Marquise DU CHASTELLET" [Mathematical Principles of Natural Philosophy, by the late Madame Du Châtelet], suggesting that she, not he, was the author.[27]

Du Châtelet used a variety of images to describe herself in her clandestine and published works. She was the "entrepreneur [*négociant*]" between cultures in her preface to Mandeville's *The Fable of the Bees*, the "surveyor of the whole edifice" of physics in the *Foundations*. No author's preface is known for her *Examinations of the Bible* or her *Principia*. Perhaps it was no longer necessary. In the *Foundations* she validated herself as a "philosopher," "geometrician," and "physicist" through the authorities she cited, her demonstrated knowledge of physics and mathematics, and by the analogies she constructed.[28] The fact of her sex seemed only important in demonstrating that she, though a woman, could claim these titles. She wrote to the executive secretary of the Bologna Academy describing the honor of her election to membership: it was "an encouragement . . . to persons of my sex: for what more flattering motive could one propose to them, in order to engage them to cultivate the sciences."[29]

27. The capitalization is in the text. Du Châtelet usually signed her letters with similar spelling, "Du Chastelet." In French the circumflex came to replace *s*, hence the variant spelling. Notice also, that the *d* is capitalized. This indicates that the "Du" is an integral part of the name. "De" with a lowercase *d* in a French name or title usually signified association with a particular place, a feudal landholding that carried a title with it. For example, when Louis XV wanted to ennoble his mistress, Jeanne-Antoinette Poisson, Mme le Normant d'Etoiles, he procured the property of Pompadour with its accompanying title for her. Thus, she became "la marquise de Pompadour."

28. See Judith P. Zinsser, "The Many Representations of the Marquise Du Châtelet," in *Men, Women, and the Birthing of Modern Science*, ed. Judith P. Zinsser (DeKalb: Northern Illinois University Press, 2005), 48–67; Julie Candler Hayes, *Reading the Enlightenment: System and Subversion* (New York: Cambridge University Press, 1999), 86–110. Erica Harth puzzles over women's ways of establishing their authority as well, and concludes: "It is as if women were allowed neither to be ignorant nor to be learned." Erica Harth, *Cartesian Women: Version and Subversions of Rational Discourse in the Old Regime* (Ithaca, NY: Cornell University Press, 1992), 38.

29. Du Châtelet to F. M. Zanotti, 1 June 1746, in Mauro De Zan, "Voltaire e Mme Du Châtelet, membri e correspondenti dell' Accademia delle Scienze di Bologna," *Studi e memorie dell' Instituto per la Storia dell' Università di Bologna* 6 (1987): 156–57. Elizabeth Goldsmith and Joan DeJean make much of women publishing anonymously. In writings on natural philosophy, many men did as well. They feared that their memoir or essay would not be read because they were associated with a particular position, for example, a Cartesian vs. a Newtonian view of the universe. See Elizabeth C. Goldsmith, "Authority, Authenticity, and the Publication of Letters by Women," in *Writing the Female Voice: Essays on Epistolary Literature*, ed. Elizabeth Goldsmith (Boston: Northeastern University Press, 1989), 48; on the practice among men, see Mary Terrall, *The Man Who Flattened the Earth: Maupertuis and the Sciences in the Enlightenment* (Chicago: University of Chicago Press, 2002), 12–13.

Without ever explicitly acknowledging it, she had accepted without question Descartes' separation of the mind or soul from the body. Thus she could be a physical woman, and even enjoy that fact, but also a thinking person, able to observe with her senses and reflect with her reason in the same way as a man.[30] She explained in the preface to the *Foundations of Physics* that Descartes had made it possible for "him who is endowed with reason" to "make his own examination" of the sciences, to "take nobody on his word alone," not even the most famous. Like her first mentor, Maupertuis, she deplored the division of "the thinking world" into adherents of the systems of Descartes and Newton. She called it "the inevitable obstinacy to which the spirit of partisanship carries one: this frame of mind is dangerous on all occasions of life; . . . [but] ridiculous in Physics." What drove her, and she assumed would drive others, was "the search for the truth." The criteria of judgment were self-evident to her: "about a book of Physics one must ask if it is good, not if the author is English, German, or French."[31]

In the twenty-first century the writings of the Marquise Du Châtelet are significant for three reasons. First, they show the capabilities of an elite woman of France's Enlightenment. Second, they give a unique view of its first decades when reason and passion, sentiment and reflection, imagination and observation, complemented rather than opposed each other. This was the era before 1750, when an individual wrote for a limited, educated, cosmopolitan group, those Du Châtelet referred to as the *gens de lettres* [men of letters]. Members of royal academies and learned societies, of university faculties, and of the cafés in Europe's capitals, men and the few exceptional women, corresponded with one another, exchanged essays, treatises, and book-length manuscripts, including those that circulated clandestinely to avoid royal censors. The works they published they reviewed for one another in the journals they edited, such as the French *Journal des savants* and the *Mémoires des Trévoux*, the London *Transactions of the Royal Society* and the Berlin *Acta Eruditorum*. This was Europe's intellectual elite, the Republic of Letters. Third, Du Châtelet's writings, taken together, give a wondrous picture of the richness of "natural philosophy," the kind of science that began with rules governing the actions of the human mind, considered theology and metaphysics integral to any understanding of causation, and refused to

---

30. Du Châtelet used the latter distinctions as defined by Locke in his *Essay concerning Human Understanding* though, as with Descartes, she never explicitly referred to him or this progression of thought, from the senses to reflection.

31. Emilie Du Châtelet, *Institutions de physique* (Paris: Prault fils, 1740), preface; x, vi, vii.

accept mere descriptions of the universe. This was "the other path," to use the phrase coined by the cultural historians of science, Steven Shapin and Simon Schaffer, the path rejected by the fixed trajectories in our subsequent histories of both the Enlightenment with its leaps to the French Revolution and of modern science with its narrow, teleological definition of "progress" from Galileo to Newton to Einstein.[32]

For Du Châtelet, like cosmologists, the most advanced physicists of the twenty-first century, believed in a "unified theory," one system of natural laws that would explain all in the cosmos from the smallest imaginable particles of matter to the vast universe. She assumed this universe consisted of multiple planetary systems, with other suns that would create spectra of light composed of completely different colors. In her search for this unified theory, she created an original synthesis of explanations and descriptions drawn from the ideas of Descartes, Leibniz, and Newton. Descartes and Leibniz furnished the belief in the exercise of her own mind and the strict rules of reasoning that must govern all thinking. She then constructed her own system of the world based on intentionally posed hypotheses substantiated by observation, demonstration, and analogies provided by experiment and mathematics.

Du Châtelet even assumed that God the Creator must follow the strict rules of reasoning she had formulated. Otherwise the cosmos would be controlled by an erratic, intervening god, who could at any moment alter reality. It was this need for stability and predictability, her belief in the fact of immutable natural laws—anything else negated the very concept of a nature governed by laws—that led Du Châtelet to her critique of the Bible. She could not accept the cruel, vengeful, apparently irrational God of the Old Testament, nor the miracles performed by the Jesus of the New Testament. Instead, she imagined a rational, benevolent Creator who "engraved in

---

32. The historian of science Thomas Kuhn identified "paradigms" of thought and method that hampered advances in modern science; similar paradigms governed the historiography of science and made it seem as if science had always been separate from metaphysics and philosophy, governed by apparently self-evident rules of reasoning and evidence. Those, like Du Châtelet, who rejected this narrow definition, were vilified, forgotten, or like Descartes, narrowly classified by subsequent commentators. Descartes became only a "philosopher," while Newton reigned as the "physicist," though both were mathematicians, both experimental physicists, and both sought not only to describe the cosmos but also to understand its relationship to God and to humanity. On the concept of the "path not taken," see Steven Shapin and Simon Schaffer, *Leviathan and the Air-Pump: Hobbes, Boyle and the Experimental Life* (Princeton, NJ: Princeton University Press, 1985); William Clark, Jan Golinski, and Simon Schaffer, eds., *The Sciences in Enlightened Europe* (Chicago: University of Chicago Press, 1999).

[human] hearts" the golden rule: "do not do unto others what you would not have done unto you."[33] In her *Discourse on Happiness* she expanded on her philosophy of human behavior, and specifically addressed women's situation. Her most important dictum was "to be resolute about what one wants to be and about what one wants to do," and to banish remorse and repentance for those errors that inevitably were part of life. She accepted that women were limited by their sex; it was, for her, as fixed as her rank in the complicated social hierarchy of her culture. Yet women like men could aspire to glory and fame. She listed the numerous paths open to men: government, diplomacy, and war. For a woman, she concluded, only study offered such a future, and a consolation "for all the exclusions and all the dependencies to which she finds herself condemned by her place in society."[34]

DU CHÂTELET'S CONTEMPORARY AND SUBSEQUENT REPUTATION

Du Châtelet is a model of both the rewards and hazards of exceptionality for a woman. In her own day, her published works brought her the learned reputation she sought. The *Mercure de France* published her critical review of Voltaire's *Elements of the Philosophy of Newton*, the Royal Academy of Sciences her *Dissertation* on fire. The *Journal des savants* praised her *Foundations of Physics* and its explication of Leibniz, as did many *philosophes*, such as Maupertuis and Clairaut, and the Swiss Cartesian, Gabriel Cramer. In the entry on "Newtonianism" in Diderot and d'Alembert's *Encyclopedia* [Encyclopédie] she was listed among the seven learned individuals who had "tried to make Newtonian philosophy easier to understand."[35] Eighteenth- and nineteenth-century biographical collections of the learned always included her as an example of the prodigy, the exception.

And then the truisms of women's history overtook the memory of her past. Beginning with the nineteenth-century journalist, Charles Saint-Beuve (*Causeries du Lundi*, 1850), she was admired for her intelligence and her accomplishments in mathematics, but the focus of the story shifted to questions of morality and sexual appetite. Saint-Beuve described the morality of her *Discourse on Happiness* as immodest, crude, "dry [*sec*]" in the sense of calculating

---

33. Du Châtelet, "Chapter 1," *The Fable of the Bees*, in Wade, ed., *Studies*, 145.

34. Du Châtelet, *Réflexions sur le bonheur*, 16–17, 21.

35. She was listed with the premier Newtonians of the eighteenth century: Henry Pemberton, Jacquier and Le Seur, Willem Jacob 's Gravesande, William Whiston, and Colin Maclauren. *Encyclopédie ou Dictionnaire Raisonné des arts et des métiers* (Lausanne: Les Sociétés Typographiques, 1778–82 edition), 22: 414.

and unfeeling. Subsequent authors added to this emerging legend. Edmond and Jules Goncourt (*La femme au dix-huitième siècle*, 1862) used Du Châtelet to personify the amoral, pleasure-seeking, capricious noblewoman of ancien régime France, the "learned woman [*femme savante*]," diverted by her reading, but incapable of serious rational thought.[36] The local historian of Cirey, Abbé Piot, portrayed her as learned, but never to the extent that she sacrificed her love of jewels, amateur theatricals, and an ever-increasing number of amorous entanglements. Her sexual exploits multiplied with each new twentieth-century biography. Popular writers such as Nancy Mitford, Samuel Edwards, and most recently, David Bodanis focused on her fifteen-year liaison with Voltaire and made a melodrama of her death at forty-two after the birth of her daughter by the younger soldier-poet, Jean-François de Saint-Lambert. It is not surprising that the first critical edition of any of her works was of the *Discourse on Happiness*, by implication the true representation of her interests and concerns.[37] Elisabeth Badinter, author of *Émilie, Émilie: l'ambition féminine au XVIIIe siècle* (1983), edited a recent publication of the *Discourse* and used it for her exploration of the psychology of a privileged eighteenth-century woman, the conflict between the noblewoman's responsibilities to her intellectual activities and to her family, especially her children. René Vaillot in his full-length biography (1978) and in the revision for René Pomeau's two-volume *Voltaire en son temps* (vol. 1, 1985–95) suggested a similar conflict within Du Châtelet's psyche, the woman of reason and the woman of passion. Only with Judith P. Zinsser's biography, *Émilie Du Châ-*

36. Nanette Le Coat, "'Le Génie de la Sécheresse': Mme Du Châtelet in the Eyes of Her Second Empire Critics," in *Émilie Du Châtelet: Rewriting Enlightenment Philosophy and Science*, ed. Judith P. Zinsser and Julie Candler Hayes, *SVEC* 1 (2006): 292–307.

37. Du Châtelet died in 1749, just before the preferred role for elite, intellectual women narrowed to that of *salonnière*, hostess and facilitator for her learned male guests. From 1750 on, even that authority was eclipsed by the virtual explosion of publishing by men of every class and persuasion initiated with the first volumes of Diderot and d'Alembert's *Encyclopedia*. No women wrote for this multivolume work, though authors freely borrowed from Du Châtelet's *Foundations* for their articles on philosophy and science. In addition, the *Encyclopedia's* entry on "Femmes [Women]" reformulated the old denigrating views of women's nature and capabilities and gave them new authority, clothing the prejudices of previous centuries in the apparent authority of the new "science" of anatomy. See the *Encyclopedia* at http:/humanities.uchicago.edu/orgs/ARTFL and the translation project at the University of Michigan, http://quod.lib.umich.edu/d/did/intro.html.

Robert Mauzi's critical edition of Du Châtelet's *Discourse on Happiness* was valuable in that it meant the modern publication of this manuscript and the first assessment of its originality within this genre of eighteenth-century literature.

Note that a version of this historiographical description previously appeared as "Madame Du Châtelet et les historiens," *Madame Du Châtelet: La Femme des Lumières*, ed. Elisabeth Badinter and Danielle Muzerelle (Paris: Bibliothèque nationale, 2006), 118–21.

telet: Daring Genius of the Enlightenment (2006), have the marquise and her writings been presented in the context of all of her worlds: her family, the royal courts, Parisian society, and the Republic of Letters.[38]

Writings of Du Châtelet's, other than the *Discourse on Happiness*, if not forgotten altogether, were attributed to her male contemporaries, dismissed as insignificant, or worse, as not "original." Even scholars ignored or minimized the significance of the *Foundations*, calling it a mere synthesis of others' ideas. Newtonian experts attributed her *Commentary* on and translation of the *Principia* to her mentor, Clairaut, when in fact he had neither the time nor the inclination to be its author.[39] The less well known writings of Du Châtelet's that circulated only in manuscript were simply forgotten, most notably her more than seven-hundred-page exegesis of the Old and New Testament.

The "rediscovery" of the Marquise Du Châtelet and modern appreciation of the many facets of her character and activities began with Theodore Besterman's collection of her letters in the late 1950s.[40] Although undoubtedly only a small part of Du Châtelet's correspondence, the letters that Besterman, the noted Voltaire scholar, found and published suggest the complexity of her life and interests. She wrote to the Duke de Richelieu and the Count d'Argental about activities in Paris and at the courts of Versailles and Fontainebleau. The letters to Maupertuis, Johann Bernoulli and his son, and other mathematicians, physicists, and *philosophes* in France, England, Italy, Switzerland, and Germany give a glimpse of her scientific studies. She is the perfect courtier in her careful letters to Frederick of Prussia. The short notes and long letters in the last year before her death to Saint-Lambert show her efforts to complete "my Newton," as she referred to it, and to be with the man she perceived to be the great passion of her life.

Initially, scholars in literature and the history of science turned to Du Châtelet because of their interest in Voltaire. Ira O. Wade's descriptions of

---

38. All of the biographies mentioned here are listed in the Bibliography. Note that Esther Ehrman did a short balanced study of Du Châtelet's life and writings that preceded Zinsser's: Esther Ehrman, *Mme Du Châtelet: Scientist, Philosopher, and Feminist of the Enlightenment* (Berg: Leamington Spa, 1986).

39. On the dismissal of women's works of synthesis as unoriginal, see Berenice A. Carroll, "The Politics of 'Originality': Women and the Class System of the Intellect," *Journal of Women's History* 2 (1990): 136–63. On Clairaut's other commitments and interests, see Judith P. Zinsser and Olivier Courtelle, "A Remarkable Collaboration: The Marquise Du Châtelet and Alexis Clairaut," *SVEC* 12 (2003): 107–20.

40. Besterman published these letters in a two-volume collection and also included them (with some additions) in his edition of Voltaire's correspondence, part of Theodore Besterman et al., eds., *Oeuvres complètes de Voltaire* (Geneva and Oxford: Institut et Musée Voltaire, and Voltaire Foundation, 1968–).

Du Châtelet and Voltaire's activities at Cirey from 1734 to 1740, and his discussions of her unpublished writings: the free translation of Bernard Mandeville's *The Fable of the Bees*, the essay "On Liberty," and the chapter-by-chapter commentary on the Bible, had already appeared even before Besterman's selection of her letters. Wade's discoveries documented the rich collaboration between Du Châtelet and Voltaire and the range of their interests. Raymond Walters did a similar study of Voltaire and Du Châtelet's submissions to the Royal Academy of Sciences competition of 1738. William H. Barber, though skeptical of Du Châtelet's originality, established her significance in bringing Leibnizian metaphysics to a French audience.[41]

Since the 1970s there has been an ongoing reassessment of Du Châtelet's life and writings, an effort to place her interests and accomplishments not only within the framework of Voltaire's circle and his activities but also in the broader world of the early decades of the European Enlightenment. Historians of science such as Carolyn Merchant, Linda Gardiner, David Bodanis, and Sarah Hutton have described the uniqueness of her synthesis in the *Foundations of Physics* and dismissed any suggestion that it was a redaction of another's work, or simply derivative.[42] Elisabeth Badinter and Mary Terrall have explained the context in which Du Châtelet wrote, and thus how the men of the Republic of Letters viewed her exceptionality.[43] Scholars of the history of mathematics, such as René Taton, R. Debever, and Antoinette Emch-Dériaz and Gérard Emch have attested to the excellence of her translation of Newton's *Principia* and her mathematical expertise.[44] I. Bernard

41. Ira O. Wade, *Voltaire and Mme. Du Châtelet: An Essay on the Intellectual Activity at Cirey* (Princeton, NJ: Princeton University Press, 1941) and *Studies*; Raymond L. Walters, "Chemistry at Cirey," *SVEC* 58 (1967): 1807–27; R. Mauzi, ed., *Réflexions sur la bonheur*; William H. Barber, "Mme Du Châtelet and Leibnizianism: The Genesis of the *Institutions de physique*," in *The Age of Enlightenment: Studies Presented to Theodore Besterman*, ed. W. H. Barber et al. (Edinburgh: Oliver and Boyd, 1967), 200–222.

42. Carolyn Merchant Iltis, "Madame Du Châtelet's Metaphysics and Mechanics," *Studies in History and Philosophy of Science* 8 (1977): 29–48; Gardiner Janik, "Searching for the Metaphysics of Science"; David Bodanis, *E=mc2: A Biography of the World's Most Famous Equation* (New York: Berkley Books, 2000); Sarah Hutton, "Emile Du Châtelet's *Institutions de physique* as a Document in the History of French Newtonianism," *Studies in History and Philosophy of Science* 35 (2004): 515–31.

43. Elisabeth Badinter, *Les Passions intellectuelles: Désirs de gloire (1735–1751)*, vol. 1 (Paris: Fayard, 1999); Mary Terrall, "Emile Du Châtelet and the Gendering of Science," *History of Science* 33 (1995): 283–310; and Terrall, "Gendered Spaces, Gendered Audiences: Inside and Outside the Paris Academy of Sciences," *Configurations* 2 (1995): 207–32; Keiko Kawashima was the first to write on reviews of Du Châtelet's works: see "Madame Du Châtelet dans le journalisme," *LLULL* 18 (1995): 471–91. See also, Francois De Gandt, ed., *Cirey dans la vie intellectuelle: La réception de Newton en France*, *SVEC* (2001): 11.

44. René Taton, "Madame Du Châtelet, traductrice de Newton," *Archives internationales d'histoire des sciences* 22 (1969): 185–209; R. Debever, "La Marquise Du Châtelet traduit et commente

Cohen, though convinced that Clairaut had written the commentary, found that hers, of all the translations of Newton's work from the original Latin, was the only one useful to his own modern rendering of the classic text into English. John Iverson has documented her place in German letters, and Sonia Carboncini has described the uses made of her *Foundations* for the Leibnizian entries in the *Encyclopedia*.[45]

These scholars, along with historians of philosophy such as Mary Ellen Waithe and Paul Veatch Moriarty, have highlighted Du Châtelet's significance in maintaining the connections between metaphysical causes and mechanical explanations of the universe, the marquise's formulation of a coherent, unified system, a project in which modern cosmologists of "string theory" are now engaged. This is most evident in her *Foundations of Physics*. Coupled with the speculations on human behavior in her version of Mandeville's *The Fable of the Bees*, particularly the free translation of the section titled "Enquiry into the Origins of Moral Virtue," and her commentary on the Bible, it has been possible to study what for her must have been an underlying, interconnected view of the relationship between man and the universe, from God's initial creation to nature's laws and the evolution of early human society.[46]

Each year new documents surface. With the discovery of a draft of Voltaire's *Treatise of Metaphysics* in the Collection of Occidental Manuscripts in St. Petersburg, Andrew Brown and Ulla Kölving have been able to analyze in greater depth the dialogue between Voltaire and Du Châtelet on ques-

les *Principia* de Newton," *Bulletin de la Classe des sciences* 73, 5th series (1987): 509–27; Antoinette Emch-Dériaz and Gérard G. Emch, "On Newton's French Translator: How Faithful was Madame Du Châtelet?" in Zinsser and Hayes, eds., *Émilie Du Châtelet*.

45. John R. Iverson, "A Female Member of the Republic of Letters: Du Châtelet's Portrait in the *Bilder-Sal . . . berümhter Schriffsteller*," in Zinsser and Hayes eds., *Émilie Du Châtelet*; see Sonia Carboncini "L'Encyclopédie et Christian Wolff: A propos de quelques articles anonymes," in *Autour de la philosophie wolffienne*, ed. Jean Ecole (New York: Georg Olms Verlag, 2001), 201–16. I am grateful to John Iverson for sending me information about German and Italian scholarship on Du Châtelet.

46. Mary Ellen Waithe, "Gabrielle Émilie le Tonnelier de Breteuil Du Châtelet-Lomont," in *A History of Women Philosophers*, ed. M. E. Waithe, vol. 3 (Boston: Klewer Academic Publishers, 1991); Paul Veatch Moriarty, "The Principal of Sufficient Reason in Du Châtelet's *Institutions*," in Zinsser and Hayes, eds., *Émilie Du Châtelet*. As Du Châtelet attempted to reconcile Leibniz's metaphysics with Newton's mechanics, so modern cosmologists work to unite Einstein's theory of relativity with quantum mechanics. For a preliminary study of Du Châtelet's overall views of human behavior and humans' place in the cosmos, see: Judith P. Zinsser, "La marquise Du Châtelet: Sa morale et sa métaphysique," in *Colloque: Tricentennaire de la naissance de la marquise Du Châtelet; éclairages et documents nouveaux*, ed. Ulla Kölving and Olivier Courcelle (Ferney-Voltaire: Centre International d'Études du XVIIIe, 2008), 217–27.

tions of metaphysics and ethics.[47] Bertram E. Schwarzbach incorporated information from William Trapnell's edition of Du Châtelet's precis of Thomas Woolston's *Discourses on the Miracles of Our Savior* (1727–29), also found in St. Petersburg, into the annotation to his critical edition of her *Examinations of the Bible*. Clorinda Donato is studying the references to Du Châtelet's *Examinations* in the work of a late nineteenth-century Catholic apologist, Nicolas Sylvestre Bergier. Literary critics have related Du Châtelet's version of Mandeville to other women's emancipatory use of this kind of writing and to that of male *philosophes* who found translation to be their entrée to the Republic of Letters and to membership in the French Royal Academy.[48] Even the favorite story of Du Châtelet's previous biographers, her final affair with Saint-Lambert, has been subject to revision. The discovery of Voltaire's liaison beginning in 1745 with his niece, Mme Denis, makes him, not Du Châtelet, the first to have been disloyal. Studies of the correspondence of Mme de Graffigny suggest that Saint-Lambert was truly enamored of Du Châtelet and affected by her death in much the same way as Voltaire. Thus the previous images of the middle-aged woman importuning a reluctant younger man have to be reassessed. Anne Soprani's new edition of Du Châtelet's love letters reveals a different story: that of a thoughtful woman careful not to rush into a liaison with a young man presumed to be the Lunéville court's favorite gigolo.[49]

Each of these specialized studies enhances the possibilities for overall assessments of the marquise's contributions to the first decades of the Enlightenment and to the history of European women: that "other voice" silenced for so long. There will always be those who prefer to see Du Châtelet as Voltaire described her in his letters to intimates, in his incidental poetry, and finally in the "eulogy [*éloge*]" he wrote about her in 1752. In his words she was a genius, "mme Newton-pom pom," "his best friend," "a tyrant," "the di-

---

47. Andrew Brown and Ulla Kölving, "Qui est l'auteur du *Traité de métaphysique?*" *Cahiers Voltaire* 2 (2003): 85–93.

48. John R. Iverson and Marie-Pascale Pieretti, "Une gloire réfléchie: Du Châtelet et les stratégies de la traductrice," in *Dans les miroirs de l'écriture: La réflexivité chez les femmes écrivains d'Ancien Régime*, ed. Jean-Philippe Beaulieu and Diane Desrosiers-Bonin (Montreal: QC, 1998), 135–44; Adrienne Mason, "'L'air du climat et le goût du terroir': Translation as Cultural Capital in the Writings of Madame Du Châtelet," in Zinsser and Hayes, eds., *Émilie Du Châtelet.* J. B. A. Suard is the best known of such translators.

49. Françoise d'Issembourg d'Happoncourt, *Correspondance de madame de Graffigny*, ed. J. A. Dainard et al. (Oxford, UK: Voltaire Foundation, 1985–); see especially volumes 1, 9, 10. For the new edition of her love letters, see Anne Soprani, ed., *Émilie Du Châtelet: Lettres d'amour au marquis de Saint-Lambert* (Paris: Éditions Paris-Méditerranée, 1997).

vine Émilie," "a soul passionate for the truth," his "muse," "Urania, idol of my heart."[50] It is to be hoped, however, that this new scholarship, her inclusion in a NOVA special on Einstein, a movie of her life for French television, her own Web site, and now the translation of many of her writings will alter the historiography and lead future generations of scholars to chronicle, analyze, and represent her life and her writings in new ways.

Judith P. Zinsser

NOTE ON THE TRANSLATIONS

These translations endeavor to balance the need for fluency and accuracy in English with the desire to preserve as much of Du Châtelet's style and phrasing as possible. Inevitably, the balance tilted now one way, now another, depending on the particular text. The marquise wrote on so many varied subjects—ethics, science, metaphysics, morality—and in so many different genres. Her writings in science are extremely Latinate, with long sentences made up of many subordinate clauses. The style seems ponderous, but is characteristic of all scientific writing in the late seventeenth and early eighteenth centuries.[51] Du Châtelet intended to give clear, detailed exposition, even if that meant being painstakingly didactic and repetitious. Her syntax, then, mimics the style of reasoning, but can be difficult to follow for a twenty-first-century reader. Adding punctuation gave clarity, and having the resulting sequence of sentences begin with conjunctions such as *and* and *but*, the preposition *for* and the adverb *thus*, retained a sense of the method of argumentation.[52]

In contrast, the *Discourse on Happiness* is closer to the standard literary French of the period. Du Châtelet seems to have looked for the striking, memorable phrase and had the opportunity over the course of years to carefully craft each sentence. This was even true of her first literary production, her "Translator's Preface" for Mandeville's *The Fable of the Bees*. The challenge

---

50. In French, "la divine Émilie," and "Uranie, idole de mon coeur"; "mme Newton-pom pom" mocks her knowledge of Newton by referring to the decorative touches made to a dress and thus to her love of beautiful clothes.

51. Du Châtelet used this same style in her biblical critique, in all likelihood to give it added authority.

52. Similarly, realizing that eighteenth-century French weights and measures have no exact modern equivalents, words such as *pied* and *livre* were not translated, but their values in modern-day weights and measures were given in the footnotes. In this case, the original language meant greater accuracy.

for us was to convey the subtleties of her choices. For example, the multiple layers of meaning or association of some French words that cannot be translated by a simple English equivalent: *âme* for mind or soul, used in the Cartesian sense as the opposite of *corps*, body; the word *esprit*, which in French carries the connotations of mind, genius, spirit, energy.[53]

Du Châtelet's letters, particularly to her friends, though often equally as complex syntactically as her scientific treatises and books, are fluid, almost conversational, and the most lyrical of her writings. Others, to patrons, to members of the Republic of Letters, follow the formal conventions of her day with elaborate opening salutations and closing phrases. Only in some of her notes to Saint-Lambert does she use short, breathless, urgent clauses; but then the pace of the language suggests that she wants to convince him of her feelings by this very rhythm and not just by the meaning of her words. The task for her translators in each instance was to preserve the style occasioned by the different circumstances surrounding each missive, for the letters reflect the many facets of this remarkable woman's life and character better than any other source.[54]

Gregory Rabassa, the translator of many contemporary Latin American authors, sees the process as imitation in contrast to cloning; Umberto Eco, the Italian novelist, translator, and scholar, as creating "linguistic tokens," "the interpretation of two texts in two different languages." Susan Bassnett, the English scholar, offers the concept of translation as a form of "refraction" which, with Du Châtelet's own interest in optics, makes an appealing metaphor, the meaning crossing from one culture to another but with a slight skewing caused by the passage from one language and cultural context to another.[55]

53. Du Châtelet considered such questions in the chapter on grammar intended for the second volume of her *Foundations of Physics*. She discussed the translation of the French word *on* as *man* in English, and the use of *je* (I) in its place by French authors. The latter practice she saw as "more polite to readers with whom an author always creates a kind of conversation." The usage of *on*, she suggested, was worth a treatise. See Du Châtelet, "Grammaire raisonnée [Reasoned Grammar]," in Wade, ed., *Studies*, 219.

54. We were fortunate to have access to a draft of the revised critical edition of Du Châtelet's letters being prepared by Ulla Kölving and Anne Soprani to be published in 2009. For purposes of easy reference, the numbers for the Besterman 1958 collection (and the *Oeuvres complètes de Voltaire*), and the more recent edition of Du Châtelet's love letters to Saint-Lambert are indicated in the footnotes.

55. For discussion of these issues, see, for example, Lawrence Venturi, *The Translator's Invisibility: A History of Translation* (New York: Routledge, 1995), chap. 1, especially 1–8, 17–24; Luise von Flotow, *Translation and Gender: Translating in the "Era of Feminism"* (Ottawa: University of Ottawa Press, 1997), 20, 41; Gregory Rabassa, *If This Be Treason: Translation and Its Discontents, A Memoir* (New York: New Directions Books, 2005), 20; Umberto Eco, *Experience in Translation*, trans.

We were aware of ourselves as both readers of Du Châtelet's texts and then as conscious writers, however "invisible" (to use the literary historian Lawrence Venturi's term) we wanted to remain. Like Du Châtelet, we hoped to "remove the thorns," the difficulties that might discourage those encountering her writings for the first time. For this reason we omitted some of the most complicated mathematical sections of her scientific works even though they are the incontrovertible proof of her mastery of physics and calculus. We did not want to overwhelm her new audience and so selected carefully from her longer works and her letters. In the final analysis, perhaps Du Châtelet described our translator's role best: we chose to be "intermediaries" or *négociants*, to use her own term, of the texts between our two cultures.

Isabelle Bour and Judith P. Zinsser

Alastair McEwen (Buffalo and Toronto: University of Toronto Press, 2001), 14, 22–23; Susan Bassnatt, *Translation Studies* (New York: Routledge, 1991, rev. ed.), see, for example, 78–80, 116–20; Clifford E. Landers, *Literary Translation: A Practical Guide* (Buffalo: Multilingual Matters, 2001); Mary Ann Caws, *Surprised in Translation* (Chicago: University of Chicago Press, 2006).

VOLUME EDITOR'S
BIBLIOGRAPHY

PRIMARY SOURCES: THE WORKS OF GABRIELLE EMILIE LE TONNELIER DE BRETEUIL, MARQUISE DU CHÂTELET

Manuscripts in the Voltaire Collection, St. Petersburg, National Library of Russia, vol. 9, 122–285. The following texts published in *Studies on Voltaire, with Some Unpublished Papers of Mme. du Châtelet*. Edited by Ira O. Wade. Princeton, NJ: Princeton University Press, 1947: *The Fable of the Bees* trans., version A; Essai sur l'optique; Grammaire raisonnée, chapters 4 and 8. De la liberté, chapter 5. For a printed version of "De la liberté," see the critical edition of the *Traité de métaphysique* and Appendix I. Edited by William H. Barber in *Oeuvres complètes de Voltaire*. Vol. 14. Oxford: Voltaire Foundation, 1989. Also see from the same collection now published as: *Six discours sur les miracles de notre sauveur* by Thomas Woolston, translated by Mme Du Châtelet. Edited by William Trapnell. Paris: Honoré Champion, 2004.

The Occidental Manuscript Collection of the St. Petersburg National Library has a version of Voltaire's *Traité de métaphysique* with Du Châtelet's annotations.

Lettres de la marquise Du Châtelet. Introduction and notes by Theodore Besterman. 2 vols. Geneva: Institut et Musée de Voltaire, 1958. This edition incorporates letters published previously in 1806 and 1878, and in the article by Louise Colet, "Mme Du Châtelet: Lettres inédites au maréchal de Richelieu et à Saint-Lambert," *Revue des deux mondes* 3 (1845): 564–96. See also: *Oeuvres complètes de Voltaire: Correspondance*. Note that assorted letters are scattered in collections with her correspondents, such as Clairaut. See also the manuscript versions of letters to Maupertuis at the Bibliothèque nationale, and to d'Argental, Jurin, and Saint-Lambert at the Morgan Library. MA2287. See the new edition: *Lettres d'amour au marquis de Saint-Lambert*. Edited by Anne Soprani. Paris, Éditions Paris-Méditerranée, 1997. A new critical edition of her letters is being prepared by Ulla Kölving and Anne Soprani.

Dissertation sur la nature et la propagation du feu published in two different editions: the first version published in *Pièces qui on remporté le prix de l'Académie Royale des Sciences en MDCCXXXVIII*. Paris, 1739: 85–168, 218–19; Recueil des pièces qui on remporté les prix de l'Académie Royale des Sciences. Paris. Vol. 4 (1752): 17, 87–170, 220–21; the second by the marquise with her interchange with Dortous de Mairan, *Dissertation sur la nature et la propagation du feu*. Paris: Chez Prault Fils, 1744. The edition of 1744 was reprinted in the Landmarks of Science microprints.

"Reply to the *Voltairomanie*." 1738, in *Oeuvres complètes de Voltaire*. Vol. 89. Oxford: Voltaire Foundation, 1969: D. app. 51, 508–12.

Examen de la Genèse and *Examen des Livres du Nouveau Testament*, MSS. 2376 and 2377, Bibliothèque de Troyes. See also less polished versions: Bénitez, no. 58, Bibliothèque de l'Académie des Sciences; Brussels National Library, ms. 15188 and 15189. Critical edition by Betram Eugene Schwarzbach, *Examens de la Bible*. Paris: Honoré Champion, forthcoming, 2009.

"Lettre sur les 'Eléments de la philosophie de Newton,'" *Journal des savants* (September 1738): 534–41. Reprinted in *Oeuvres complètes de Voltaire*. Vol. 84.

Institutions de physique published in two editions: Paris: Chez Prault Fils, 1740 (with two reprints, Amsterdam and London, 1741); Amsterdam: Aux Depens de la Compagnie, 1742. The edition of 1740 was reprinted in the Landmarks of Science microprints. Chapter II appeared as "De l'existence de Dieu" in *Lettres inédites de Madame la Marquise Du Châstelet à M. le Comte D'Argental*. Paris: Xhrouet, 1806. A facsimile of the 1742 edition was reprinted in Christian Wolff, *Gesammelte Werke: Materialien und Dokumente*. Edited by J. École, H. W. Arndt, Charles. A. Corr, J. E. Hofmann, and M. Thormann. Vol. 28. Hildesheim: G. Olms, 1988. Translations were made of the 1742 edition: Italian, Venice: Giambattista Pasquali,1743; German translation by Wolf Balthasar Adolph von Steinwehr. Leipzig; in der Rengerischen Buchhandlung, 1743.

Réponse de madame la marquise du Châtelet à la lettre que M. de Mairan lui a ecrite le 18 fevrier 1741 sur la question des forces vives. Bruxelles: Foppens, 1741. Reprinted with the *Dissertation sur la nature et la propagation du feu*, 1744 ed. Included with the *Dissertation* in the Landmarks of Science microprints. Translated into Italian to accompany the Venetian edition of the *Institutions*; translated into German by Louise A. Gottsched. Leipzig, 1741.

Réflexions sur le bonheur in five reprints with only minor variations in the text: *Huitième Recueil philosophique et littéraire* (Bouillon: Société typographique de Bouillon, 1779); *Opuscules philosophiques et littéraires* (Paris: Chenet, 1796); in *Lettres inédites de madame la marquise Du Châtelet à m. le comte D'Argental* (Paris: Xhrouet, 1806); a critical edition by Robert Mauzi. Paris: Société d'édition Belles Lettres, 1961; and a reprint with an introduction by Elisabeth Badinter. Paris: Rivages poche, 1997.

Principes mathématiques de la philosophie naturelle par feue madame la marquise Du Chastelet. 2 vols. Paris: Desaint et Saillant, 1759. Facsimile reprint of incomplete 1756 edition, Paris: Blanchard, 1966; of complete 1759 edition, Sceaux: Éditions Jacques Gabay, 1990.

OTHER PRIMARY SOURCES

Bertrand, Jean. *La Fable des Abeilles*. London, 1740.

The New Oxford Annotated Bible with the Apocrypha. Edited by Herbert G. May and Bruce M. Metzger. New York: Oxford University Press, 1977.

Breteuil, Baron de. *Mémoires*. Edited by Evelyne Lever. Paris: F. Bourin, 1992.

Créqui, Marquise de. *Souvenirs de la marquise de Créqui, 1710–1803*. 5 vols. Paris: Michel Lévy Frères, 1867.

Clairaut, Alexis-Claude. Correspondance in (Prince Baldassare de) Boncompagni, "Lettere di Alessio Claudio Clairaut." *Atti dell' Accademia Pontificia de' Nuovi Lincei* 45 (1892): 233–91.

————. *Elémens d'algèbre*. Paris: Chez les Frères Guerin, 1746.

————. *Elémens de géometrie*. Paris: Chez David Fils, 1741.

Descartes, René. *Discourse on Method*. Translated by John Veitch. In *The Rationalists*. Garden City, NY: Dolphin Books, 1962.

Diderot, Denis. *Encyclopédie; ou, Dictionnaire raisonné des sciences, des arts et des métiers par une société de gens de lettres; mis en ordre & publié par M. Diderot . . . ; & quant á la partie mathématique par M. D'Alembert*. 35 vols. Lausanne: Les Societés Typographiques, 1778–82.

Du Deffand, Marie, Marquise. *Correspondance complète de la marquise Du Deffand avec ses amis*. 2 vols. Paris, 1865.

Fontenelle, Bernard le Bovier de. *Conversations on the Plurality of Worlds*. With an introduction by Nina Rattner Gelbart. Translated by H. A. Hargreaves. Berkeley: University of California Press, 1990.

Graffigny, Françoise d'Issembourg d'Happoncourt. *Correspondance de madame de Graffigny*. Edited by J. A. Dainard et al. 12 vols. Oxford: Voltaire Foundation, 1985– .

Hénault, Charles Jean François. *Mémoires du Président Hénault*. Edited by François Rousseau. Paris: Hachette, 1911.

Jordan, Charles-Etienne. *Histoire d'un voyage littéraire fait en MDCCXXXIII*. The Hague: Adorren Hoetiers, 1735.

Longchamps, Sébastien G., and Jean Louis Wagnière. *Mémoires sur Voltaire, et sur ses ouvrages*. Paris: Aimie André, 1826.

Luynes, Charles Philippe d'Albert, Duc de. *Mémoires du duc de Luynes sur la cour de Louis XV, 1735–1758*. Edited by L. Dussieux and E. Soulié. 17 vols. Paris: Firmin Didot, 1860–65.

Mandeville, Bernard. *The Fable of the Bees; or, Private Vices, Publick Benefits*. London: J. Roberts, 1714.

Maupertuis, Pierre-Louis Moreau de. *Discours sur les différentes figures des astres*. Paris: De L'Imprimerie royale, 1732.

Mercier, Louis Sébastien. *Le Paris de Louis Sébastien Mercier: Cartes et index toponymique*. Edited by Jean-Claude Bonnet. Paris: Mercure de France, 1994.

Missy, Jean Rousset de. "Treatise of the Three Imposters." In *The Enlightenment: A Brief History in Documents*, edited by Margaret C. Jacob. New York: Bedford/St. Martin's, 2001.

Montaigne, Michel de. *The Complete Works of Montaigne*. Translated by Donald Frame. Stanford, CA: Stanford University Press, 1957.

Newton, Isaac. *Opticks: A Treatise of the Reflections, Refractions, Inflections & Colors of Light*. (4th edition, London, 1730.) New York: Dover Publications, 1952.

————. *The Principia: Mathematical Principles of Natural Philosophy*. Translated by I. Bernard Cohen and Anne Whitman, assisted by Julia Budenz. Berkeley: University of California Press, 1999.

Pope, Alexander. *An Essay on Man*. In *The Best of Pope*, edited by George Sherburn, 114–54. New York: Ronald Press Company, 1929.

Sainte-Beuve, Charles Augustin. *Causeries du lundi: Portraits de femmes et portraits littéraires*. 14 vols. Paris: Garnier Frères, 1852–62.

Voltaire. "Eloge historique de Madame du Chastelet." *Bibliothèque impartiale* 5 (January–February 1752): 136–46.

————. *Mémoires*. Edited by Jacqueline Hellegouarc'h. Paris: Livre de Poche, 1998.

————. *Oeuvres complètes de Voltaire*. Edited by Theodore Besterman et al. Geneva and Oxford: Institut et Musée Voltaire and Voltaire Foundation, 1968–.

SECONDARY SOURCES

Backer, Dorothy Anne Liot. *Precious Women: A Feminist Phenomenon in the Age of Louis XIV*. New York: Basic Books, 1974.

Badinter, Elisabeth. *Émilie, Émilie: L'ambition féminine au XVIIIe siècle*. Paris: Flammarion, 1983.

————. *Les Passions intellectuelles*. 2 vols. Paris: Fayard, 1999.

Barber, W. H. "Mme. Du Châtelet and Leibnizianism: The Genesis of the *Institutions de physique*." In *The Age of Enlightenment: Studies Presented to Theodore Besterman*, edited W. H. Barber et al., 200–222. Edinburgh: Oliver and Boyd, 1967.

Bassnett, Susan. *Translation Studies*. New York: Routledge, 1991.

Bodanis, David. *E=MC2: A Biography of the World's Most Famous Equation*. New York: Berkley Books, 2000.

Bonnel, Roland. "La correspondance scientifique de la marquise Du Châtelet: La 'lettre-laboratoire.'" *SVEC* (Studies on Voltaire and the Eighteenth Century) 4 (2000): 79–95.

Brown, Andrew, and Ulla Kölving. "Qui est l'auteur du *Traité de metaphysique?*" *Cahiers Voltaire* 2 (2003): 85–93.

Carboncini, Sonia. "L'Encyclopédie et Christian Wolff: À propos de quelques articles anonymes." In *Autour de la philosophie wolffienne*, edited by Jean Ecole, 210–16. New York: Georg Olms Verlag, 2001.

Carroll, Berenice A. "The Politics of 'Originality': Women and the Class System of the Intellect." *Journal of Women's History* 2, no. 2 (1990): 136–63.

Chapman, Sara. "Patronage as Family Economy: The Role of Women in the Patron-Client Network of the Phélypeaux de Pontchartrain Family, 1670–1715." *French Historical Studies* 24, no. 1 (2001): 11–35.

Charles-Gaffiot, Jacques. *Lunéville: Fastes du Versailles Lorrain*. N.p.: Editions Didier Carpentier, 2003.

Chaussinand-Nogaret, Guy. *The French Nobility in the Eighteenth Century: From Feudalism to Enlightenment*. Translated by William Doyle. Cambridge: Cambridge University Press, 1995.

Cohen, Bernard. "A Guide to Isaac Newton's *Principia*." In Isaac Newton's *The Principia: Mathematical Principles of Natural Philosophy*. Translated by I. Bernard Cohen and Anne Whitman, assisted by Julia Budenz. Berkeley: University of California Press, 1999.

Craveri, Benedetta. *The Age of Conversation*. Translated by Teresa Waugh. New York: New York Review Books, 2005.

Debever, R. "La marquise Du Châtelet traduit et commente les *Principia* de Newton." *Bulletin de la classe des sciences* 73, 5th series (1987): 509–27.

De Clercq, Peter. *The Leiden Cabinet of Physics: A Descriptive Catalogue*. Leiden: Museum Boerhaave Communication 271, 1997.

De Gandt, François, ed. *Cirey dans la vie intellectuelle: La réception de Newton en France*. *SVEC* 11 (2001).

DeJean, Joan. *Tender Geographies: Women and the Origins of the Novel in France*. New York: Columbia University Press, 1991.

Delisle, Jean. *Translation: An Interpretive Approach*. Translated by Patricia Logan and Monica Creery. London, Ottawa: University of Ottawa Press, 1988.

Delpierre, Madeleine. *Dress in France in the Eighteenth Century*. Translated by Caroline Beamish. New Haven, CT: Yale University Press, 1997.

De Zan, Mauro. "Voltaire e mme Du Châtelet membri e correspondenti dell' Academia delle Scienze di Bologna." *Studi e memorie dell' Instituto per la Storia dell' Università di Bologna* 6 (1987): 141–58.

Eco, Umberto. *Experiences in Translation*. Translated by Alastair McEwen. Toronto, Buffalo: University of Toronto Press, 2001.

Edwards, Samuel. *The Divine Mistress*. New York: David McKay, 1970.

Ehrman, Esther. *Mme Du Châtelet: Scientist, Philosopher, and Feminist of the Enlightenment*. Leamington Spa, UK: Berg, 1986.

Emilie Du Châtelet: Rewriting Enlightenment Philosophy and Science. Edited by Judith P. Zinsser and Julie Candler Hayes. *SVEC* 1 (2006).

Fairchilds, Cissie. *Domestic Enemies: Servants and their Masters in Old Regime France*. Baltimore: Johns Hopkins University Press, 1984.

Fara, Patricia. *Pandora's Breeches: Women, Science, and Power in the Enlightenment*. London: Pimlico, 2004.

Feingold, Mordechai. *The Newtonian Moment: Isaac Newton and the Making of Modern Culture*. New York: Oxford University Press, 2004.

Goldgar, Anne. *Impolite Learning: Conduct and Community in the Republic of Letters, 1680–1750*. New Haven, CT: Yale University Press, 1995.

Goncourt, Edmond, and Jules de Goncourt. *La femme au dix-huitième siècle*. Preface by Elisabeth Badminter. Paris: Flammarion, 1982.

Goodman, Dena. *The Republic of Letters: A Cultural History of the French Enlightenment*. Ithaca, NY: Cornell University Press, 1994.

Gordon, Daniel. *Citizens without Sovereignty: Equality and Sociability in French Thought, 1670–1780*. Princeton NJ: Princeton University Press, 1994.

Greene, Brian. *The Elegant Universe: Superstrings, Hidden Dimensions, and the Quest for the Ultimate Theory*. New York: W. W. Norton, 1999.

Hahn, Roger. *The Anatomy of a Scientific Institution: The Paris Academy of Science, 1666–1803*. Berkeley: University of California Press, 1971.

Harth, Erica. *Cartesian Women: Versions and Subversions of Rational Discourse in the Old Regime*. Ithaca, NY: Cornell University Press, 1992.

Hayes, Julie Candler. *Reading the Enlightenment: System and Subversion*. New York: Cambridge University Press, 1999.

Hecht, Helmut. "Leibniz's Concepts of Possible Worlds and the Analysis of Motion in Eighteenth-Century Physics." In *Between Leibniz, Newton, and Kant: Philosophy and Science in the Eighteenth Century*, edited by Wolfgang Lefevre, 27–45. Boston: Kluwer Academic Publishers, 2001.

Hutton, Sarah. "Émilie Du Châtelet's *Institutions de physique* as a Document in the History of French Newtonianism." *Studies in the History and Philosophy of Science* 35 (2004): 515–31.

Iltis, Carolyn Merchant. "Madame du Châtelet's Metaphysics and Mechanics." *Studies in the History and Philosophy of Science* 8, no. 1 (1977): 28–48.

Iverson, John, and Marie-Pascale Pieretti. "Une gloire réfléchie: Du Châtelet et les stratégies de la traductrice." In *Dans les miroirs de l'écriture; La réflexivité chez les femmes écrivains d'Ancien Régime*, edited by Jean-Philippe Beaulieu and Diane Desrosiers, 135–44. Montreal: Département d'études françaises, Université de Montreal, 1998.

———. "'Toutes personnes . . . seront admises à concourir': La participation des femmes aux concours académiques." *Dix-huitième Siècle* 36 (2004): 313–32.

Janik, Linda Gardiner. "Searching for the Metaphysics of Science: The Structure and Composition of Madame Du Châtelet's *Institutions de physique*, 1737–1740." *SVEC* 201 (1982): 85–113.

Joly, Bernard. "Les théories du feu de Voltaire et de mme Du Châtelet." *SVEC* 11 (2001): 212–37.

Kawashima, Keiko. "Madame Du Châtelet dans le journalisme." *LLULL* 18 (1995): 471–91.

Kronturis, Tina. *Oppositional Voices: Women as Writers and Translators of Literature in the English Renaissance*. New York: Routledge, 1992.

Lancaster, Henry Carrington. *The Comédie française, 1701–1774: Plays, Actors, Spectators, Finances*. Transactions of the American Philosophical Society Vol. 41. Philadelphia: American Philosophical Society, 1951.

Landers, Clifford E. *Literary Translation: A Practical Guide*. Buffalo: Multilingual Matters, 2001.

Ledeuil-d'Enquin, Justin. *La marquise Du Châtelet à Sémur et le passage de Voltaire*. Sémur: Millon, 1892.

Lilti, Antoine. *Le Monde des salons: Sociabilité et mondanité à Paris au XVIIIe siècle*. Paris: Fayard, 2005.

Lynn, Michael R. "Enlightenment in the Republic of Science: The Popularization of Natural Philosophy in Eighteenth-Century Paris." PhD dissertation, University of Wisconsin, Madison, 1997.

Maclean, Ian. *Woman Triumphant: Feminism in French Literature, 1610–1652*. Oxford: Clarendon Press, 1977.

Mangeot, George. *La famille de Saint-Lambert, 1596–1795*. Paris: Libraire Croville-Morant, 1913.

Masson, Pierre-Maurice. *Madame de Tencin (1682–1749): Une Vie de femme au XVIIIe siècle*. Geneva: Slatkin Reprints, 1970.

Maugras, Gaston. *La Cour de Lunéville au XVIIIe siècle*. Paris: Plon-Nourrit, 1904.

Mitford, Nancy. *Voltaire in Love*. London: Hamish Hamilton, 1957.

Modern Women Philosophers, 1600–1900. Edited by Mary Ellen Waithe. Boston: Kluwer Academic Publishers, 1991.

Muratori-Philip, Anne. *Le Roi Stanislas*. Paris: Fayard, 2000.

Neuschal, Kristen Brooke. *Word of Honor: Interpreting Noble Culture in Sixteenth-Century France*. Ithaca, NY: Cornell University Press, 1989.

Newton, William R. *L'Espace du roi: La Cour de France au château de Versailles, 1682–1789*. Paris: Fayard, 2000.

O'Keefe, Cyril B. *Contemporary Reactions to the Enlightenment (1728–1762): A Study of Three Critical Journals, the Jesuit Journal de Trévoux, the Jansenist Nouvelles Écclésiastiques, and the Secular Journal des Savants*. Geneva: Slatkin, 1974.

Petrovich, Vesna C. "Women and the Paris Academy of Sciences." *Eighteenth-Century Studies* 32, no. 3 (1999): 383–90.

Picard, Roger. *Les Salons littéraires et la société française, 1610–1789.* Paris, New York: Brentano's, 1943.

Piot, l'abbé. *Cirey-le-Château; La marquise Du Châtelet: Sa liaison avec Voltaire.* Saint-Dizier: O. Godard, 1894.

Pomeau, René. *Voltaire en son temps.* 2 vols. Oxford: Voltaire Foundation, 1992.

Rabassa, Gregory. *If This Be Treason: Translation and Its Discontents; A Memoir.* New York: New Directions Books, 2005.

The Rise of Modern Philosophy: The Tension between the New and Traditional Philosophies from Machiavelli to Leibniz. Edited by Tom Sorrell. Oxford: Clarendon Press, 1993.

Rizzo, Tracey. *A Certain Emancipation of Women: Gender, Citizenship, and the Causes Célèbres of Eighteenth-Century France.* Selinsgrove, PA: Susquehanna University Press, 2004.

Roche, Daniel. *The Culture of Clothing: Dress and Fashion in the Ancien Régime.* Translated by Jean Birrell. Cambridge: Cambridge University Press, 1994.

Saget, Hubert. *Voltaire à Cirey.* Chaumont: Le Pythagore, 2005.

Sargentson, Carolyn. *Merchants and Luxury Markets: The Marchands Merciers of Eighteenth-Century Paris.* London: Victoria and Albert Museum in association with J. Paul Getty Museum, 1996.

Schaffer, Simon. "Glass Works." In *The Uses of Experiment: Studies in the Natural Sciences,* edited by David Gooding, Trevor Pinch, and Simon Schaffer, 67–104. New York: Cambridge University Press, 1989.

———. "Halley, Delisle, and the Making of the Comet." In *Standing on the Shoulders of Giants: A Longer View of Newton and Halley,* edited by Norman J. W. Thrower, 254–92. Berkeley: University of California Press, 1990.

———. "Self-Evidence." In *Questions of Evidence: Proof, Practice, and Persuasion across the Disciplines,* edited by James Chandler, Arnold I. Davidson, and Harry Harootunian, 56–91. Chicago: University of Chicago Press, 1994.

Schiebinger, Londa. *The Mind Has No Sex? Women in the Origins of Modern Science.* Cambridge MA: Harvard University Press, 1989.

The Sciences in Enlightened Europe. Edited by William Clark, Jan Golinski, and Simon Schaffer. Chicago: University of Chicago Press, 1999.

Sgard, Jean. *Dictionnaire des Journaux, 1600–1785.* 2 vols. Oxford: Voltaire Foundation, 1991.

Shank, John Bennett. *Before Voltaire: Newtonianism and the Origins of the Enlightenment in France, 1687–1734.* Ann Arbor, MI: UMI, 2000.

Shapin, Steven. "The House of Experiment in Seventeenth-Century England." In *The Scientific Enterprise in Early Modern Europe,* edited by Peter Dear, 273–304. Chicago: University of Chicago Press, 1997.

Shapin, Steven, and Simon Schaffer. *Leviathan and the Air-Pump: Hobbes, Boyle, and the Experimental Life.* Princeton, NJ: Princeton University Press, 1985.

Simonin, Charlotte, and David W. Smith. "Du nouveau sur Mme Denis: Les apports de la correspondence de Mme de Graffigny." *Cahiers Voltaire* 4 (2005): 25–56.

Smith, David W. "Nouveaux regards sur la brève rencontre entre mme Du Châtelet et Saint-Lambert." In *The Enterprise of Enlightenment: A Tribute to David Williams,* edited by Terry Pratt and David McCallam, 329–43. Oxford: Peter Lang, 2004.

Smith, George E. "The Newtonian Style in Book II of the *Principia.*" In *Isaac New-ton's Natural Philosophy*, edited by Jed Z. Buchwald and I. Bernard Cohen, 249–98. Cambridge, MA: MIT Press, 2001.

Solnon, Jean-François. *La Cour de France.* Paris: Fayard, 1987.

Sonnet, Martine. *L'Education des filles au temps des Lumières.* Preface by Daniel Roche. Paris: Les Éditions du Cerf, 1987.

Spink, J. S. *French Free-Thought from Gassendi to Voltaire.* London: University of London, Athlone Press, 1960.

Steinbrugge, Lieselotte. *The Moral Sex: Woman's Nature in the French World.* Translated by Pamela E. Selwyn. New York: Oxford University Press, 1995.

Sturdy, David J. *Science and Social Status: The Members of the Académie des Sciences, 1666– 1750.* Woodbridge, Suffolk: Boydell Press, 1995.

Sutton, Geoffrey V. *Science for a Polite Society: Gender, Culture, and the Demonstration of En-lightenment.* Boulder, CO: Westview Press, 1995.

Taton, René. "Mme Du Châtelet, traductrice de Newton." *Archives internationales d'histoire des sciences* 22 (1969): 185–209.

Terrall, Mary. "Émilie du Châtelet and the Gendering of Science." *History of Science* 33 (1995): 283–310.

———. "Gendered Spaces, Gendered Audiences: Inside and Outside the Paris Acad-emy of Sciences." *Configurations* 2 (1994): 207–32.

———. *The Man Who Flattened the Earth: Maupertuis and the Sciences in the Enlightenment.* Chicago: University of Chicago Press, 2002.

———. "Vis Viva." *History of Science* 42 (2004): 189–209.

Tronquart, Martine. *Les Châteaux de Lunéville.* Metz: Editions Serpenoise, 1991.

Vaillot, René. *Madame Du Châtelet.* Preface by René Pomeau. Paris: Albin Michel, 1978.

Venuti, Lawrence. *The Translator's Invisibility: A History of Translation.* New York: Rout-ledge, 1995.

von Flotow, Luise. *Translation and Gender: Translating in the "Era of Feminism."* Ontario: Uni-versity of Ottawa Press, 1997.

Wade, Ira O. *The Clandestine Organization and Diffusion of Philosophic Ideas in France from 1700 to 1750.* Princeton, NJ: Princeton University Press, 1938.

———. *Voltaire and Madame du Châtelet: An Essay on the Intellectual Activity at Cirey.* Princeton, NJ: Princeton University Press, 1941.

———, ed. *Studies on Voltaire, with Some Unpublished Papers of Mme. Du Châtelet.* Princeton NJ: Princeton University Press, 1947.

Waff, Craig B. "Comet Halley's First Expected Return: English Public Apprehensions, 1755–58." *Journal for the History of Astronomy* 17, no. 1 (1986): 1–37.

Walters, Robert L. "The Allegorical Engravings in the Ledet-Desbordes Edition of the *Eléments de la philosophie de Newton.*" In *Voltaire and His World: Studies Presented to W. H. Barber*, edited by R. J. Howells, A. Mason, H. T. Mason, and D. Williams. Oxford: Voltaire Foundation, 1985.

———. "Chemistry at Cirey." *SVEC* 58 (1967): 1807–27.

Walters, Robert L., and W. H. Barber. "Introduction." *Éléments de la philosophie de Newton.* Vol. 15. Oxford: Voltaire Foundation, 1992.

Writing the Female Voice: Essays on Epistolary Literature. Edited by Elizabeth C. Goldsmith. Boston: Northeastern University Press, 1989.

Zinsser, Judith P. *Emilie Du Châtelet: Daring Genius of the Enlightenment* (originally published as *La Dame d'Esprit: A Biography of the Marquise Du Châtelet*). New York: Penguin, 2006.

———. "Emilie Du Châtelet: Genius, Gender, and Intellectual Authority." In *Women Writers and the Early Modern British Political Tradition*, edited by Hilda L. Smith, 168–90. New York: Cambridge University Press, 1998.

———. "Entrepreneur of the "Republic of Letters": Emilie de Breteuil, Marquise Du Châtelet, and Bernard Mandeville's *The Fable of the Bees*." *French Historical Studies* 25, no. 4 (2002): 595–624.

———. "The Many Representations of the Marquise Du Châtelet." In *Men, Women, and the Birthing of Modern Science*, edited by Judith P. Zinsser, 48–70. DeKalb: Northern Illinois University Press, 2005.

———. "La Marquise Du Châtelet: Sa morale et sa métaphysique." *Colloque: Tricentennaire de la naissance de la marquise Du Châtelet*. Geneva: Center for Eighteenth-Century Studies, 2008.

———. "Mentors, the Marquise Du Châtelet, and Historical Memory." *Notes and Records of the Royal Society* 61 (2007): 89–108.

———. "Translating Newton's *Principia*: The Marquise Du Châtelet's Revisions and Additions for a French Audience." *Notes and Records of the Royal Society* 55, no. 2 (2001): 227–45.

Zinsser, Judith P., and Olivier Courcelle. "A Remarkable Collaboration: The Marquise Du Châtelet and Alexis Clairant." *SVEC* 13 (2003): 107–20.

I

BERNARD MANDEVILLE'S
THE FABLE OF THE BEES

VOLUME EDITOR'S INTRODUCTION

The first selection is the preface Du Châtelet wrote for her translation of selections from Bernard Mandeville's *The Fable of the Bees*. She was then living at Cirey, the Du Châtelet family estate, more or less continuously from 1735 until 1739. Mandeville was a physician, a Dutch immigrant to London in the 1690s, who became famous for biting, satirical critiques of society. The work had been reprinted many times before Du Châtelet decided to translate parts of it. *The Fable of the Bees*, in fact, consisted of three parts: "The Grumbling Hive," a long verse allegory; an essay, "Enquiry into the Origins of Moral Virtue"; and explanatory "Remarks," which Mandeville used to expand upon the meaning of selected couplets from the opening poem.

This was her first extended effort at translation from English into French, though notes from other reading in English show her interest in a number of British authors. She chose this particular work because, as she explained in her preface, it was "the best book of morality ever done." As such, her very free translation forms part of her explorations of "la morale" (the rules of conduct permitted and practiced in a society) and "la métaphysique" (the study of first causes and the nature of being). Du Châtelet agreed with other translators of her era, such as Anne Dacier and Pierre Coste, that a text could be altered to suit a French audience. As she explained to her readers, she believed verse too difficult and so did not translate the verse allegory. She condensed the "Enquiry" and made it her chapter one. Of the "Remarks," she only chose to do some, and those that she included show deletions and additions.[1] Most noticeable is her incorporation of the golden rule despite

---

1. A table of contents for Du Châtelet's translation appears at the end of this selection with the corresponding section in *The Fable of the Bees* indicated.

the fact that it contradicts Mandeville's central thesis—that public good results from private vice—which otherwise she never questions and for which she adds her own examples. Unlike her contemporaries, however, she did, in most instances, indicate when she had included her own views on a particular point by setting off the paragraphs with quotation marks. In the end, she created a book of ethics that is very much her work, a transposition of Mandeville's thought, rather than a translation as understood in the twenty-first century.[2]

Du Châtelet's "Translator's Preface" is remarkable in that it gives not only her philosophy of translation but also her thoughts on elite women's education and the obstacles placed in the way of their intellectual advancement. At twenty-eight, a young, privileged, married woman with two children, she was not a feminist, but she was eager and intense about her desire to read, study, and participate on equal terms in what she and her contemporaries called the Republic of Letters. It is evident from comments in her "Translator's Preface" that Du Châtelet intended her work to be published. She perhaps conceived of it as her first contribution to these cosmopolitan, transnational interchanges. That it was not remains a mystery.[3] Instead, it only exists in multiple drafts, part of the collection of Voltaire's papers and books sold to Catherine the Great of Russia at the time of his death.

The "Translator's Preface" presented here is what was probably the last of four drafts now in the Voltaire Collection at the St. Petersburg National Library. Ira O. Wade first published it as an appendix to *Voltaire and Madame Du Châtelet: An Essay on the Intellectual Activity at Cirey* (1941) and again with minimal annotation in his *Studies on Voltaire, with Some Unpublished Papers of Mme Du Châtelet* (1947).

The two letters that precede the "Preface" present a picture of Du Châtelet's activities as a young, married woman in Paris, of "les choses frivoles," the frivolous things that she renounced when the attentions of "men of letters" convinced her that she too was "a thinking creature." These were the times when she frequented supper parties at the Villars-Brancas after going to the theater or the opera with her older friend, Marguerite Thérèse Colbert de Croissy, Duchess de Saint-Pierre. Louis de Brancas de Forcalquier was a member of this family and the duchess's younger lover. He and Louis

---

2. For a discussion of eighteenth-century translation and of this text, see Judith P. Zinsser, "Entrepreneur of the 'Republic of Letters': Emilie de Breteuil, Marquise Du Châtelet, and Bernard Mandeville's *The Fable of the Bees," French Historical Studies* 24, no. 4 (2002): 595–624.

3. The work was controversial, and this may explain why it circulated only in manuscript, if at all. When a literal French translation by Jean Bertrand was published in 1740 the French royal authorities condemned it.

François Armand Du Plessis, Duke de Richelieu, Du Châtelet's relative by marriage, were then on active service in the Royal Army, thus the reason for her letters to them. Keeping one's friends abreast of the news was a common practice when they had to be away from Paris and Versailles. It is also evident that Du Châtelet had sought Richelieu's advice about her decision to join Voltaire at Cirey. This would make public her liaison with the controversial poet, and only if her husband accepted him officially into their household would the proprieties be observed.

Note the formal phrasing of the opening and closing sentences of Du Châtelet's letters, especially in contrast to other teasing, affectionate remarks, the references to acquaintances by nicknames, and the concern about health in the centuries before modern medicine. All were characteristic of correspondence in this era between women and men of Du Châtelet's world.

The letter to Forcalquier will be published for the first time in the new edition of Du Châtelet's letters edited by Ulla Kölving and Anne Soprani and was translated for this volume with their kind permission. The letter to Richelieu is from the Besterman collection.

⁂

RELATED LETTERS

To Louis de Brancas de Forcalquier, Marquis de Céreste[4]

I have not yet found, monsieur, the imagination required to answer you; but I have no need of it to show you my extreme anxiety for your health. My affection and the interest I take in your concerns will serve me in place of wit.

The people who tell me about you have been at the court of *la fée Boscotte* for a week. I arrived from Burgundy precisely at the time when Mme de Feriol left to go there.[5] I give myself the pleasure of filling the void left in their

---

4. This letter will be included in the new edition of Du Châtelet's letters edited by Ulla Kölving and Anne Soprani. It is translated here by their kind permission.

5. Probably Jeanne Grâce Bosc Du Bouchet, wife of Charles-Augustin de Feriol, Count d'Argental, a councilor to the Paris court, the *parlement*. His aunt was Mme de Tencin, and so he and his wife would have been well connected in the social worlds of Paris and Versailles. "La fée Boscotte" is a nickname for one of their contemporaries, possibly taken from a well-known fairy tale. I am grateful to Penny Arthur of the Graffigny Project for this suggestion.

circle by her absence. We constantly speak of you. We miss you in our little *loge*, that is scarcely inhabited since we lost hope of receiving you there. And if I do not speak nonsense in as good faith as the good Feriol, I surely do not surrender to her anything of the pleasure that I find speaking of you with a friend with whom I share very fondly the anxieties her affection for you occasions. This occupation has up to now stood me in stead of the pleasure of writing to you. But since she has left, I found that it was better to write a dull letter to you and to have your news than to live in an ignorance so disquieting; and that it would be better to tell you in vile prose how much your health concerns me than to let you doubt my affection. On this occasion, I have sacrificed my pride; it is necessary to shed it when one lives with you, but I find that one benefits enough from hearing from you, and that it would be too presumptuous to pretend to answer someone like you.

In the evening I was witness to the reception of Mme de Richelieu at her beautiful *hôtel*.[6] It looks like the palace of Psyche, but love was lacking there. Twenty servants, none of whom she knows, passed before her, and gave her a charming supper, accompanied by the most elegant lighting; the most beautiful apartment in the world, lit with one hundred candles and filled with music. This had all the brilliance of the gayest fetes but without the stiffness or the crush, for there were only seven of us, she included.

You undoubtedly know of Voltaire's trip to the camp. He tells me how distressed he is not to have been able to see you. If you had stayed at Spire, he intended to pay his court to you there. I do not know where things stand with his affairs. It seems to me, from M. de Richelieu's letters, that he hopes that he will be done with them soon. It is said here that he was taken for a spy and that he feared he might be hanged. It would be sad to be burned in Paris and hanged in Philippsbourg.[7]

At Paris, this 12 July [1734]

Mme d'Autrey is dying, as you know, but in such an affecting struggle that your uncle could not resist her charms. He spends his life here. I believe her

6. The word *hôtel* has no easy equivalent in English; it signifies an elegant residence, a townhouse, consisting of a number of apartments, or suites of rooms. The *hôtel* Richelieu in the Place Royale (now the Place des Vosges) made the Richelieus neighbors of Du Châtelet's family when she was growing up. The Duke de Richelieu had just married Elisabeth Sophie, Princess de Guise.

7. France was then engaged in the War of Polish Succession (1733–38). Voltaire visited the military camp at Philippsbourg to see the Duke de Richelieu when he was fleeing from a government order of arrest occasioned by the unauthorized publication in Rouen of his *Philosophical Letters*. Condemned by the government and the Paris *parlement*, the book was ritually burned in the Place de Grèves by the public executioner.

pride is quite satisfied to win him without rouge and in the horrors of convulsions, a victory that had eluded all the splendor of her charms and the
roses with which she concealed them.[8]

A short piece is playing called *La Pupille* [The Ward], said to be by Pont
de Veyle and which is a great success; I have not seen it.[9] I will surely not
speak to you either of Italy or of Danzig; I am not political and Germany
is the only object of my curiosity. If something may console me for the de-
rangement of your health, it is that at least it shelters you from cannon balls
and puts a curb on your warrior's ardor. Keep yourself safe for your friends,
monsieur, and please believe that no one takes a truer interest than I.

I am delighted by the good you tell me of my friend Beauvau.[10] I like
saying pleasant things to him too much, not to take advantage of the first
opportunity I find to tell it to him. It is people like him who are worthy of
loving and pleasing you, because he can feel all that you are worth, without
being jealous; very few people deserve this praise.

To Louis François Armand Du Plessis, Duke de Richelieu[11]

[about the 15th of June 1735]

I love the gossip of the heart as much as I hate that of the mind. So, since you
give me carte blanche, I believe that my letters will become in-folio volumes.
Yours came at just the right time. I was going to write to you, first in order
to write to you, and next to tell you that this is what you are like: that you
love people eight days, that you have carried on a flirtation with me, but that
I who take affectionate friendship[12] as the most serious thing in the world,
and who love you truly, have been worried by your silence, have been distressed by it. I have said to myself: "one should," it is said, "love one's friends
with their faults. M. de R. is flighty and changeable, one must love him such
as he is." I felt that my heart could gain nothing from this bargain. I well
knew that I might have been happy, and believed you incapable of friendship, but I could not without chagrin give up this beautiful dream of making

8. Marie-Thérèse Fleuriau, Countess d'Autrey, was another member of the Villars-Brancas
circle. At court, both women and men rouged their cheeks.

9. Du Châtelet did subsequently see the play written by d'Argental's brother, Antoine Feriol,
Count de Pont-de-Veyle.

10. Charles Just de Beauvau, Prince de Beauvau-Craon, a courtier from Lorraine, but well connected to the royal court at Versailles.

11. No. 38 in the Besterman collection, vol. 1, 72–78; D876 of *Oeuvres complètes*.

12. Du Châtelet uses the word "amitié," which conveys the ideas of affection and friendship.

you a friend, you who are believed to be only made for flirtation, you whom I never would have taken it into my head to love, but whose friendship I can no longer do without. These are the ideas that occupied my mind while you were, as you claim, *blocked*. You give me such a comical description of the state you were in that, if I were not concerned about your health, I would tell you that your letters are at once tender and pleasing, two things which ordinarily do not go together. But let me warn you that, though you may describe your melancholy with all the grace and gaiety in the world, this is no excuse for not writing to me. It is the privilege of affection to see a friend in all the situations of his soul. I love you sad, gay, lively, blocked; I want my friendly feelings to add to your pleasures and diminish your troubles, and I want to share them. There is no need for this to experience true misfortunes or great pleasures; I need no happenings, and I am as interested in your vapors and your flirtations as others are interested in the happiness or sadness of the people whom they call friends. I acknowledge that it is best to see one's lover with rouge on than not to see him at all. I, for example, will substantiate my claim: I have extremely confused ideas this evening; I feel that I am not eloquent, but the desire to communicate my ideas to you, however confused they are, makes me suppress my pride.

Here is a story designed to save you a sentence that I was about to begin and which would have had no end. . . .

[Du Châtelet quickly changes the subject and gives the duke amusing gossip of common acquaintances. She returns, however, to the topic she said she would avoid.]

It is impossible that you should make these disastrous predictions; otherwise, I would be all too afraid of the one you dwelt on. There are on this point things I have never told to anyone, not you, not even V.[13] There is heroism or perhaps madness in my shutting myself up at Cirey, with the three of us. However, the decision has been made. I believe myself yet more able to destroy the suspicions of my husband than to stop V.'s imagination. In Paris, I would lose him irretrievably and without remedy; at Cirey, I can at least hope that love will thicken the veil that should, for his happiness and for ours, cover my husband's eyes. I ask you, as a favor, to convey nothing of this to V. His head would spin with anxiety and I fear nothing as much as distressing him, above all pointlessly. Do not for-

13. Du Châtelet commonly refers to Voltaire as "V" in her letters. Richelieu seems to have suggested that she was not serious in her feelings for Voltaire and that her husband would never approve of the poet joining their household as more than an occasional guest seeking refuge from the authorities.

get your eloquence with my husband,[14] and prepare yourself to love me unhappy if ever I become so. To make sure that I am not wholly so, I am about to spend the three happiest months of my life. I leave in four days and it is in the midst of the inconveniences of departure that I dare write to you. My mind is overwhelmed, but my heart swims in joy. The hope that this step will persuade him that I love him obscures all other ideas, and I can only see the extreme happiness of curing all his fears and of spending my life with him.

You see that you are wrong; I am certainly infatuated, and yet I confess to you that his anxieties and his suspicions distress me. I know that this makes the torment of his life; this in turn poisons mine, but we might well both be right. There is much difference between jealousy and the fear of not being loved enough: one can brave the one when one feels that one does not merit it, but one cannot help being touched and distressed by the other. Jealousy is an annoying feeling, and the fear of it a delicate anxiety, against which there are fewer weapons and fewer remedies, other than to go to be happy at Cirey. There, in truth, is the metaphysics of love, and this is where the excess of this passion leads. All this appears to me as the clearest and most natural thing in the world. Only by comparison do I perceive that I would condemn it in Marivaux.[15] But do not be surprised: this evening I spent two hours walking in the Tuileries with his father Fontenelle.[16]

Speaking about Fontenelle, Maupertuis is going to the Pole to measure the earth on that side;[17] he states that he cannot stay in Paris after I leave. There is an anxiety in his mind that makes him unhappy and proves that it is more necessary to occupy one's heart than one's mind. But unfortunately, it is much easier to make algebraic calculations than to be in love—I say in love like me, for it is necessary with you to define words and to be exact.

---

14. Du Châtelet had asked Richelieu to speak to the marquis about her relationship with Voltaire and to quiet his suspicions that it might be sexual rather than intellectual. The problem was not adultery, but rather the lower social rank of the poet.

15. Pierre Marivaux, a contemporary writer known particularly for his comedies. Du Châtelet is suggesting that this was like a scene in one of his comedies.

16. Bernard le Bovier de Fontenelle was a family friend of Du Châtelet's and a member of both the intellectual and social elite of Paris. Walking in the extensive gardens of the palace of the Tuileries was a common activity for Du Châtelet and her circle.

17. Maupertuis had gained sponsorship from the Royal Academy of Sciences for an expedition to the North Pole (Lapland) to determine whether the earth was flattened in the middle or at its ends. Newton had hypothesized that attraction caused a flattening at the poles. Verification of this would be a proof of his theory of universal attraction.

I took Du Fay to Saint-Maur where I spent eight days.[18] On the way we argued about the existence of you know what. In truth, I almost made a proselyte of him, and he would have been a good one, but we were not able to stop the sun in its course, thus, not giving me enough time to convince him. If ever I spend more time with you, I want to undertake your conversion, but I will never be happy enough to achieve that, except at Richelieu.

I spoke of you with your *président* the only time I met him; I spoke of you with Mme de Boufflers, who makes miracles for me. I would have a liking for her if I were not so frightened of her.[19] She says that she believes you made a frightful mess with him; she asked me to tell you of it and would not tell me what the matter was. I assured her that it was impossible that you should ever cause any trouble.

The new Polienos is more enamored than ever of Circe. He keeps talking about her always in this connection. She is now an arbiter on works of the mind, but she has more pimples than Pierrot had; it is even said that he gave her his. If you do not understand me, read the satire of Petronius.[20]

I have been to *Abensaïd*. The madness of the parterre for this piece can only be compared with that of the parterre at the Opera for Mlle Lebreton.[21] As for me, I believe that the public is being charitable to the author, and that this will make all the bad pieces sprout which otherwise would not dare show themselves in broad daylight. I regret to have to tell you that there are many incidents, but nearly as many different interests. Assuredly, *Héraclius* is not in that style: I have never found Corneille so sublime, but in *Héraclius* he astonished my soul.[22] The feeling of admiration is so rarely excited that it

18. Charles François de Cisternay Du Fay was the keeper of the *Jardin du roi*, the royal botanical garden; he and Du Châtelet were staying at a Condé family château. The discussion was probably about Newton and attraction.

19. Charles Jean François Hénault was presiding officer of one of the royal courts. Usually referred to as *Président* Hénault, he was a very successful courtier and member of the important Parisian *salons*. Madeleine Angélique de Neuville-Villeroy, Duchess de Boufflers, was also a prominent courtier Du Châtelet cultivated at this time.

20. Courtiers often wrote about one another in code names with references to stories that their reader would understand.

21. This was a tragedy by Jean Bernard Le Blanc; Mlle Lebreton, was a singer at the Royal Opera. Du Châtelet expresses her annoyance with the role of the parterre, the men who bought cheap tickets to stand in the large area in front of the stage. They showed their feelings about the performers and the play or opera very openly and could be responsible for the success or failure of a presentation. Rival playwrights often paid young men to heckle each other's opening nights.

22. Pierre Corneille (1606–1684), the great French seventeenth-century playwright. *Héraclius* was first presented in 1647.

seems to me that it is one of those which give the greatest pleasure. I do not think it impossible that if V. maintains what he said to you, we will quarrel for two or three hours at Cirey over *Héraclius,* for I excluded my admiration for Corneille from the complete sway I have given him over my soul. My love, and even my admiration for him, demand reasons, and one would be wrong not to ask some of him, for he often has good ones.

M. De Modena stays here for another month. The king told him that he relied on him to make Mme de Modena keep her word. Mme d'Orléans sees her no longer and forbade her daughters to see her; one cannot behave (leaving aside what she did at Versailles) with more wisdom and moderation. Not one word of ill humor has escaped him. That is very uncommon and very respectable, so she is generally pitied.[23]

Ah, how much I agree with you! I do not believe that a retinue of mediocre beauties is the currency of a sublime thought. There is a feeling of the sublime in Maurice's lines that you cite; it is at the same time the sublime of the heart and of the mind, and Racine surely did not have the currency of that.[24] You can see from what I have just said about V. that taste, kindness, not even superiority can carry away my feeling. Thus, I have the merit of having thought completely as you did and I cherish that fact.

Oh, do read this *Vie de Turenne* [Life of Turenne] since you so like to be bored in a learned way! It is not to bother you that Bécherand wrote to you, it was to have your answer to read in public. The bother was surely only an accessory. My aversion to her made me see la Pierotte one more time before my departure; in truth, propriety and not the summer have extracted this platitude from me.[25]

The last words of your letter, which make me fear some trouble, are a promise not to leave me for long without giving me your news. Think how much I love you, since I ask your news when about to leave for Cirey. Address that letter to Paris still.

જ

23. Charlotte Aglaé d'Orléans, wife of Francesco III of Modena and notorious for her quarrels with her husband. Du Châtelet refers here also to King Louis XV and her mother the Duchess d'Orléans.

24. Jean Racine (1639–1699) was the other great seventeenth-century French playwright. Maurice was a character in Corneille's *Héraclius.*

25. Du Châtelet is probably referring to the Marie Hortense Victoire de La Tour d'Auvergne, Duchess de La Trémoille who was interested in the Abbé Bécherand, a cleric given to convulsions, and thus, may have acquired this nickname.

TRANSLATOR'S PREFACE FOR *THE FABLE OF THE BEES*

Since I began to live with myself, and to pay attention to the price of time, to the brevity of life, to the uselessness of the things one spends one's time with in the world, I have wondered at my former behavior: at taking extreme care of my teeth, of my hair and at neglecting my mind and my understanding. I have observed that the mind rusts more easily than iron, and that it is even more difficult to restore to its first polish.

Such sensible reflections do not, however, give the soul back that flexibility it lost from lack of exercise when one is no longer in the first flush of youth. The fakirs of the East Indies lose the use of the muscles in their arms, because those are always in the same position and are not used at all. Thus do we lose our own ideas when we neglect to cultivate them. It is a fire that dies if one does not continually give it the wood needed to maintain it. So, wishing if it is possible to make up for such a great mistake, and to make it bear the fruits that I can still look forward to, I sought for some kind of occupation that could, in focusing my mind, give it that firmness (if I can put it that way) that can never be acquired unless one has chosen a goal for one's studies. One must conduct oneself as in everyday life; one must know what one wants to be. In the latter endeavors irresolution produces false steps, and in the life of the mind confused ideas.

Those who have received very decided talent from nature can give themselves up to the force that impels their genius, but there are few such souls[26] which nature leads by the hand through the field that they must clear for cultivation or improvement. Even fewer are sublime geniuses, who have in them the seeds of all talents and whose superiority can embrace and perform everything. Those, however, who have most claim to this universal monarchy of the fine arts attain perfection in one particular field with more facility and make it their favorite. M. de Voltaire, for example, although a great metaphysician, great historian, great philosopher, etc. has given preference to poetry. And the epithet of France's greatest poet, as well as that of "universal man," will be his distinctive characteristic.

It sometimes happens that work and study force genius to declare itself, like the fruits that art produces in a soil where nature did not intend it, but these efforts of art are nearly as rare as natural genius itself. The vast majority of thinking men—the others, the geniuses, are in a class of their own—

---

26. Du Châtelet is using the word *soul* in the Cartesian sense of *mind* or *spirit* as separate from the body. René Descartes (1596–1650) was the principal authority in France on philosophy in its broad eighteenth-century meaning.

need to search within themselves for their talent. They know the difficulties of each art, and the mistakes of those who engage in each one, but they lack the courage that is not disheartened by such reflections, and the superiority that would enable them to overcome such difficulties. Mediocrity is, even among the elect,[27] the lot of the greatest number. Some are busy removing the thorns that slow the true geniuses in their course, and it is they who supply us with so many dictionaries, and similar works which are so useful in literature. The colors of the great painters must be ground by someone. Others periodically report to the public all that happens in the Republic of Letters. Lastly, others convey from one country to another the discoveries and thoughts of great men, and remedy, to the best of their abilities, the misfortune of the multiplicity of languages, so often deplored by true lovers of learning.

I know that obtaining for one's country riches drawn from its own resources does it a greater service than informing it of foreign discoveries; Van Robés has been more useful to France than the man who first brought imported cloth from England.[28] But one must cultivate the portion one has received and not give in to despair, because one has only two *arpents* of land while others have ten *lieues* of land.[29]

This passage from the Gospels can be applied to the arts; according to the Evangelists: *sunt plures mansiones in domo patris mei*.[30] It is certainly more valuable to give a good translation of an esteemed English or Italian book than to write a bad French one.

27. *Élus* in French means the *elect*. Du Châtelet uses the word here to designate those "chosen" by blood, the elite, her privileged circle.

28. Joseph Van Robais (ca. 1630–1685), born in Courtrai, came to France from what is now Belgium as part of Louis XIV's principal minister Colbert's policy to encourage French manufactures. Von Robais established a cloth factory at Abbeville that became competitive with English cloth manufacturers and ended French dependence on English imports. The king awarded him and his workers naturalization and dispensation as Protestants.

29. French measurements in Du Châtelet's day can only be approximated. The basic length was a *pied du roi* (roughly 32.48 cm). Twelve *pouces* made up one *pied* and were roughly 27.07 mm. A *toise*, like the English fathom, was six *pieds* (1.95 m). A *lieue*, or "league," was 2,000 *toises*, or 3.9 km. An *arpent* could indicate length and was equivalent to 200 *pieds* (64.97 m) or area (4,221 m^2) when it was roughly equivalent to an English acre. Available at http://en.wikipedia .org/wiki/French_units_of_measurement#Length, accessed 19 November 2007.

30. *In domo Patris mei mansiones multae sunt;* "In my Father's house there are many mansions," John 14:2. The Bible used by Du Châtelet was the Latin Vulgate, St. Jerome's translation from Greek. The English translation of the Latin Vulgate was the standard translation into English until the mid-twentieth century for Roman Catholics. The translation above is the one used in that translation. See *The Holy Bible*, translated from the Latin Vulgate (Rockford, IL: Tan Books and Publishers, 1899, repr. 1971).

Translators are the entrepreneurs of the Republic of Letters,[31] and they should at least be praised for perceiving and knowing their limitations and for not undertaking to produce works themselves, and thus attempting to carry a burden under which they would succumb. Besides, if their work does not require creative genius, which no doubt holds the first rank in the empire of the fine arts, it calls for an application for which they must be even more grateful, as they can expect less glory from it.

Of all works, those of discursive thought seem to me the most susceptible to a good translation. Reason and morals know no country. The genius of a language, the curse of translators, is less to be felt in works where ideas are the only things to be conveyed, and where the graces of style are not the main merit. By contrast, works of imagination can rarely be transmitted from people to people, because to translate a good poet, one must be almost as good a poet as the author.

But while it is impossible to have faithful records of men's poetic imagination in many languages, it is not impossible to have those of their reason; for this we are indebted to translators. Thus, if human nature, in general, is indebted to the wise M. Locke for enabling it to know its own noblest part, namely, its understanding, the French are without doubt indebted to M. Coste, for having made known to them this great philosopher.[32] For so many men, even among the readers of Locke, cannot master the English language. Furthermore, very few among those who have learned the language of modern philosophy would be able to understand M. Locke in English, and thus to overcome both the difficulties of the language and those of the subject matter.

No doubt, before deciding to translate one must convince oneself that it is commentators, not translators, who are made to say in the "Temple of Taste":

> Taste is nothing, we are accustomed
> To write out at length, idea to idea,
> What others think, but we don't think at all.[33]

---

31. Du Châtelet uses the word "négociant," which in her day meant a merchant who introduced exotic products from one culture to another, such as a member of the Company of the Indies; thus "entrepreneur" rather than merchant seems a good translation. See *Le Dictionnaire de la l'Académie française* (Paris, 1694), 2.115.

32. She is referring to John Locke (1632–1704), the English philosopher and author of *An Essay concerning Human Understanding* (1690), translated into French by Pierre Coste (1668–1747) as *L'Essai sur l'entendement humain* (1700).

33. "Le Temple du goût" is a long satiric poem by Voltaire, published in 1733. Du Châtelet has chosen a section mocking commentators and their lengthy tomes.

The judicious author of these charming lines of verse has appreciated the difference between writing thick volumes on a passage from Dictys of Crete[34] that one does not understand at all, and is of no use to us, and making available for one's country the works and the discoveries of others.

But as any opportunity can be misused, the desire to make money and to be published has produced almost as many bad translations as bad books.

If making a good translation is not without some difficulty, at least, it ought to be easy to choose a good book as the object of one's work. However, translations often appear when the original work has already been forgotten. The English fall into this kind of error more often than we do. There are scarcely any bad French books that they do not translate, witness Sethos and so many others.[35] Yet the profound genius of the English ought to make them less avid for our books, which for the most part, are frivolous in comparison to theirs. It seems to me that one could apply to French books what the count of Roscomon said of our poetry: that all the gold of an English line made into a French wire would fill several pages.

> The weighti [sic] bullion of one sterling line
> Drawn to a French wire would through all pages shine.
> (The word *line* in English means both *ligne* [line] and
>   *vers* [line of poetry])[36]

I believe that translation is more widespread in England because, since French is part of the education of Englishmen, more among them are able to translate.

There are many unfaithful translators, some translating word for word because they are afraid to be unfaithful to the original. Others, because their author's meaning is difficult to grasp, just miss it and obscure a brilliant thought of which their mind has had only a glimpse. As for those who

Le goût n'est rien, nous avons l'habitude
De rédiger au long, de point en point
Ce qu'on pensa, mais nous ne pensons point.

34. Dictys of Crete was the supposed author of a Latin narrative of the Trojan War that probably was a source of Greek legends for medieval Europe.

35. Sethos is perhaps a reference to Abbé Jean Terrasson (1670–1750), a classicist who translated a book of Egyptian myths and legends into French. The book was titled *Sethos* (1731). I am grateful to Penny Arthur of the Graffigny Project at the University of Toronto for this reference.

36. Wentworth Dillon, Earl of Roscommon (1633–1685), an Irish nobleman, minor poet, and translator. Du Châtelet is quoting from his "An Essay on Translated Verse," lines 53–54, available at http://www.hn.psu.edu/Faculty/KKemmerer/poets/roscommon/default.html, accessed 13 September 2003.

substitute their absurdities for those of the author they translate, I look on them as similar to the travelers who take advantage of the proverb: *long ways, long lies.*[37] I believe that only translators of works in oriental languages have fallen into this excess.

The difficulties of each art are for artists what the circumstances of the smallest events are for their contemporaries. The interest that both groups take in their endeavors and the point of view from which they consider events magnify the objects for the former and the latter. Posterity and the public judge very differently. Thus, while it is true to say that a good translation requires application and labor, it is, nonetheless, at best, a very mediocre work.

However mediocre this kind of work, it may be thought that it is audacious for a woman to aspire to do it.

I feel the full weight of prejudice that excludes us [women] so universally from the sciences,[38] this being one of the contradictions of this world, which has always astonished me, as there are great countries whose laws allow us to decide their destiny, but none where we are brought up to think.

Another observation that one can make about this prejudice, which is odd enough, is that acting is the only occupation requiring some study and a trained mind to which women are admitted, and it is at the same time the only one that regards its professionals as infamous.[39]

Let us reflect briefly on why for so many centuries, not one good tragedy, one good poem, one esteemed history, one beautiful painting, one good book of physics, has come from the hands of women. Why do these creatures whose understanding appears in all things equal to that of men, seem, for all that, to be stopped by an invincible force on this side of a barrier;[40] let someone give me some explanation, if there is one. I leave it to naturalists to find a physical explanation, but until that happens, women will be entitled to protest against their education. As for me, I confess that if I were king I would wish to make this scientific experiment. I would reform an abuse that

37. "To lie well goes a long way" would be a more complete version of this proverb. Du Châtelet often liked to use proverbs to make a point, even in her writings on natural philosophy.

38. "Sciences" had a broad meaning in the eighteenth century. However, Du Châtelet and her contemporaries gave it the Aristotelian connotation of "certain knowledge," similar to our idea of quantifiable, demonstrable fact ascertained through the subjective observation of our senses.

39. In her era, actors were excommunicated by definition.

40. Du Châtelet often uses words that had meanings in mathematics and mechanics; for example, "semblable" is the word she uses for "equal," a geometric term to express equivalency; and "force," used in eighteenth-century physics to mean the impetus for motion.

cuts out, so to speak, half of humanity. I would allow women to share in all the rights of humanity, and most of all those of the mind. Women seem to have been born to deceive, and their soul is scarcely allowed any other exercise. This new system of education that I propose would in all respects be beneficial to the human species. Women would be more valuable beings, men would thereby gain a new object of emulation, and our social interchanges which, in refining women's minds in the past, too often weakened and narrowed them, would now only serve to extend their knowledge. Some will probably recommend that I ask M. the abbé of St. Pierre to combine this project with his.[41] Mine will perhaps seem as difficult to put into practice, even though it may be more reasonable.

I am convinced that many women are either ignorant of their talents, because of the flaws in their education, or bury them out of prejudice and for lack of a bold spirit. What I have experienced myself confirms me in this opinion. Chance led me to become acquainted with men of letters, I gained their friendship, and I saw with extreme surprise that they valued this amity.[42] I began to believe that I was a thinking creature. But I only glimpsed this, and the world, the dissipation, for which alone I believed I had been born, carried away all my time and all my soul. I only believed in earnest in my capacity to think at an age when there was still time to become reasonable, but when it was too late to acquire talents.

Being aware of that has not discouraged me at all. I hold myself quite fortunate to have renounced in mid-course frivolous things that occupy most women all their lives, and I want to use what time remains to cultivate my soul. Feeling that nature has refused me the creative genius that discovers new truths, I have done justice to myself, and I am content to render with clarity the truths others have discovered, and which the diversity of languages renders useless for most readers.

Having decided upon this kind of work, my esteem for the English, and the taste that I have always had for the free and virile way of thinking and expressing themselves of this philosophical people, caused me to prefer their books to those of other nations. I have chosen this book titled *The Fable of the Bees*, among all those I could have translated, because it appears to be, of all the books in the world, one of those most designed for humanity

---

41. Charles-Irénée Castel, Abbé de St. Pierre (1658–1743), wrote a number of well-known treatises on education, including one that advocated *collèges* for girls with a course of study similar to that for boys.

42. "L'amitié," the French word used here, means more than simple friendship and carries the connotation of affection and concern.

in general. It is, I believe, the best book of ethics ever written, that is to say, the one that most leads men to the true source of the feelings to which they abandon themselves almost all without examining them. Mandeville,*[43] who is the author of it, could be called the Montaigne of the English if it were not for the fact that he has more method and healthier ideas about things than Montaigne.[44]

I do not, however, have for my author the idolatrous respect of all translators. I acknowledge that his work is written badly enough in English. Parts of it are too long, and it sometimes goes too far, as when it says, for example, *that a thief is as useful to society as a bishop who gives alms, and that there is no merit in saving from a fire an infant about to be engulfed.* And in many other places it advances several things which are not true and could be dangerous. I have taken care to add a corrective statement in those places in order to prevent them from having dangerous consequences. I have taken the liberty of pruning the author's style in a number of places, and of cutting out all that was directed to the English and which is therefore too specific to their customs.

I also took the liberty of adding my own reflections, when the material on which I was working suggested them to me, reflections that I believed deserved to be included. But, in order that the reader could discern them, I have taken care to indicate them with quotation marks.[45]

The reader will find in this book thoughts that may seem a little bold, but before making this judgment it is only a matter, I believe, of examining whether they are accurate. For, if they are true, and if they teach men how to know themselves, they cannot fail to be useful to those thinking men, and it is for those only that this book is destined. *Odi prophanum vulgus et arceo.*[46]

I admit that, having had the temerity to undertake this work, I have the temerity to wish to succeed. I believe myself all the more obligated to give it all my most careful attention as success alone can justify the undertaking. The unfairness of men in excluding us women from the sciences should at least be of use in preventing us from writing bad books. Let us try to enjoy this advantage over them, so that this tyranny will be a happy necessity for us, leaving nothing for them to condemn in our works but our names.

43. Du Châtelet adds her own note here: "*He was the son of a French refugee. He proves by his example that French minds need to be transplanted to England to acquire power." In fact, Du Châtelet was mistaken about Mandeville's origins. He was a Dutch refugee.

44. Du Châtelet is referring to Michel de Montaigne (1533–1592), and his *Essais* (1588); Montaigne was also a favorite of Mandeville. When she suggests that Mandeville had more "method," she means that his reasoning was more systematic.

45. French translators of the eighteenth century often made additions, but Du Châtelet is unusual in her desire to set them apart in this way.

46. "I hate the vulgar rabble and drive them away," Horace *Odes*, III, 1,1.

TABLE OF DU CHÂTELET'S CHOICES FOR TRANSLATION FROM BERNARD MANDEVILLE'S *THE FABLE OF THE BEES*

Du Châtelet's Translation	Mandeville's *The Fable of the Bees*
Preface du Traducteur	[No equivalent in Mandeville]
Avertissement du Traducteur	[No equivalent in Mandeville]
[Not translated]	The Preface
[Not translated]	The Grumbling Hive: or, Knaves Turn'd Honest
Preface de l'Autheur	The Introduction
Chapitre 1er: De l'Origine des Vertus Morales	An Inquiry into the Origin of Moral Virtue
Chapitre 2e: Du Choix des Différentes Professions	A. Whilst others follow'd Mysteries To which few Folks bind 'Prentices;
Chapitre 3e: Des Négocians	B. These were call'd Knaves, But bar the Name, The grave Industrious were the same.
Chapitre 4e: De l'Honneur et de la Honte	C. The Soldiers that were forc'd to fight, If they surviv'd, got Honour by't.
Chapitre 5e: Des Marchands en Detail	D. For there was not a Bee but would Get more, I won't say, than he should; But than, etc.
Chapitre 6e: Des Jouëurs	E. —As your Gamesters do, Who, tho' at fair Play, ne'er will own Before the Losers what they've won.
Chapitre 7e: Des Professions Fondées sur les Vices des Hommes	F. And Virtue, who from Politicks Had learn'd a thousand cunning Tricks, Was, by their happy Influence, Made Friends with Vice.
Chapitre 8e: Que les Plus Méchants Sont Encore Utiles à la Societé, et Principalement les Cabaretiers	G. The worst of all the Multitude Did something for the Common Good.
Chapitre 9e: Des Musicos d'Hollande	H. Parties directly opposite, Assist each other, as 'twere for spight.
Chapitre 10e: De l'Avarice	I. The Root of Evil, Avarice, That damn'd ill-natur'd baneful Vice, Was Slave to Prodigality. [No J. exists in the Mandeville]
Chapitre 11e: De la Prodigalité	K. That noble Sin—
Chapitre 12e: Du Luxe	L. —While Luxury Employ'd a Million of the Poor, &c. [Remarks M.-T., V., X., and Y. not translated; no U or Z exist.]

II

DISSERTATION ON THE NATURE
AND PROPAGATION OF FIRE

VOLUME EDITOR'S INTRODUCTION

By the summer of 1737 Du Châtelet had already turned most of her attention to readings in natural philosophy, particularly Newton's *Opticks* and his *Principia*. Voltaire, back at Cirey from his brief exile in Holland and the opportunity to watch the Dutch physicist, Willem Jacob 'sGravesande's demonstration lectures, was excited by the prospect of conducting his own experiments at Cirey. His man of business in Paris was instructed to send the necessary instruments, including a reflecting telescope, prisms and lenses, a vacuum pump, and a burning glass [*verre ardent*]. Eager to advance his claims to membership in the Royal Academy of Sciences after the publication in 1737 of his *Elements of the Philosophy of Newton*, Voltaire proposed to enter the Academy prize competition for 1738. The essay was due in September, 1737 and so, with Du Châtelet's assistance, they began their second scientific collaboration, a study of "the nature and propagation of fire," the topic set by the Academy.[1]

In addition to reading in Boerhaave's *Elementa chemiae* [Elements of Chemistry], the text Du Châtelet favored, and Musschenbroek's *Elementa physicae* [Elements of Physics], they consulted Descartes, Newton, and the publications of Europe's scientific academies, such as the *Transactions* of London's Royal Society. Determined to add more proof to Newton's theory of attraction, Voltaire concluded from their experiments with burning iron at the local forge, wood piles set fire on the estate, and numerous substances heated

---

1. Much of this account comes from Judith P. Zinsser, *Émilie Du Châtelet: Daring Genius of the Enlightenment* (New York: Penguin, 2006), esp., 152–62, 168–69. See on their previous collaboration on the *Elements*, 145–50, and also Robert L. Walters and W. H. Barber, "Introduction," *Eléments de la philosophie de Newton* (Oxford: Voltaire Foundation, 1992).

in the simple laboratory in the gallery of the château, that fire consisted of particles that, having mass and weight, obeyed Newton's laws of attraction. Du Châtelet disagreed. She believed that the experiments were inconclusive and that any changes in the weight of the substances they heated or burned could be explained without attributing weight to fire itself. Perhaps angry, certainly determined to present, what to her, was a true explanation of the phenomenon of fire, she secretly wrote and submitted her own essay for the competition. Only her husband, the marquis, knew of this audacious act, the first entry by a woman to these competitions, and this to a scientific academy that in the twentieth century still would not admit women members, not even the twice-honored Nobel Laureate Marie Curie.[2]

In her answer Du Châtelet wrote far more extensively and thoughtfully than Voltaire. In fact, she came very close to describing fire in much the same terms as our modern concept of "energy," an entity animating all substances. Similar to Descartes and Herman Boerhaave, she saw it as neither matter nor spirit. It could penetrate all in nature. In addition to considering the presence of fire, she also considered its absence. As with flammability, she concluded that freezing came from the addition of something to the substance. In particular, from her observation of how one made ices to eat in the summer, she deduced that nitrates must join with water particles to create snow, as nitrates around a vase froze the water it contained. The *Dissertation*, then, reflects her vast reading in the authorities and is a combination of experiment, reflection, and hypothesis. In addition, here and in her *Foundations of Physics*, she demonstrated her mastery of the rhetorical style of eighteenth-century science. For example, she outlines all of the objections to a hypothesis and then answers them one by one; her paragraphs go from the presentation of a phenomenon through observed consequences or variations, to a conclusion prefaced by "thus."

The announcement of the winners of the competition came in April 1738, three respected Cartesian authors. Voltaire, never discouraged by such defeats, worked to have his entry published by the Academy along with those of the victors, a not uncommon practice. When Du Châtelet admitted to him that she had also submitted a dissertation, he agitated for publication of both of their essays. In a surprising gesture to "one of the best of our poets" and "a young lady of high rank," as characterized by the Academy, it agreed to publish all five competitors in a limited 1739 edition of three

2. See Vesna C. Petrovich, "Women and the Paris Academy of Sciences," *Eighteenth-Century Studies* 32, no. 3 (1999): 383–90.

hundred copies, complete with the "errata" she insisted on including to make clear her new adherence to Leibnizian views on the nature of collisions. It appeared again in the formal series of collected prize essays published by the Academy later in1752.[3]

Perhaps because of the restricted printing of the *Dissertation* and the usual delay in the appearance of the *Recueil*, Du Châtelet undertook its revision and arranged its publication in 1744 with all of the changes incorporated, the language made clearer and more concise. It is this revised edition that has been chosen for translation. This version is also included in the Landmarks of Science microprint series, and can be accessed in an edited form online.[4]

The letters presented before the *Dissertation*, letters to Pierre Robert le Cornier de Cideville, a longtime friend of Voltaire's from his days in Rouen studying law, and to Maupertuis, Du Châtelet's mathematics tutor and mentor in philosophy, show something of her and Voltaire's life at Cirey. They describe her problems with the abbé Michel Linant, a protégée of Cideville and Voltaire, brought to the château to be her son's *précepteur* (a companion and tutor expected to stay with the boy until he was old enough to continue his education in Paris); her negotiations for inclusion of the "errata" for her *Dissertation*; and suggest her excitement at the collaboration with Voltaire on the *Elements of the Philosophy of Newton* and with mathematics and science more generally.

The letter to Charles Augustin Feriol, Count d'Argental shows her ongoing concern for Voltaire, who spent a number of months in exile, in the winter of 1736/37, and her continuing hope that she could return him to royal favor. D'Argental, now a friend to both of them, was someone to whom she could confide her fears and describe her efforts with her family and court patrons such as the Duke and Duchess de Richelieu.

All three letters are from the Besterman collection.

⁂

---

3. *Recueil des pièces qui ont remporté les prix de l'Académie des Sciences* [Collection of Pieces Which Won the Prizes of the Academy of Sciences], vol. 4 (Paris: Gabriel Martin, 1752). On the eighteenth-century publication history of the *Dissertation*, see Walters, "Chemistry at Cirey," 1819–20.

4. Available at http://www.fontesdart.org/index2.php?option=com_contentandtask=view.and id=25andItemid=33. R. K. Smeltzer alerted me to this Web site. Voltaire's and Leonhard Euler's dissertations are also included. Euler, along with Louis-Antoine Du Fech, and Jean-Antoine de Créquy, count de Canaples, were the winners.

RELATED LETTERS

To Pierre Robert le Cornier de Cideville[5]

Cirey, this 27 February [1736]

I have up to now, monsieur, only been M. de Voltaire's secretary. I want to be my own today and tell you myself the interest that I take in your situation, and how much I would like it to end in a manner that permitted you to come see a hermitage where our affection would do you honor. Your friend's health is a little better; a diet and moderate work would render it more stable, if he were capable of this self-discipline. I understand that the love of work is an all-consuming passion, when one produces *Alzires*. You have shared his success with such sensibility that I love to speak to you of it.[6]

You will have seen from the last letter of M. de V. what he thinks of M. Linant. You must believe that, as you had recommended him to me, I would have told you of his behavior if I had not been pleased with him. M. de V. says that it is your kindnesses that have spoiled him. Be that as it may, one cannot have a more unbearable and more ill-grounded pride than his. Such vanity would be unjustifiable even if he were all he believes himself to be. I believe him to have a very mediocre talent, no imagination, no invention, an extraordinary ignorance. He will always be a very mediocre man. The only evil in this is not to believe it. But if you were mistaken about his mind, I fear that you were even more mistaken about his heart. The kindness and the unvarying sweetness of M. de V. has made him conceal and pass over the traits of ingratitude that, without the protection he honors him with, would a long time ago have forbidden him my house. The wrongs toward me count for little. He only owes me respect, and lack of education may explain his not showing it. But for M. de V., to whom he owes all that he is, and who showered him with kindnesses, that is altogether different. Only a heart is required for one to sense what one owes in gratitude. If you still feel, monsieur, some kindness toward him, make him sense that he owes, in all sorts of instances, the greatest respect and the most lively gratitude to M. de V., that to stray from that for a moment, and not constantly to be

---

5. No. 55 in the Besterman collection, vol. 1, 100–101; D1025 of the *Oeuvres complètes*.

6. *Alzire* was the play by Voltaire that Du Châtelet first helped with in 1734 and 1735. It was presented at the Comédie française in January of 1736. As with so many of Voltaire's plays it was a great success both in performance and when published. Cideville was a friend of Voltaire's from his time in Rouen when his father expected him to study the law. They exchanged writings throughout this period, poetry and plays.

concerned with it, is to render himself unworthy of your kindnesses and his, and that his sad lack of fortune (which was all that held me back until now) will no longer affect me, no more than the protection of M. de Voltaire, if I see him put pride and malignity in the place of respect, of the gratitude that he owes him. M. de V. has much contributed to spoiling him, but he must never abuse the marks of kindness that his position in my household, and the politeness that is an inseparable part of your friend's character, have brought him. As for me, I have nothing to reproach myself with; I treated him harshly enough, I sensed that he needed it, and, moreover, I have no taste for his wit, although I admit that he has some.[7] I believe in the end that he has done justice to himself by the kind of work he has undertaken, and he is no longer thinking of a tragedy. He strives for the French Academy, and I believe him much more capable of that. This is surely enough talk of him, but the interest that you take in him led me to it.

You believe, given what I think of the brother, that I am not inclined to take on the sister. In what capacity would I employ her, anyway? As a lady's maid? This would demean my son's tutor too much, who must enjoy some respect among the servants; and as for having him in my company, I do not believe that you had that in mind.[8]

Adieu, monsieur, I am ashamed at the length of this letter and the few things that it contains. I hope that the assurances of my esteem and of my wish to have the honor of seeing you here will make you tolerate it. Your friend speaks a thousand tender things to you.

To Pierre-Louis Moreau de Maupertuis[9]

At Cirey, the 1st December [1736]

Is it possible that one should still write to you at the Pole? I would not have thought that there were passions that enjoyment could increase. I was delighted to receive your news. The gazette said you ran the risk of being eaten by flies. I was very glad to learn that they respected you. You may be

---

7. Du Châtelet discovered that Abbé Michel Linant did not know Latin well enough to teach her son, nor science or mathematics. After leaving her service he did finally complete a play and eventually was admitted to the French Academy.

8. Du Châtelet did take Linant's sister on as her *femme de chambre* [lady's maid], and also made provision in her household for Linant's mother. The sister, Linant, and the mother left precipitously in December 1737 when Du Châtelet discovered that the young woman had been writing negatively to friends about her mistress.

9. No. 73 in the Besterman collection, vol. 1, 125–26; D1216 of *Oeuvres complètes*.

indebted for that to the protection of M. de Réaumur,[10] for there is no indication that they sense your value as much as female Lapps. All the letters you write to Paris are said to be full of praises for them. It is apparently for one of them that your companion has left me.[11] You can tell me of it without being indiscreet. It seems to me that there cannot be women from the Pole here, but, seriously, what I would like you to let me know is when you are coming back.

We have used your absence to render the people who inhabit Cirey worthy of you, for one does not lose hope of seeing you here one day. We have become real philosophers. The companion of my solitude has written an introduction to the philosophy of M. Newton, which he has dedicated to me and the frontispiece of which I send you.[12] I believe that you will find the verses worthy of the philosopher of whom they speak, and of the poet who made them. You will find this almost printed on your return. If you had been in this part of the world, one would have asked for your advice. You have for a very long time wanted to make a philosopher of the first of our poets and you have succeeded, for your advice contributed to his determination to give himself up to his thirst for knowledge. As for me, you know more or less the dose of physics and mathematics I can take. I enjoy a great advantage over the greatest philosophers: that of having had you as my master. I am yet more proud, if possible, to see that you have not forgotten me.

I beg you to continue to give me your news, for the descriptions in your last letter make me anxious about your health. You owe to my affection for you the justice of being certain that no one takes a truer interest in it than me. The philosopher of Cirey, who is one of the people in the world who admire you, love you, and who wish you here the most, asks me to tell you all that on his behalf. The epistle in verse is his letter. Answer me promptly, or better still come and give us yourself news of the shape of the Earth and above all of you. You will see from the verses which ellipsoid we

10. Réaumur, aside from being one of the principal leaders of the Academy of Sciences, was known for his study of insects. See chapter 2, note 29.

11. Rumors circulated in Paris that one of the party had taken a Laplander as a mistress; Du Châtelet is suggesting that it was Clairaut. In fact, they did bring back two women, but both were Swedish, not Lapp. Du Châtelet was one of the Parisian elite who assisted them finding placement, one in marriage and the other in a nunnery.

12. The frontispiece for Voltaire's *Elements of the Philosophy of Newton* portrayed the author as a Virgil-like figure seated writing at a table. Newton is in the heavens above him and the light from the Englishman's mind reflects down onto the poet's pages from a mirror held by a classically dressed female figure also aloft on some clouds.

favor. It is for you to conform your observations to it, for it would be hard to sacrifice the two lines:

> Earth, change shape, and let attraction,
> Flattening its ends, raise the equator.[13]

It is much easier for you to change the shape of the Earth. I implore you, leave, in the changes that you will make, Cirey as it is, and above all, never forget how much you are loved here.

To Charles Augustin Feriol, Count d'Argental[14]

Tuesday 22 [January 1737]

I find a safe opportunity to write to you and you may be sure that I will not miss it. Have you received my packet by the mail coach? If they searched it, then I am lost. I hope soon to be reassured and that you will let me know how things stand. It is ages since I had your news, but I can imagine that you have nothing to tell me, that you are waiting for me to inform you about the response of the bailly, and that your heart is still watching over us.[15]

I await from your response by the mail coach my good or my ill, my life or my death. My courier for Holland always has his boots greased. I wrote to you yesterday to tell you that I had news on the 8th. One left for Amsterdam the 13th, still with the intention of having the *Philosophy* [Elements of the Philosophy of Newton] printed; it is even announced in the gazette as being in press.[16] I hope that the letters he will receive from me, and those that you wrote to him on this topic at my request, will make him change his mind. I would regard this as a false step. Above all, there is a chapter on metaphysics that is very out of place and very dangerous. He would be forced to remove

---

13. Voltaire wrote a preface to his *Elements* that extolled Du Châtelet's role in its writing and her exceptionality as a learned woman. In the poem that also preceded the text, he included this acknowledgment of Maupertuis' role in providing one of the proofs hypothesized by Newton for the role of attraction in the universe, that it would cause the Earth to bulge at the equator and be flattened.

14. No. 93 in the Besterman collection, vol. 1, 166–70; D1265 in the *Oeuvres complètes*.

15. Louis Gabriel Froullay, Bailly de Froullay, was Du Châtelet's uncle, a royal magistrate, and a successful connection of hers at court.

16. While Voltaire was in exile in Holland, he often sent his own notices to the news journals, such as the *Gazette d'Amsterdam*, in theory to confuse the government spies. Du Châtelet refers to Voltaire in this letter as "one" as she assumes these spies will be opening the mail from Cirey.

it in Paris to have the *approbation;*[17] but in Holland he will let it stand. Finally, I regard it as a blow for his happiness to prevent this Amsterdam edition of his *Philosophy* from not preceding that of Paris. I have spared no effort to dissuade him from it; I hope that you will have done as much. I have let you know my reasons for it, as well as my entreaties that he should be extremely prudent in this new edition of his works; it is announced in the *Gazette* as being revised by himself. He must sense what this announcement commits him to. And, above all, let him not include *Le Mondain* there.[18] At every point he must be saved from himself, and I use more politics to guide him than all the Vatican uses to retain Christendom in its fetters. I count on you to second me. All my letters are sermons, but one is on guard against them; one says that I am afraid of my own shadow, and that I do not see things as they are. There are no such prejudices against you, and your advice will sway him.

One sent me, by the letter of the 8th, the copy of a letter to the prince royal,[19] which is very good and very wise in all ways. But this is what I find there: "I will have the temerity to send to your royal highness a manuscript that I would never dare show except to a mind as unconstrained by prejudices as yours, and to a prince who, among so many homages, merits that of unlimited confidence."

I know this manuscript. It is a metaphysics so much too reasonable that it would have this man burned, and this book is a thousand times more dangerous and certainly more punishable than *La Pucelle*.[20] Imagine how I shuddered. I am not yet recovered from the astonishment and, also, I confess, from the anger. I have written a fulminating and threatening letter. But they take such a long time to arrive that the manuscript could have left before they do. Or at the least, one will tell me so, for one is sometimes obstinate, and this demon of a reputation (which I think is misconstrued) does not desist. I must confess that I could not refrain from lamenting my fate when I

---

17. Every book published in France needed the approval, or *approbation*, of the royal censors before going to press. Du Châtelet is referring to a chapter of Voltaire's *Treatise on Metaphysics* where he speculates on animals having souls, an idea he took from John Locke's *Essay concerning Human Understanding*. He planned to include this chapter in his book on Newton. Such a concept brought immediate censorship from the French authorities as its possibility negated what was presumed to be the key difference between animals and humans.

18. Voltaire continually revised and reprinted his writings as a way to earn money, if not to enhance his fame. Authors were paid a lump sum at the time of publication; no royalties or copyright existed. "Le Mondain" was a poem that had already caused an uproar with the authorities because of disrespectful references to Adam, the first man.

19. Frederick of Prussia, who would become king in 1740 on the death of his father, had initiated the correspondence with Voltaire and wished him to visit Prussia.

20. Du Châtelet is referring to Voltaire's satirical epic on Joan of Arc. He had to seek asylum in Cirey when stanzas circulated in Paris.

saw how little I could rely on the tranquility of my life. I will spend it fighting against him for himself, without saving him, trembling for him, or lamenting his mistakes and his absence. But in the end such is my destiny, and it is still dearer to me than the most happy one. You must help me ward off this blow to the extent that it is possible, for you sense that this recklessness will be his undoing sooner or later and irretrievable. The prince royal will not keep his secret any better than he kept it himself, and sooner or later it will be leaked. Further, the manuscript will pass through the hands of the king of Prussia and of his ministers before reaching the prince, all of whose packets, as you will easily believe, are opened by his father. You will easily believe that M. de la Chétardie, who, anyway, is quite idle, is advised to know what passes between the prince royal and V. as much as it is possible.[21] Finally, if there was only the incongruity of such conduct, confiding the secret of his life, his tranquility, and that of the people who have linked their life to his, to a *prince* of twenty-four years, whose heart and mind are not yet formed, whom a disease can render devout, whom he does not know, in truth, he ought not do that. If a friend of twenty years asked him for this manuscript, he ought to refuse him, and he sends it to an unknown man and a prince! Besides, why make his tranquility depend on another, and that unnecessarily, for the vain gratification (for vanity is the correct word) of showing to someone, who is no judge of it, a work in which he will see only recklessness? He who confides his secret so lightly deserves to be betrayed. But what have I done to him, to have the happiness of my life depend on the prince royal? I confess that I am outraged, as you can see; and I cannot believe that you disapprove of me. I feel that when this mistake has been made, if all it took to right it was to give my life, I would do it; but I cannot see without a very bitter sadness that a creature so amiable in every way wants to make himself miserable by unnecessary recklessness, for which there is not even the least pretext.

What you can do, and what I implore you to do is to write to him *that you know that the king of Prussia opens all his son's letters, that M. de la Chétardie spies on all that concerns him in Prussia, and that he cannot be too cautious in all that he will send and all that he will write to the prince royal, and that this is an opinion that you believe you owe to him*, without entering into any detail, for he would never forgive me this letter if he came to know of it, and yet it is necessary to parry this blow or renounce him forever.

Mme de Richelieu has not spoken to the Keeper of the Seals and I am quite pleased about that, for he might have caused a breach between her and me. But she still maintains, in the most affirmative tone, and M. de Richelieu

21. Joachim Jacques Trotti, Marquis de la Chétardie, the French ambassador to Prussia.

too, that they have the word of the Keeper of the Seals never to do anything against M. de Voltaire without warning them of it, and thanks to these assurances we ought to sleep peacefully. I do not know what to believe. But one thing is sure, that this promise is the only thing I have asked of them since I came to live at Cirey, and that they only told me they had it two weeks ago. Tell me what you think. Apparently writing to M. Du Ch. was not part of the bargain.[22]

In case my letters had been intercepted, I should tell you that a big packet of very important letters should have reached you on Sunday the 20th, by the mail coach from Bar-sur-Aube, in a little box in the care of a coachman and that I counted on your response to leave by the same means this Saturday, the 26th.

Did you receive a chevreuil, which may have arrived spoiled?[23]

Adieu. Write to me. Your letters are the solace and the medicine of my soul. I have missed them for two weeks. Give them back to me and let me keep your pity and your affection. The man who will convey this letter stays in Paris.

꒰

FROM THE *DISSERTATION*

Ignea convexi vis, et sine pondere coeli
Emicuit, fummaque locum sibi legit in arce.

—Ovid[24]

FIRST PART: *OF THE NATURE OF FIRE*

How difficult it is to define Fire

Fire appears to us as so many different phenomena that it is nearly as difficult to define it by its effects as to know its nature entirely. Every moment it

22. One of Du Châtelet's relatives, François Victor le Tonnelier de Breteuil, the titular head of her family, was threatening to write a condemnatory letter to the marquise's husband in hopes of precipitating the end of his protection of Voltaire and his willingness to sequester the poet on his estate at Cirey. The Keeper of the Seals was Germain Louis Chauvelin (1685–1762).

23. A small deer common in that part of France and certainly from the Du Châtelet lands. It was delayed and the marquise sent a second that arrived in time to be eaten.

24. Ovid, "On the Creation of the World," *Metamorphoses*, I: 26–27: "The force of fire ascended first on high / And took its dwelling in the vaulted sky," trans. Sir Samuel Garth, et al., available at http://classics.mit.edu/Ovid/metam.1.first.html, accessed 17 September 2007.

defies the ability of our mind to grasp it, even though it is part of ourselves and of all the bodies that surround us.

I

That Fire is not always hot and luminous

Of all the effects of fire, heat and light most strike our senses. Thus, it is by these two signs that one customarily recognizes it, but giving sustained thought to the phenomena of nature will make us wonder if fire does not have a more universal effect on bodies, by which it might be defined.

One must never conclude from the particular to the general, so, though heat and light are often united, it does not follow that they always are. These are two effects of the being that we call *fire*, but do these two properties,*[25] to light and to heat, constitute its essence? Can they be removed from it? In other words, is fire always hot and luminous?

IF FIRE IS ALWAYS HOT AND LUMINOUS.
Many experiments decide in the negative.

LIGHT WITHOUT HEAT IN THE RAYS OF THE MOON.
1. There are bodies that give us a great deal of light without heat. Such are the rays of the Moon, united at the center of a burning glass [*verre ardent*][26] (this makes us see in passing the absurdity of astrology). It cannot be said that it is because the Moon sends so few rays; for these rays are thicker, denser, when united in the focus of a burning glass than those which come from a candle. And yet, not only this candle, but even the smallest spark, burns us, while, united at the same distance, the Moon's rays have no effect on us.

25. Du Châtelet adds a note at this point: "*I use here interchangeably the words *modes* and *properties*, to avoid the frequent return of the same word, because strictly fire is not always hot and luminous, heat and light are *modes* and not the *properties* of the being that we call fire." See Chapter 3, "Of Essence, Attributes, and Modes," of Du Châtelet's *Foundations of Physics* for her explanation. By "properties" Du Châtelet means primary constituent aspects of a being; "modes," then, are secondary manifestations or characteristics. These distinctions come from her reading in Gottfried Wilhelm Leibniz (1646–1716) and represent an addition to the 1744 version.

26. The eighteenth-century *Encyclopédie* (1750–72) compiled by Denis Diderot (1713–1784) and his circle gives a long entry on "ardent" in which a *verre ardent* is described as a convex lens that, through refraction, joins rays of light together to a point, or *foyer*. Du Châtelet and Voltaire, like other experimenters of their day, placed many different substances at this focal point to see how quickly they would ignite and burn. See the ARTFL online edition of the *Encyclopédie ou dictionnaire raisonné des sciences, des arts et des métiers*.

Neither is this because these rays are reflected, for the rays of the Sun reflected by a flat mirror, and reflected again on a concave mirror, have more or less the same effects as if the concave mirror received them directly.

Finally, it cannot be because of the space that it traverses from the Moon to here; 90,000 *lieues* extra cannot make the rays lose a property that they conserve for 33 million *lieues* coming from the Sun.[27] This effect may have to be attributed to the particular nature of the body of the Moon, and perhaps the satellites of Jupiter and Saturn give some heat to these planets, even though our Moon gives us none.

The rays heat less as one rises above the Atmosphere, even though they give there the same light as near the surface of the Earth; yet they are purer higher up where the atmosphere is lighter. Thus, heat is not essential to elementary fire.

WATER DOES NOT EXTINGUISH THE LUMINOSITY OF GLOWWORMS.

Dails[28] and glowworms are luminous without giving off any heat, and water does not extinguish their light. M. Réaumur even reports that water, far from extinguishing it, revives the light of dails.[29] I have verified this on glowworms, I have plunged some in very cold water, and their light was not affected.

From these experiments it would seem that water only acts on the property of fire that we call heat, since it destroys the heat, and does not alter the light when the property of lighting is separated from that of heating.

HEAT WITHOUT LIGHT IN IRON ON THE POINT OF CATCHING FIRE.

2. Some bodies would burn the hand that comes close to them and give off no light: such as iron on the point of catching fire. Thus, fire can be deprived of light as well as heat.

Thus, heat and light would seem to be to fire as mode is to substance; light being nothing other than fire transmitted in a straight line to our eyes, and heat, the agitation in every direction that this same fire excites in us when it insinuates itself into our pores.

27. See chapter 1, note 29.

28. A kind of shellfish found on the coast of Poitou in France containing a liquid that is luminous in the dark.

29. René Antoine Ferchault de Réaumur (1683–1757) was most famous for his extensive study of insects; Du Châtelet is here referring to his 1723 memoir for the Royal Academy of Sciences in which he reported a number of experiments with dails. He also often served as an officer of the Academy and was a close friend of its perpetual secretary, the mathematician and natural philosopher Bernard le Bovier de Fontenelle (1657–1757).

DIFFERENT PROPAGATION OF LIGHT AND HEAT.

3. Heat and light propagate differently. Light always acts in a straight line, and heat insinuates itself into our bodies from all sorts of directions. In addition, the speed of light is infinitely greater than that of heat; but the ratio is not known, for it would be necessary to know the different degrees of speed with which fire penetrates different bodies. This is very difficult.

OTHER DIFFERENCE BETWEEN HEAT AND LIGHT.

4. Another very remarkable difference between heat and light is that a body can lose its light in an instant, but it never loses its heat except successively over time. This difference follows from the way in which heat and light act. To put an end to light, it suffices to interrupt the fire in a straight line; but since fire, to create heat, must penetrate a body from all directions, this action must be more difficult to stop. Thus, if you cover a burning glass with an opaque veil, the light disappears in a moment at its center, and yet a solid body exposed to it would conserve for a long time the heat that it acquired. This is also why bodies cool slowly in Boyle's void, even though they go out very promptly.[30]

DESCARTES' VIEW, WHICH JUSTIFIES THIS OPINION.

5. If one wished to rely on authority, one would say that Descartes made light his second element, and fire his first. In truth he did not give any reason for this idea, and I do not claim to examine it here, but it could only be founded on the fact that this great man thought light and heat were two modes of the being that we call fire.[31]

6. Light and heat are the objects of two of our senses, touch and sight, and for this reason they do not appear suited to constitute the essence of

30. Du Châtelet and Voltaire had a vacuum pump [*machine de vuide*, literally "machine for making a void"] made by France's premier craftsman and entrepreneur of scientific experimental instruments, Abbé Jean-Antoine Nollet (1700–1770); this was one of the scientific instruments Voltaire ordered while at Cirey. Here, Du Châtelet is referring to the Englishman, Robert Boyle (1627–1691), famous for countless experiments to study the effects of a vacuum. Note that neither in the eighteenth century nor today have scientists achieved a complete void, the absence of any atmosphere.

31. René Descartes (1596–1650) was the principal authority in France on philosophy in its broad eighteenth-century meaning. In *Les Principes de la philosophie* [The Principles of Philosophy] (1644), the *Dioptriques* [Dioptrics] (1637), and *Le Monde* [The World], he identified fire and light as primary elements. Note that Du Châtelet considers all that is animate a "being," a concept familiar to all natural philosophers. On Descartes, see, for example the Internet Encyclopedia of Philosophy, http://www.iep.utm.edu/d/descarte.htm, accessed 23 December 2007.

a being as universal as fire. These are sensations, modifications of our soul, that seem to depend on our existence and of the manner in which we exist; for a blind man will define fire as *that which heats*, and a man deprived of universal touch, *that which lights*. So, they would have two different ideas of the same being, and one deprived of these two senses, would not have any idea of it.[32] Now, suppose that it pleased God to create in Sirius, for example, a globe whose beings had none of our senses (and it is very possible that there are such beings in the immensity of the universe), fire would certainly be neither hot, nor luminous on this globe, and yet it would not be destroyed. So it seems that some more universal effect must be sought in fire, and one whose existence does not depend on our senses.

HOW OUR SENSES DECEIVE US ABOUT HEAT.

7. The necessity of such a sign for us to judge with certainty the presence of fire is evident from the way in which our senses make us judge the heat of bodies. For one and the same body will appear to be a different temperature, according to the circumstances in which we find ourselves; thus, when we touch a body with two hands, one of which has just come out of cold water and the other out of hot water, this body seems cold and hot at the same time. The alterations to our health also change the heat of bodies for us; a man in the heat of a fever will find cold the very body that, when he was shivering, seemed hot to him. Thus, the heat that bodies make us feel cannot enable us to judge with certainty the amount of fire they contain.

II

What is the most universal effect of Fire?

What then is the most universal effect of fire? By what sign can we recognize it? I mean, recognize it as *Philosophers*, for there are two ways to know bodies, and those who study nature see it with a different eye from the common people.[33]

---

32. Du Châtelet and her contemporaries were fascinated with the effects of being deprived of one of the senses. Given her belief in John Locke's ideas that sensations and experience define our knowledge, deprivation of one sense could have significant effects on what and how one acquired knowledge. See chapter 1, note 32 for Locke. Diderot explored this idea of sense deprivation in his essay *Lettre sur les aveugles* (1749), a copy of which he sent to Du Châtelet.

33. Du Châtelet is using the term "Philosopher" in its early modern meaning, when "philosophy" included all that could be known. In England, the specific study of nature was called "natural philosophy." In French the equivalent term was "Physique [physics]."

THE MOST UNIVERSAL EFFECT OF FIRE IS TO INCREASE THE VOLUME OF ALL BODIES. RAREFACTION WITHOUT HEAT.[34]

This certain sign of the presence of fire, this effect it produces in all bodies, which one sees, touches, and measures, which operates in the void as easily as in the air, is to increase the volume of bodies before consuming their particles, spreading them in all their dimensions, and when its action is prolonged, to separate them down to their basic elements. This effect does not depend on the light and heat of fire, for the air is very rarefied on the top of mountains where the heat is imperceptible, and this rarefaction of the air, which is very much greater at the summit of mountains, such that the inverse ratio of weights suggests gravity must in part be attributed to fire, which at that height rarefies the air without perceptibly heating it.

Water, which boils at about 212 degrees on a Mercury thermometer,[35] and which over that temperature does not acquire any more heat from the most violent fire, evaporates if it goes on boiling; for it can only evaporate, its rarefaction only increases, and its particles only become more and more separated.

AND WITHOUT LIGHT.

Finally, a candle that you extinguish, and that ceases to illuminate, evaporates and is further rarefied by the smoke it produces. Thus, the rarefaction depends neither on the light nor on the heat of the fire, since it subsists in the bodies that the fire penetrates independently of their heat, and their light. . . .[36]

However, several objections can be raised about this definition, which makes rarefaction the distinctive property of fire.

OBJECTIONS TO THE UNIVERSAL RAREFACTION OF FIRE, AND ANSWERS TO THESE OBJECTIONS.[37]

1. It may be said that the rarefaction caused by fire is not always evident to us.

---

34. *Rarefaction* in the eighteenth century had a very general meaning: a decrease in density and pressure, as in the air. With fire this decrease would cause the spreading of the material being burned.

35. Daniel Fahrenheit (1686–1736), the instrument maker, invented the mercury thermometer, a great improvement over that of Réaumur, which could not withstand the same extreme heats as it was filled with ethyl, or distilled, alcohol (*esprit de vin*). Fahrenheit also discovered the changes in the boiling point of water occasioned by changes in atmospheric pressure.

36. In the eighteenth century, terms like *evaporation* were applied to all kinds of substances and had a more general meaning than ascribed to them today. In this context it means *disappearance*.

37. This is a clear example of the way in which Du Châtelet uses classical rhetorical forms to advance her views; in this case, presenting objections and then answering each one.

But it is the nature of fire that it is thus. Fire is spread equally in all bodies (as I will prove in what follows), and we cannot perceive these effects when they are the same throughout. We need differences as our *criterium* [criterion] and to guide us in our judgments. For we have no sign for knowing fire when it is enclosed between the pores of bodies; it is like the air they all contain and is only revealed to us when some cause releases it.

2. Fire, it will be said, rarefies bodies in increasing their heat.

This is true in general, but I do not believe that one could conclude from this that heat is the cause of the rarefaction, for I have just shown, for example, through the example of water that boils, that there are circumstances in which rarefaction continues increasing, although the heat no longer increases. Since heat does not always accompany rarefaction, it must be conceded that rarefaction does not depend on heat.

3. It will perhaps be said that air and water also increase the volume of bodies, and that it follows that one cannot make rarefaction the distinctive property of fire.

It cannot be denied that air and water have this effect on bodies; but in augmenting their volume, they do not separate them into their constituent parts, they do not cause them to evaporate, to move away from each other as fire does. Thus, the kind of rarefaction they sometimes cause in bodies is essentially different from that which operates because of fire. Perhaps it may even be that the kind of rarefaction that operates with air and water is caused by fire itself, for it is by movement that air and water penetrate bodies, and the internal movement of bodies probably only comes from the fire they contain.

In truth, frozen water increases in volume and floats on liquid water, although it contains much less fire when it is frozen than when it is in its fluid state, but this phenomenon must be attributed to a particular cause, which I will discuss in the second part of this work.

4. It can also be said that fire does not rarefy all bodies, that horn, animal dung, and many other bodies harden in fire, diminishing in volume. Now, these effects are precisely the opposite of rarefaction, so, rarefaction cannot be the universal property of fire, since there are bodies in which it produces opposite effects.

This objection collapses if one reflects that fire only hardens these bodies, and makes them smaller because it has really rarefied them, because it has made the water between their particles evaporate, and that then the particles that resisted its action are all the more compact, occupy all the less volume, as the fire removed more of the aqueous material from between their pores.

5. Finally, it can be objected that the rays of the Moon that are fire do not rarefy the bodies exposed to them. But the limits of our senses are so narrow that it is scarcely justified to state anything on the basis of what they report to us; thus, although the rays of the Moon, however gathered together they are, may have no effect on the thermometer, we cannot conclude from this that they are entirely deprived of the power to rarefy, we are only certain that they are incapable of exciting in us the sensation that we have called *heat*. But perhaps, some instrument will be invented sensitive enough to reveal to us in the rays of the Moon this rarefying power that appears to be inseparable from fire.

THE RAREFACTION OF BODIES BY FIRE APPEARS TO BE ONE OF THE PRIMARY LAWS OF NATURE.

The rarefaction that fire effects in all bodies it penetrates appears to be one of the primary laws of Nature, one of the mainsprings of the Creator, and the end for which fire was created. Without this property of fire all would be dense in nature; all fluidity, and perhaps all elasticity, comes from fire, and without this universal agent, without this breath of life that God has spread over his work, nature would languish at rest, and the universe could not subsist a moment as it is.

Thus, far from motion being the cause of fire, as some philosophers thought, fire is on the contrary the cause of the internal motion of the particles of all bodies.

This is the place to examine the reasons that prove that fire is not the result of motion.

III

Whether motion produces Fire

MOTION DOES NOT PRODUCE FIRE.

1. If fire were the result of motion, all violent motion would produce fire, but very strong winds like the east or the north wind, far from producing inflammation of the air and the atmosphere that they act on, produce, on the contrary, a cold that all of nature feels, and that is often fatal to animals and to the produce of the Earth.

2. There are, in chemistry, fermentations[38] which lower the thermom-

---

38. In the eighteenth century, chemists identified *fermentation* as the transformation of substances by adding an agent that caused effervescence.

eter; it is true that in these fermentations the igneous particles evaporate, since the vapor the mixture gives off is hot, thus, these very fermentations are caused by the fire that withdraws from the pores of the liquids, but it is no less true that the quantity of fire is diminished in the bodies that ferment, the parts of which are, however, in a very violent motion; so, the motion of these liquids has deprived them of the fire they contained, far from having produced any.

Finally, in these fermentations, the mixture coagulates in some places, which proves what I said above, that without fire all would be compact in nature.

3. The rays of the Moon, which are in very great motion, give off no heat.

TENTAMINA FLORENTINA.[39]

4. A mixture of ammonium chloride [*sel ammoniac*] and sulfuric acid [*huile de vitriol*] produces a fermentation that lowers a thermometer, but if one throws in a few drops of ethyl alcohol [*esprit de vin*] the effervescence stops, and the mixture heats, and then makes the thermometer rise. So this is a case in which the motion being diminished, the heat has increased: so motion does not produce fire.[40]

IV

Whether Fire has all the properties of matter?

But what is this being we call fire? Has it all the properties of matter? This is what the sagacity of Boyle, Musschenbroek, Boerhaave, Homberg, Lémery, 'sGravesande, etc. has not yet been able to determine.[41]

39. *Tentamina Florentina* refers to papers presented to the Florentine Academy, or Academia del Cimento [Academy of Experiment], which only lasted from 1657 to 1667. Nevertheless, its publication, *Saggi di naturali esperienze fatte nell Academia del Cimento*, was very influential and republished at least three times in the eighteenth century. I am grateful to R. K. Smeltzer for this identification. One of Du Châtelet's authorities, van Musschenbroek, translated the *Saggi* into Latin (1731). On van Musschenbroek, see note 41, below.

40. For translations of eighteenth-century chemical terms, see http://dbhs.wvusd.k12.ca.us/webdocs/Chem-History/Obsolete-Chem-Terms1.html, accessed 18 November 2007.

41. Du Châtelet here refers to the major experimental natural philosophers of the era: Petrus van Musschenbroek (1692–1761), Herman Boerhaave (1668–1738), Wilhelm Homberg (1652–1715), Nicolas Lémery (1645–1715), and Willem Jacob 'sGravesande (1688–1742). This was Voltaire's position, that fire had the properties of matter, in particular that it obeyed the law of gravity.

Non nostrum inter vos tantas componere lites[42]

It seems that a truth that so many competent physicists have not been able to discover is not to be known by humanity. With regard to first principles, only conjectures and probabilities are within our reach. Fire appears to be one of the mainsprings of the action of the Creator, but this power is so subtle that it escapes us.

FIRE IS EXTENDED, DIVISIBLE, ETC.

We can see clearly in fire some of the properties of matter, extension, divisibility, etc. It is not the same for two other properties of matter: there are doubts about whether fire possesses impenetrability and the tendency toward a center.

BUT IT HAS PERHAPS NEITHER WEIGHT NOR IMPENETRABILITY.

All these properties that we perceive in matter being only phenomena,*[43] there is no contradiction in supposing that there were combinations in which these phenomena do not occur. For it cannot be denied that "simple beings," which, combined, make up all sentient beings, could be combined in such a way that there would not result from their union any of the phenomena we regard as inseparable properties of the being we call *matter*, so it is for experiment to teach us whether fire is heavy and impenetrable.

V

Is Fire impenetrable?

REASONS THAT CAN RAISE DOUBTS ABOUT THE
IMPENETRABILITY OF FIRE.

It seems equally difficult to deny and to accept this property in fire. Here are some of the reasons that can create doubt about its impenetrability.

    1. We see through a hole made in a card by a pin, the fourth part of the

42. From Virgil's *Eclogue* III: "It is not for us to end such great disputes." A humble note in theory, but as it was taken from Voltaire's *Philosophical Letters* (1734), "Letter XIV" on Descartes and Newton and his description of their disagreement over the nature of matter, it has a clear satirical purpose, for she believes that one particular view is true, not that it can never be decided.

43. Du Châtelet added this comment to her text for this 1744 edition: "*It will easily be seen here that the principles of Leibnizian philosophy underlie this statement." She added the asterisk and the reference to Leibniz to indicate her continued adherence to Leibniz's ideas on matter, first demonstrated in the corrections she insisted on making when the Academy of Sciences printed the first version of her *Dissertation* in 1739.

sky, and all the objects between the horizon and us in this space. Now we cannot see an object unless each visible point of it sends rays to our eyes, so, the prodigious quantity of rays that passes through this pinhole and crosses there without blending and without producing any confusion in our view, astonishes the imagination, and one is very tempted to believe that a being that appears to be so easily penetrated is not impenetrable.

2. The most powerful fire that men have created so far is that in the focal point of the great mirror of the Palais Royal, or the mirror of Lyons.[44] And yet, the smallest discernible object can be seen through the luminous cone that melts gold in this focus, without this gathering together of rays between the object and the eye in any way detracting from the image of this object.

3. A candle carries its light in a sphere of a half-league in radius; how incredibly tiny must the particles that illuminate all this space be, since they all are contained in this candle? It is difficult to conceive of them there, if they cannot be penetrated.

4. M. Newton demonstrated to our eyes and mind that colors are nothing other than the different colored rays.[45] So in order for us to see objects, all elementary rays must cross, in passing through the pupil, without ever blending and without the blue ray taking the place of the green or the red that of the indigo, etc. This seems almost impossible if the rays are impenetrable.

5. Glass, which transmits light, has many fewer pores than muslin, which reflects it almost entirely. The pores of oiled paper that transmit rays are much less big than those of dry paper through which they find no point of passage. Thus it is not size or the quantity of pores of a body that make it permeable to light, since the way to make bodies transparent is to fill their

---

44. Du Châtelet undoubtedly means not a simple mirror, but a *miroir ardent*. This was a highly polished concave mirror that created a focal point of the light rays it reflected and that could be used to burn substances as described here. Villette, a seventeenth-century craftsman from Lyons, made four of these large reflecting mirrors, one of which he gave to Louis XIV who in turn gave it to the Academy of Sciences. This last mirror was thirty-six inches wide, and may be the one Du Châtelet is referring to at the Palais Royal. Or she is referring to Homberg's experiments with the large burning lens that belonged to the Duke d'Orléans who lived at the Palais Royal. See Voltaire, *Essai sur la nature du feu*, ed. W. A. Smeaton and Robert L. Walters, in *Oeuvres complètes de Voltaire* (Oxford: Voltaire Foundation, 1991), 17.40 n. 15.

45. Du Châtelet added the following statement to this 1744 edition of her *Dissertation*: "*The reader will undoubtedly understand that I mean by *colored ray* the ray that has the power to excite in us the sensation of such color." She and Voltaire had been studying Sir Isaac Newton's writings. Newton (1642–1727), most famous as a mathematician and physicist, described all of his experiments with lenses and his conclusions about the nature of light in *Opticks* (1704).

pores: so then it is very probable that fire is not impenetrable, since it penetrates bodies independently of the nature of their pores.

But these reasons, which can cause doubt about the impenetrability of fire, can be challenged by other very strong reasons.

REASONS IN FAVOR OF THE IMPENETRABILITY OF FIRE.

1. The rays of the Sun make smoke change direction, and reunited by a burning glass, they melt gold and stones, and cause vibrations in the spring of a watch placed half exposed in the focus of this glass. One cannot see how it would be possible for fire to act on bodies, nor how it could make this watch spring vibrate, if it did not resist the effort made by these bodies to oppose its action.

One can respond that the soul is not impenetrable and that, nonetheless, it makes our body, which is composed of resistant particles, move. And that finally all that acts on bodies is not corporeal, since God certainly is not matter, and he acts nonetheless on matter.

2. Rays reflect up from bodies to come to our eyes. The reflection necessarily carries the elasticity of the reflected body: thus, since the rays reflect, they must be composed of resistant particles.

But one can still respond that M. Newton demonstrated that it is not in rebounding off of the solid particles of bodies that light reflects, and that consequently the reflection of the light does not prove the impenetrability of fire, that even this phenomenon of reflectiveness could make one believe that light is not impenetrable. For how will the perpendicular ray return after the reflection, following the same line by which it fell, if in this line it encounters a continuation of itself, which will resist it by its solid particles and consequently prevents it from returning by the line already described? If one says that this ray will not describe an axiom of optics, considered incontestable, I ask what would be the reason for this deflection of the ray, and what would determine deflection more to the left than to the right? Finally, I am told that the extreme porosity the microscope discovers in bodies subjected to our research leads us to believe that the fineness of the constituent particles of fire can suffice: to cause the reflection of the perpendicular ray, and all the phenomena of light that most astonish our mind and that could make us doubt the impenetrability of fire. I ask how it can be conceived that a ray composed of a thousand pores separating its solid particles could come from the Sun to us in a straight line, without being interrupted and without merging with the millions of other rays of different colors that emanate at the same time from the Sun?

So one is obliged to acknowledge that there is some ground for regarding the impenetrability of fire as doubtful.

VI

Does Fire tend to the center of the Earth?[46]

Philosophers will probably agree that there can be several bodies that do not tend toward the center of the Earth; such must be, for example, the matter that makes gravity and that drives bodies toward the center of the Earth.[47] Let us see then if this is the case with fire, or whether it tends toward the Earth like other bodies.

It is again experiment, this great master of philosophy, that teaches us whether fire has this property.

I will be content to examine here M. Homberg's experiment on the weight of pure antimony [*régule d'antimoine*] calcinated in a burning glass, and that of M. Boerhaave on the weight of flaming iron.[48]

M. Homberg reports that 4 *onces* of pure antimony exposed at a *pied* and a half from the true focal point of the Palais Royal mirror, increased their weight by 3 drams and some grains during their calcination, that is to say, by about a tenth. But that, having next been brought to the melting point at the true focal point, they lost this acquired tenth, and an eighth of their own weight.

M. Boerhaave, by contrast, having weighed 8 *livres* of iron, found no difference in weight between this flaming iron and this iron when absolutely cold.[49]

Several remarks can be made about these two experiments.

---

46. This section has significant additions in this 1744 version to make more explicit her disagreement with Voltaire about the nature of fire. He insisted that fire is matter, subject to the laws of gravity, Newton's position.

47. Like most of her contemporaries, including Newton, Du Châtelet assumes a mechanical explanation for gravity, some sort of particles or a substance that make it operate on bodies.

48. Du Châtelet is referring to Homberg's *Essais de chimie* [Essays on Chemistry] (1702–09) and Boerhaave's *Elementa chemiae* [Elements of Chemistry] (1733). *Calcination* is the process of turning a substance into powder through heating it.

49. If a grain is the basic apothecary's unit, roughly 0.065 g, a dram is 60 grains, or roughly, 3.89 g. A *livre* was the basic French measurement for weight, roughly 489.59 g or a bit more than an English pound. Sixteen *onces* made up one *livre*, 30.59 g. Available at http://en.wikipedia .org/wiki/Apothecaries27_system, accessed 19 November 2007.

EXAMINATION OF THE EXPERIMENT OF M. HOMBERG ON THE WEIGHT
OF CALCINATED PURE ANTIMONY ON A BURNING GLASS.

1. During all the time of the calcination of pure antimony by M.
Homberg, it had to be stirred with an iron spatula; now the heat may well
have detached some particles of this instrument, which being added to the
pure antimony will have increased its weight. The salts and sulfurs, with
which the air is always laden, may also have become fused with the pure antimony by the action of the fire, and with the help of the continual movement of the spatula used to stir it. Thus, one is very far from being sure that
it was the fire that increased its weight; for if fire is the most subtle solvent
in nature, it is also the most powerful agent for uniting bodies.

2. What confirms this conjecture is that the bodies that most increase
their weight by fire are those that are stirred during their calcination, and
that lose all the weight acquired and even their own substance when they
are melted again. Boyle himself acknowledges that the continual agitation
during calcination is what most contributes to the increase in the action of
fire on bodies.[50]

3. M. Homberg's pure antimony having been melted at the true focal
point, lost all the weight acquired, as well as an eighth of its own weight.
Now if some particles of fire had increased its weight in calcination, how
could it have lost this weight at the true focal point? Should not, on the contrary, a new fire have produced a new increase, and since the weight is diminished in melting instead of increasing, is it not likely that the fire at the
focal point, being more violent than that which had calcined it, separated
the heterogeneous particles that were united in pure antimony and that had
increased its weight during the calcination?

4. All melting metals lose in weight, and yet melting is the state in
which they receive the greatest quantity of fire. Thus if fire increased the
weight of bodies, it should considerably increase that of melting metals, but,
on the contrary, their weight diminishes. So it is certain that the greatest
quantity of fire these metals can receive does not increase their weight.

It is easy to see that the diminution in weight of melting metals must be
attributed to the particles that this violent fire causes to evaporate from between their pores, and to the increase in their volume.

---

50. Du Châtelet probably read of Boyle's experiment in the *Philosophical Transactions* of the London Royal Society. He wrote on the actions of fire in the 1670s. Later in this *Dissertation* she
identifies one such treatise.

EXAMINATION AND CONFIRMATION OF M. BOERHAAVE'S EXPERIMENT
ON THE WEIGHT OF FLAMING IRON.[51]

5. The iron of M. Boerhaave, while it was all sparkling with fire, must
have contained very many more igneous particles than the pure antimony of
M. Homberg, which had been calcinated at 18 *pouces* from the true focus of
the mirror; and yet this iron, all permeated with fire, did not weigh a grain
more than when it was completely cold. However, I do not see any reason
why, if fire had weight, it would not always increase the weight of the bodies
it penetrates. I can certify that this equality of weight was found in masses of
iron weighing from one *livre* to 2,000 *livres*, which I had weighed before me
all aflame, and after that, entirely cold.

OTHER EXPERIMENTS ON THE WEIGHT OF FIRE. . . .

6–13. [Du Châtelet describes experiments conducted by Boyle, Hartsoeker, and Boerhaave with a variety of materials, in the open air and in a
void.[52] She concludes that all prove that increases in weight are caused by
the different circumstances of operation, not by fire.]

That is why, of all the repeated experiments on the weight of bodies exposed to fire, none is entirely the same. The increase the same fire causes in
bodies is now greater, now less, as one can be convinced by reading about
Boyle's experiments or conducting them oneself; this proves that the increase in the weight of bodies must not be attributed to a cause as invariable as fire. . . .

Now, in supposing the emission of light, all the fire that the Sun sends
over our hemisphere during one hour of the hottest day of summer must
scarcely weigh what M. Homberg supposes had penetrated into his pure
antimony. Here is the demonstration, if I am not mistaken.

The speed of the rays of the Sun has been known since the observations that Messrs. Huygens and Rømer made on the eclipses of the satellites

---

51. Du Châtelet and Voltaire duplicated many of these experiments at a forge near Cirey
with Voltaire insisting that they proved fire had weight despite the inconclusive results they
achieved. In this essay, Du Châtelet concludes the opposite.

52. She identifies Boyle's treatise as *De flammæ ponderabilitate* [On the Weight of Fire] (1673),
which described experiments with iron, zinc, and charcoal. She mentions Gilles-François Boul-
duc (1728–1769), a chemist and royal pharmacist, whose experiments with pure antimony she
describes. Also noted are Hartsoeker's experiments with pewter and lead and Boerhaave's with
lead. Nicolaas Hartsoeker (1656–1725) was a Dutch instrument maker and natural philosopher,
known particularly for his work in optics and physics.

of Jupiter;[53] this speed is around 7 to 8 minutes from the Sun to us. Now if the Sun is at 24,000 demi-diameters from the Earth, it follows that, in coming from this star to us, the light traverses a thousand million *pieds* per second in round numbers; and a cannonball of one *livre* of shot propelled by a half *livre* of powder only covers 600 feet in a second, thus the speed of the rays of the Sun surpasses, in round numbers, 1,666,600 times that of a one-*livre* cannonball.

But, the effect of the force of bodies being the product of their mass by the square of their speed,[54] a ray that was only the 1/2777555560000th part of a one-*livre* cannonball would have the same effect as the cannon, and a single instant of light would destroy all the universe. Now I do not believe that we can determine a *minimum* for the extreme fineness of a body that, being only the 1/27775555560000th part of a one-*livre* cannonball, would have such terrible effects, and of which millions of billions pass through a pinhole, penetrate the pores of a diamond, and strike ceaselessly the most delicate organ of our body, the eye, without wounding it or even being felt.

14. . . . So I believe that it is incontrovertibly demonstrated, by the way we see, by the phenomena of light, and by the first laws of the collision of bodies, that (supposing fire has weight) we cannot perceive its weight, and that if all the rays of the Sun sent to our hemisphere during the longest day of summer weighed only 3 *livres*, our eyes would be useless, and the Universe could not sustain a moment of light.

---

53. Christiaan Huygens (1625–1695) was a Dutch mathematician and astronomer, author of *Systema Saturnium* [Saturn's System] (1659) on the rings of Saturn; Ole Christensen Rømer (1644–1710) was a Danish mathematician and astronomer who worked with French astronomers of the Royal Observatory in Paris, and was the first to determine the velocity of light.

54. Note that in the eighteenth century, the word *force* did not have its modern meaning; it only signified the impetus for motion. This is the controversial paragraph that Du Châtelet insisted must be annotated in the version published by the Academy of Sciences. She had originally given the formula as f = mv and credited the Academy *Mémoire* [Memoir] of 1728 by Jean-Jacques Dortous de Mairan (1678–1771). Dortous de Mairan became the official Academy spokesman for Cartesian physics and this formula. Although Academy policy did not allow revisions to a submission, Réaumur, perhaps at the urging of Maupertuis, whom Du Châtelet enlisted in her support, agreed to allow her to add a list of "changes," of which this was by far the most significant. She removed the reference to Dortous de Mairan and replaced it with Johann Bernoulli's formula, f = mv². Bernoulli (1667–1748) was the Swiss mathematician with whose son, Johann Bernoulli II (1710–1790), Du Châtelet corresponded. He was known for his work in calculus and as a vocal adherent of this Leibnizian formula. Thus, Du Châtelet signaled her disagreement with the official position of the Academy's most important members. For a complete description of the controversy over this formula, see Mary Terrall, "Vis Viva," *History of Science* 42 (2004): 189–209.

M. MUSSCHENBROEK'S ARGUMENT IN FAVOR OF THE WEIGHT OF FIRE.[55]

15. The learned M. Musschenbroek made a seemingly very strong argument in favor of fire having weight: *The glowing iron that you weigh*, he says, *you weigh in air which is a fluid. Now fire having increased the volume of this iron by rarefaction, it should weigh less in the air when it is hot and its volume is greater than when it contracts because of the cold and its volume is diminished. You only find the same weight in the cooled iron because the fire had really increased the weight of the flaming iron; for if [fire] had not increased it, you should have found your iron less heavy when it was glowing, than when it was cooled.*

ANSWER TO THIS ARGUMENT.

This argument would be incontrovertible if one could be sure that no other body than the fire had been introduced in the flaming iron; but we are very far from being sure of it, for if foreign bodies can mix with bodies calcinated by the rays of the Sun (the purest fire we know), will it not be all the easier for particles of wood or charcoal to penetrate the bodies exposed to ordinary fire? Thus it will be easily perceived that, in refuting M. Homberg's experiments, I have also intended to refute those of Messrs. Boyle and Lémery, in a word, all those made on bodies increased in weight by fire; this increase that fire here on Earth causes in bodies should even be perceptible because of the quantity of heterogeneous particles it must introduce to their pores, and it is only imperceptible in some because they lose much of their own substance by the action of the fire, and because their specific gravity diminishes with rarefaction.

From all these experiments it must, then, be concluded that fire has no weight, or that if it does, it is impossible for its weight ever to be perceptible to us.

VII

What the distinctive properties of Fire are

FIRE NATURALLY TENDS UPWARD.

But if after examining the experiments on the weight of fire, we come to consider its nature and to look for its properties, we cannot but acknowledge that far from tending toward the center of the Earth, as other bodies do, it flees from this center, and that its action naturally carries it upward. . . .

---

55. Du Châtelet is probably quoting from Musschenbroek's *Elementa physicae* [Elements of Physics] (1734).

[Du Châtelet describes experiments proving this from the *Tentamina Florentina* of the Florentine Academy.]

This tendency of fire to rise depends on another property specific to fire, by which it tends to equilibrium and spreads equally in space when nothing opposes it. Thus, fire always tends to disengage from the pores of bodies and to spread upward where there is no perceptible atmosphere and where it can extend equally to all sides without obstacle. For the atmosphere contributes infinitely to the heat in which we live, as the cold that prevails in the mountains proves.

A very simple experiment, which I have often repeated, proves this tendency of fire to rise.

If you put a plate or a board on one of the tall glass cylinders that serve as covers for candles in the summer, and you leave a lighted candle under the covered cylinder, there is no doubt that the heat of the flame must at every moment rarefy the air trapped in this glass. So if the flame rose only because of its specific lightness (as is claimed), it should continuously be seen to become round and lose its conical shape, since this air trapped in the cylinder rarefies at each instant, but this is not what happens. The flame keeps this conical shape up to the moment when it goes out and, when it is much diminished in height, its point can still be seen tending upward.

WHY THE FLAME RISES IN VERY RAREFIED AIR.

The cause of this phenomenon is that the flame of the candle contains enough fire for it to oppose the natural tendency of this flame toward the center of the Earth, and for the fire to make it rise by the superiority of its force, independent of the specific gravity of the air. Fire would perhaps not have the same effect on all flames, for some contain far fewer igneous particles than others. . . .

[Du Châtelet explains the similar behavior of smoke in experiments done in a void.]

M. Geoffroy[56] made an experiment in which one sees by the naked eye that fire tends to spread equally to all sides, and that it ceaselessly endeavors to separate the particles of bodies one from the other. For this clever academician reports that, having melted iron with a reflecting mirror and having collected the sparks that it threw off, he found that these sparks were each a small globe of hollow iron. So the fire had fought against the cohesion of these iron particles and their weight, and had overcome them.

---

56. Etienne-François Geoffroy (1672–1743), French chemist, most famous for his *Table des rapports* [Table of Affinities] (1718), to which Du Châtelet may be referring.

FAR FROM BEING SUBJECT TO IT, FIRE IS THE ANTAGONIST OF GRAVITY.
NO REST IN NATURE.

So, far from being subject to gravity, fire is perpetually antagonistic to it. Thus, all in nature is in perpetual oscillations of dilation and contraction caused by the action of fire on bodies and the reaction of bodies that oppose the action of fire by their weight and the cohesion of their particles. And we do not know of any perfectly hard bodies, because we do not know any that does not contain fire and the particles of which are in perfect repose. Thus, the ancient philosophers who denied absolute rest, were surely more sensible, perhaps without knowing it, than those who denied motion.

FIRE CONSERVES AND VIVIFIES EVERYTHING IN NATURE.

Without this perpetual action and reaction of fire on bodies, and bodies on fire, all fluidity, all elasticity, all softness would be banished; and if matter were for a moment deprived of this spirit of life which animates it, of this powerful agent ceaselessly opposing the adunation of bodies,[57] all would be compact in the universe, and it would soon be destroyed. Thus, not only do the experiments not demonstrate the weight of fire; but claiming that fire has weight is to destroy its nature, in a word, to take away its most essential property, that by which it is one of the mainsprings of the Creator.

FIRE IS EQUALLY SPREAD THROUGHOUT. ALL BODIES IN THE SAME AIR
CONTAIN THE ELEMENT OF FIRE.

Another attribute of fire, which also appears to be specific to it, is to be equally distributed in all bodies. Men must probably have taken a long time to accept this truth. We are inclined to believe that marble is colder than wool; our senses tell us this, and to undeceive us it was necessary for us to create, as it were, a being [device] able to measure the degree of heat spread through bodies; this being [device] is the thermometer. *It* taught us that the most compact matter and the lightest, the most volatile and the coldest, marble, hair, water, and ethyl alcohol, Boyle's void, and gold, in a word, all bodies (except animate creatures) in the same air contain the same quantity of fire.[58]

57. *Adunation* is a term from chemistry that Du Châtelet took from Boyle's writings, meaning *union* or *combination*.

58. A set of experiments conducted in the gallery of Cirey by Du Châtelet and Voltaire. Note that Du Châtelet treats a void as a "body."

FIRE IS SPREAD NOT ACCORDING TO MASSES, BUT ACCORDING
TO SPACES.

It follows from this property of fire, 1. that all bodies are equally hot in the
same air, since they all have the same effect on the thermometer. 2. that fire
is distributed not according to masses but according to spaces, since gold and
the pneumatic void contain it equally. 3. that no body is permeated with fire
more than another, or can retain a greater quantity, since in the same air ethyl
alcohol is not hotter than water, and they cool off to the same extent.

ETHYL ALCOHOL DOES NOT CONTAIN MORE FIRE THAN WATER.

If our senses tell us that wool contains more fire than marble, our reason
seems to tell us ethyl alcohol contains more than water. It refracts light
more, the least fire inflames it, it is entirely consumed by the flame, it never
freezes; lastly this liquid appears to be all igneous, especially when it becomes alcohol by distillation. However, despite all these phenomena, the
thermometer decides for equality, and one cannot see how ethyl alcohol
could contain more fire than other bodies without the thermometer's causing us to notice it. For it cannot be said that this greater quantity of fire
contained in ethyl alcohol is in equilibrium with its particles, just as a lesser
quantity is in equilibrium with that of water, and when action and reaction
are equal, it is as if there had been no action. This would be supposing a
thing entirely contrary to all that we know of the action of fire on bodies
and of the reaction of bodies on fire; bodies only resist the action of fire by
their mass, or by the coherence of their particles. Now ethyl alcohol is of all
fluids the lightest (if you make an exception of air), and its particles appear
the least coherent; alcohol, which is lighter than ethyl alcohol, is yet more
flammable. Thus, the more one considers fire as a body acting according to
the laws of collisions with other bodies, the less likely it will appear that the
lightest body of all should resist the action of fire most. Thus, since the thermometer demonstrates that ethyl alcohol contains no more fire than water, it
must be acknowledged that fire is equally distributed in all space without regard to the bodies it fills. If ethyl alcohol scatters light more than denser liquids, if it never freezes, this probably depends on the structure and the distribution of its pores, and not at all on a greater quantity of fire contained in
its substance. And if it inflames more easily, that is because it contains more
pabulum ignis, and that its particles are more easily separated.[59]

59. *Pabulum ignis* is a term from the Old Testament (Leviticus 3:11) signifying a quality that
causes matter to burst into flame. Later in the eighteenth century, the concept became associ-

Marble appears to us colder than wool because, being more compact, it touches our hand at more points, and consequently it takes more of our heat; thus, despite all appearances, we are forced to acknowledge this equal distribution of fire in all bodies.

The artificial cold that Fahrenheit found the way to produce, and that made the thermometer fall to 72° below the point of freezing, proves that in the greatest cold we experience, no body is deprived of fire, it lives in us and in all time.

FIRE TENDS BY ITS NATURE TO EQUILIBRIUM

This equal distribution of fire in all bodies, this equilibrium to which it tends by its nature and which remained so long unperceived, was, however, indicated to us by a thousand effects caused by the operations of fire ceaselessly before our eyes, though we ignored them.

PROOFS

[Du Châtelet offers eight proofs from experiments and by reasoning from observations. Three examples follow.]

2. A body all sparkling with fire, to which a cold body is applied, loses its heat until it has communicated to this other body a quantity of fire that reestablishes the equilibrium between them.

3. Potassium carbonate [*huile de tartre, huile de tartre par défaillance*], which seems so igneous to us, and distilled turpentine oil [*huile de térébenthine*], which protects our bodies from the cold and seems so hot to us, are by themselves no more so than pure water. For being mixed with water, they do not change its temperature, which proves that the effervescence caused by some liquids when added to water does not happen because these liquids contain more fire than pure water. . . .

6. The same fire that melts gold and stones in the focus of a reflecting mirror spreads a heat in the air that is scarcely perceptible to us. For air does not oppose the equilibrium of fire like gold and other bodies, which by their solidity retain it some time in their pores. This is also why the fire of the Sun rarefies the upper atmosphere without heating it perceptibly, for the pressure of the atmosphere no longer opposing resistance to the fire, it spreads without obstacle and is no longer sufficiently dense for us to notice its heat.

---

ated with the idea of "phlogiston," a property in all matter that accounted for its inflammability. Johann Joachim Becher (1635–1682) first conceived of this property. It was popularized in France after Du Châtelet's death. Its most famous proponent was the English chemist, Joseph Priestley (1733–1804).

The necessity of this atmospheric pressure for the heat of fire, is perceptibly demonstrated with water, which acquires a greater degree of heat in boiling, in proportion to the greater weight of the atmosphere. . . .

So one of the distinctive and inseparable properties of fire is to be spread equally in all space, without any regard for the bodies it fills, and to tend to reestablish the equilibrium of heat between bodies, as soon as the cause which disrupted it ceases.

VIII

Conclusion of the first part

I conclude from all that I have said in this first part:

1. That light and heat are two very different effects and very independent of each other, and that they are two ways of being, two modes, of the being we call *fire*.

2. That the most universal effect of this being which it causes in all bodies and in all places, is to rarefy bodies, to increase their volume, and to separate them down to their elementary particles, when its action is continuous.

3. That fire is not the result of motion.

4. That fire has some of the properties of matter, its extension, its divisibility, etc.

5. That the impenetrability of fire has not been demonstrated.

6. That fire is not heavy, that it does not tend toward a center like other bodies.

7. That it would be impossible (even supposing that it had weight) for us to perceive its weight.

8. That fire has several properties specific to it, over and above those it has in common with other bodies.

9. That one of its properties is not to be determined toward a point, but to spread equally through all bodies, and by its nature to tend to equilibrium.

10. That it is this property that ceaselessly counters the adunation of bodies, and lastly, that it is because of this that fire is one of the mainsprings of the action of the Creator, whose work it vivifies and conserves.

11. That fire is the cause of the internal motion of the particles of bodies.

12. That fire is capable of more or less motion, but that absolute rest is incompatible with its nature.

13. That fire is equally spread in all space, and that in the same air all bodies contain an equal quantity of it, except for the creatures that have life.

After examining the nature of fire and its properties, it remains for me to examine the laws it follows when it acts on bodies, and when its effects are perceptible.

SECOND PART: *OF THE PROPAGATION OF FIRE*

I

How Fire is distributed in bodies

Fire is distributed here on Earth in two different ways.

1. Equally in all space, whatever bodies it fills, when the temperature of the air which contains them is even.

2. In creatures endowed with life, which contain more fire than vegetable matter, and the other bodies in nature.

FIRE ACTS ON ALL NATURE.

Fire being spread all over exerts its action on all nature; it unites and dissolves everything in the universe.

But this being, whose effects are so powerful in our operations, eludes our senses in the operations of nature and very precise experiments. Very profound reflections were required for us to discover the imperceptible action that fire produces in all bodies.

If the equilibrium fire seeks were never disrupted, neither in ourselves nor in the bodies that surround us, we would not have any idea of cold, or of heat, and we would only know the light of fire.

But as it is impossible for the universe to subsist without this equilibrium being constantly disrupted, we sense almost constantly the vicissitudes of cold and of heat that the changes in our own temperature or in that of the bodies around us make us feel.

The action of fire, whether it is concealed from us or perceptible, can be compared to *force vive* [live force] and *force morte* [dead force];[60] but just as

---

60. Du Châtelet, following the ideas of Leibniz, and the Swiss mathematician Johann Bernoulli, distinguished between *force vive*, equivalent to the modern concept of kinetic energy (mv^2, now expressed as $\frac{1}{2} mv^2$), and *force morte* (mv), the concept of momentum. In this, she disagreed with Newton and with most of the prominent mathematicians and physicists in France. From Bernoulli and Leibniz she took the idea of opposing bodies using their force

the force of bodies is perceptively stopped without being destroyed, so fire conserves in this state of apparent inaction the force by which it opposes the cohesion of the particles of bodies. And the perpetual combat of this effort of fire and of the resistance bodies offer to it, produces almost all the phenomena of nature.

Thus, fire can be considered in three different states, resulting from the combination of these two forces.

1. When the action of fire on bodies, and the reaction of bodies to it, are in equilibrium, then it is as if there were no action, and the effects of fire are imperceptible to us.

2. When this equilibrium is disrupted and the resistance of the bodies prevails over the force of fire, then bodies become more compact. Part of the fire they contain is forced to leave them, and they give to us the sensation of cold.

3. Finally, when the action of fire prevails over the reaction of bodies, then the bodies heat up, rarefy, become luminous, depending on whether the quantity of fire they receive in their substance increases, or the force of the fire they naturally contain is more or less excited. If this power of fire passes certain limits, the bodies on which it acts melt or evaporate. In this case, the fire, no longer having an antagonist, forces by its tendency to move *quaquaversum* [every way] the particles of the bodies to avoid one another, move them farther and farther away from each other, until it has entirely separated them. . . .

. . . [T]hus, from these two forces combined, *the coherence of bodies and the effort that fire makes to oppose it*, result all the gathering together and all the dissolutions of the universe: cohesion uniting, compressing, connecting the particles of bodies, and fire, in contrast, moving them apart, separating them, rarefying them.

So it is necessary to examine the different effects resulting from the combination of these forces.

II

Of the causes of the heat of bodies

A body heats either because it receives fire in its pores or because that which is enclosed in it is subject to a new motion.

---

to hold each other at rest after a collision, rather than the force being completely dissipated by the collision.

It seems to me the different causes that can produce these effects on bodies can be ascribed to two principles.

The first is the presence of the Sun and the direction of the rays it sends us; the presence of the Sun gives bodies a new fire in their pores, and they receive all the more as the incidence of its rays is more perpendicular. . . .

The second cause of heat, which interrupts the equilibrium to which fire tends, is the friction of bodies, against one another. All the ways in which fire here on Earth may be excited are only modifications of this cause, just as all our senses are only variations of touch.[61]

HOW THE FIRST MEN DISCOVERED FIRE.

It is probably this friction between bodies that made fire known to the first men. Conflagrations of some forests from the agitation of their branches, or the collision of two pebbles, introduced them to this being that animated them, the existence of which they did not even suspect.

Thus, for a long time, the first men will have been able to see the light of the Sun, and to feel its heat. They will have been able to experience the vicissitudes of cold and heat caused by health and sickness without having any idea of fire, that is to say, of this being we have the power to excite, and so to speak, to create. For the first fire that men produced must have appeared to them a true creation.

Nature having allowed men to divine the secret of fire, it must have been a long time before they suspected that the rays of the Sun and the fire that they lit were of the same nature. Burning glasses had to be invented for them to know that this Sun, whose return brings health and rejuvenates all nature, is capable of destroying everything as well as vivifying everything, and that the effect of its rays, when focused, very much surpasses that of fire here on Earth.

III

Of the Fire produced by friction

The second cause, which shows the fire that bodies contain, acts all the more powerfully as the bodies that are rubbed together are in closer contact; thus, three things can increase the effects that fire produces by friction.

61. Du Châtelet is thinking of all actions on our senses as a form of touching, as light rays touching the retina.

1. The mass of the bodies.
2. Their elasticity.
3. The rapidity of motion. . . .

[Du Châtelet explains how these factors affect friction; for example, that the less dense mass of fluids explains their lack of friction. In fact, they decrease friction, as in grease on the hubs of carriage wheels.]

The friction of bodies is both the most universal and the most powerful cause for arousing the power of fire . . . [for example, stone struck against iron] is probably one of the greatest miracles of nature: that the most violent fire can be produced in a moment by the collision of apparently very cold bodies.

IV

Of the action of Fire on solids

[Du Châtelet notes the tendency of fire to cause a solid to expand, making reference to experiments described by Bernoulli and Musschenbroek, who used a pyrometer to measure the expansion of metals when heated. From these experiments she deduces a number of rules: that a body acquires a predetermined amount of heat past which even the most violent fire cannot heat it more; that the more heat a body acquires, the slower its expansion; that there is no uniform degree of temperature at which bodies melt; that mixing two metals together may result in their combination melting with less fire.[62]]

METALS NO LONGER HEAT AFTER MELTING.

7. When the expansion of bodies is in its last phase, their particles have to yield to the action of the fire and separate; then fire makes them pass from the state of solids to that of fluids, and this is the last step in the action of the fire on them. For their pores being sufficiently dilated, they give off as many of the particles of fire as they have received; thus the heat of the bodies no longer increases after melting.

If the power of fire on bodies were not limited, fire would soon destroy the universe. These limits that the Creator imposed on it, and which it never

62. These conclusions must have been the result of the experiments done with Voltaire at the forge near Cirey, or in the gallery of the château.

oversteps, are one of the great proofs of the design that reigns in this universe. . . .

[Du Châtelet concludes this section with further speculations on the effects of fire; for example, that less heat at the poles explains why "Lapps are small and robust," and that bodies heat more or less depending on their color. She suggests further experimentation to test for a correlation between a color and its heat; for example, does red give off more heat than violet?]

V

How Fire acts on liquids

BOILING WATER DOES NOT ACQUIRE MORE HEAT.
Few things would be known of the way in which fire acts on liquids without the discovery of M. Amonton. One knows that this learned man, in seeking the means to make a more perfect thermometer than that of Florence, discovered that boiling water acquires a fixed degree of heat, past which it heats no more even by the greatest fire.

The celebrated Messrs. Réaumur and Fahrenheit, this artisan philosopher, both perfected Amonton's discovery. . . .[63]

It is easy to understand the reason for what happens then under different atmospheric pressures, for when the surface of the water is pressed down by a greater weight, fire separates its parts with more difficulty, and consequently it requires a greater quantity of fire to make it boil, since it is in this separation of the particles of the liquids that boiling consists. Thus, it is probable that water, pressed by a weight equivalent to that which the atmosphere would have at 409,640 *toises* below the surface of the Earth,[64] would shine like melting metals, for at that depth the weight of the atmosphere would be equal to that of gold, according to the calculations of M. Mariotte. . . .[65]

[Du Châtelet describes numerous experiments on the varying boiling

63. Guillaume Amonton (1663–1705), a French scientist, known primarily for perfecting a variety of scientific instruments.

64. A unit of measurement, similar to the English fathom, but used on land and at sea by the French. It is six *pieds* or about 1.949 m.

65. Edmé Mariotte (1620–1684), French physicist, had his essays on the effects of changes in pressure on the volume of gases and on color published in the *Recueils* [collections of essays presented at meetings] of the Royal Academy of Sciences.

points of liquids and of mixtures of liquids, which she found particularly interesting.]

The heat that hot fermentations produce lasts until the motion of the liquids ceases, then they return to their original temperature, just as the heat that solids acquire by friction dissipates as soon as the internal motion of their particles ceases. . . .

But the most singular effect of these mixtures, which appears entirely inexplicable, is that two equal quantities of any liquid whatever, acquire by mixing, a degree of heat that is half the difference of the heat that these two portions of the same liquid had before being mixed. Thus, a *livre* of water that registers on the thermometer at 32°, when mixed with another *livre* of boiling water registered at 212°, makes the thermometer rise to 90° after the mixing. Now 90° is half the difference between 32° and 212°.

In whatever manner one explains this singular phenomenon, it is always certain that it is a new proof of the equality with which fire spreads in bodies.

In all fermentations, be they hot or cold, motion continues until liquids have regained their ordinary temperature. This combat between the action of fire on bodies and the resistance bodies offer to it by their cohesion proves to us that cold fermentations also depend on the combination of these two forces.

VI

How Fire acts in plants and animals

THE PRINCIPLE OF LIFE SEEMS TO BE IN FIRE.
The thermometer teaches us that creatures that are endowed with life contain a greater quantity of fire than the other bodies in nature, the greatest heat in the summer in our climates being 80° and rarely 84°, and that of a healthy man being 90° or 92°, and in children, it even goes up to 94°.[66] Thus, the principle of life seems to be in fire, since animate creatures have received a greater quantity of it than others, and children, in whom the principle of life is as yet undepleted, have a greater degree of heat than grown men and grown men more than old men. . . .

[Du Châtelet continues to say that animals have greater heat than men and are more vigorous.]

66. These measurements reflect the inaccuracy of even Fahrenheit's thermometer.

PAGE 148.[67] WHAT DEGREE OF HEAT CAUSES ALL ANIMALS TO PERISH.

The celebrated Boerhaave, in his excellent *Treatise on Fire*, reports that several animals having been put in a place where sugar is dried and where the heat was 146°, not only died very soon, but also their blood and all their humors became corrupted, in such a way that they produced an unbearable smell. Men cannot sustain the heat of this place, and the workers who labor there must take turns almost constantly to go and breathe fresh air. M. Boerhaave concludes from this experiment and some others that we would soon die if the air that surrounds us only caused the thermometer to rise to 90° [*sic*]. Thus, we can regard this degree of heat as roughly the point at which all animal kind would perish.

In 1709 the thermometer dropped to zero degree in Iceland, and animal-kind did not perish; thus, it is probable that we are more capable of withstanding great cold than great heat, provided, however, that it does not last long.

WHEN IT IS THAT COLD STOPS PLANT GROWTH.

Plant growth ceases at the point of freezing, for although trees and some grasses, like hay, resist it, they do not grow as long as the air is at this temperature. Thus, this point can be regarded as the limit for vegetation in relation to cold, and if it was prolonged, trees and plants, no longer growing, would soon be entirely destroyed.

The degree of heat of melted wax which, floating on hot water, begins to coagulate, can be regarded as the highest point allowing growth in relation to heat. For since a greater heat would melt wax, which is a vegetable substance, this heat would disperse and separate the nutritive matters instead of gathering and uniting them, and then plants could only decline.

VII

Of the nutriment of Fire

[Following the work of other chemists, and using the experiments that she and Voltaire conducted on the grounds of Cirey, Du Châtelet reports that different substances burn "according to their respective densities," and that air and atmosphere are necessary to sustain fire. She accepts the common

---

67. Du Châtelet, unlike her contemporaries, usually gives the sources for her points, even, as in this case, the actual page.

idea that fire's basic nutrient is *pabulum ignis*, a substance that is in all bodies and constitutes its lightest and most volatile particles.]

WHY WATER EXTINGUISHES FIRE, AND WHY WIND INFLAMES IT.
It is for this reason that water extinguishes fire, and a bellows inflames it. For water prevents the oscillations that the air communicates to fire from reaching it, and in contrast, a bellows produces stronger and more frequent vibrations of the atmosphere.

The force with which a double bellows at a forge blows air into the fire, being equal to the 30th part of the weight of the atmosphere,[68] this force must expel the air with great speed, and constantly renew it. It may be grasped from this how much a violent wind must increase the fire.

SOME CAUSES OF THE EXTINCTION OF FIRE.
Fire lasts as long as the action and reaction caused by this atmospheric pressure continue. Thus, three things can make fire cease.

1. The consumption of combustible bodies.
2. The suppression of the weight of the atmosphere.
3. The destruction of the air's elasticity.

VIII

If Fire is the cause of elasticity[69]

[Du Châtelet argues that rather than being the cause, fire destroys the elasticity of the air and in all bodies.]

IX

If electricity depends on Fire

FIRE SEEMS TO BE THE CAUSE OF ELECTRICITY.
There are good grounds for believing that fire is the cause of electricity.

68. Elsewhere in her *Dissertation* Du Châtelet identifies the weight of an atmosphere as 2,240 pounds per square *pied*, a close approximation.
69. *Elasticity* was the ability of a body to regain its shape after a collision.

Analogy, this thread that we have been given to guide us in the labyrinth of nature, makes, it seems to me, this opinion quite probable.

1. All bodies contain fire, almost all have the property to retain and to produce light, and all become electric by friction, metals and liquids being excepted. But these bodies do not become electric by themselves, but by communication.

2. There is no electricity without friction, and consequently without heat.

3. Almost all electric bodies show outwardly the cause that animates them, by the sparks they throw off, which can be perceived in darkness.

4. Their light subsists after their electricity is destroyed, just as there are bodies that give light without heat.

5. Freezing and calm weather are more favorable to electricity than a great heat, as in a reflecting mirror.

6. Fire and electrical matter need air to be active.

7. Bodies the most susceptible to electricity are the least appropriate for transmitting it, just as bodies reflect much less light as they heat up more.

8. Humidity destroys electricity in bodies without destroying their electric light, just as water cools bodies, but does not extinguish dails and glowworms, etc.

9. Homogeneous bodies become suffused with electricity in proportion to their volume, just as fire distributes itself according to volumes and not according to mass.

10. Bodies become more electric when heated before being rubbed, etc.

It seems by all these effects, that, with some probability, fire may be considered the cause of electricity. . . .

[Du Châtelet acknowledges that there are other phenomena of electricity that do not follow the analogy with fire, but assumes that future "work, application, and sagacity of the mind" will discover their cause.]

X

How Fire acts in the void

Air appears to be as necessary for fire to burn as it is for animals to live; however, the pneumatic machine [vacuum pump] has demonstrated to us that this very general rule also has its exceptions. . . .

[Du Châtelet then describes Hauksbee's and Boyle's experiments on flaming substances in a void,[70] and also those that she and Voltaire conducted at Cirey. She ascribes variations in results to the amount of the element of fire in the substances and to the purity of the substance. She believes the substances do not freeze because "the air takes all their heat." She reports that effervescence operates the same in air and in a void.]

XI

For what reason Fire acts

[Du Châtelet describes six experiments and from them reasons that light and fire do not act from the same cause: the quantity of their particles. After watching "the prodigious effects of burning glasses" she suggests that rays of fire particles, as they approach one another, acquire another force in proportion to their proximity to one another.]

The effort particles of fire ceaselessly make to avoid one another and to spread equally to all sides is visible to the naked eye when two candles are brought close together and an attempt is made to unite their flames. For the closer they are brought together, the more clearly they can be seen moving apart, well away from each other.

It does indeed appear that fire always acts in bodies for a reason compounded of these two reasons: the density of its particles and the force they acquire in approaching each other. . . .

Effervescence shows us that most of the particles of matter are to one another as little magnets and that they have an attracting and a repulsing side. The tendency the particles of bodies have to stay together by their cohesion, and the effort of the fire held in their pores ceaselessly to make them separate, are probably the cause of these phenomena. This combat between these two antagonistic powers causes effervescence, and perhaps most of the miracles of chemistry. . . .

[Du Châtelet suggests that the additional force caused by proximity may only be the result of an increase in their movement. She then gives three objections to this hypothesis and answers each.]

---

70. Francis Hauksbee (1666–1713) was famous for his experiments with electricity. Du Châtelet read of him perhaps in the *Philosophical Transactions* of the London Royal Society, or in his *Physico-Mechanical Experiments on Various Subjects* (1709).

CONJECTURE ON THE ACTION OF FIRE ON SATURN AND THE COMETS.
This increase in the force of fire by the coming of its particles closer together, is perhaps one of the means the Creator has used to make up for
the remoteness of Saturn and the comets from the Sun. The rays may act in
these globes in proportion to the cube of their proximity, and then a very
small quantity of rays can suffice to heat and to illuminate them.

XII

Of the cooling of bodies

THE DENSEST BODIES ARE THOSE WHICH COOL MOST SLOWLY.
    1. The more difficult it is for a body to receive fire in its pores, the lon-
ger it conserves it there; for this body resists, by its mass and by the coherence of its parts, the effort the fire makes to leave it. Thus the more solid a
body, the longer it takes to cool. . . .
    3. Two globes of iron heated equally conserve their heat in direct pro-
portion to their diameter; the greater the diameter, the less surface area in
proportion to their mass and the less outlet the fire finds to escape from their
pores. In addition, the exterior air that surrounds them touching at fewer
points, takes less of their heat.

CONJECTURE ON THE FORM OF THE SUN.
For the same reason, the spherical shape is more calculated to conserve heat
for a long time. For of all figures, it is that which has least surface area in relation to its mass; and the fire does not find in a globe any place that it could
more easily leave than another, for they all offer equal resistance.
    This reasoning could lead one to believe that the Sun and the fixed
stars are perfectly spherical bodies (if the effect of their centrifugal force is
ignored).
    4. The bodies that take the most heat from other bodies are considered
the coldest. This is why marble appears colder to us than silk; for the densest bodies are those which take the most of our heat, because they touch us
at more points, and marble being specifically denser than silk, must appear
colder to us.

IN WHAT PROPORTION BODIES COMMUNICATE THEIR HEAT.
    5. A cube of hot iron being placed between two cold cubes, one of
marble, and the other wood, this iron will cool more by the contact with the

marble, but it will heat the wood more in the same time. For marble heats with more difficulty than wood, more or less in relation to the specific gravity of these two bodies.

[6.] But if these three cubes are left long enough in the same place, the heat of the cube of iron will spread to the other two and to the air that surrounds them; so that after some time all three will be the same temperature as the air around them. . . .

All these rules, governing the way fire leaves bodies, are subject to exception, just as the rules governing its penetration of them are, . . .

[Du Châtelet then refers to experiments with the pyrometer measuring the contraction of bodies, but, she explains, Academy constraints on the length of the dissertation prevent their description.]

XIII

Of the causes of the freezing of water

There are three kinds of cold.

The first is that which depends on the arrangement of our organs, for our senses often make us judge a body to be colder than another, though they are both the same temperature. Through this illusion marble appears colder to us than wool, and people believe caves to be hotter in winter than in summer, etc.

The second is that of bodies that really cool and that fire leaves; this kind of cold is nothing but the diminution of fire, which I discussed in the preceding article. Thus, all nature cools and contracts in winter because of the absence of the Sun and the obliqueness of its rays.

The third is freezing.

THE ABSENCE OF FIRE IS NOT THE ONLY CAUSE OF FREEZING.
It seems, from all the circumstances that accompany this third type of cold, that it cannot be attributed solely to the absence of fire, and it is necessary to look for another cause in Nature.

PROOFS.
1. Fire rarefies all the bodies it permeates and consequently increases their volume. Thus, if ice were only caused by the absence of fire, it would be contracted water, and it should have a greater specific gravity than water; but the reverse is the case. The volume of water increases in freezing,

in the proportion of about 8 to 9, and it increases all the more as the cold is greater and the water is more contracted. Thus ice is not caused by the absence of fire only. . . .

[In 2. and 3. Du Châtelet dismisses air bubbles in the water as the cause. Knowing that the transformation of water into ice expands and then breaks the vase in which it is contained, its particles must still be active despite the loss of fire. She mentions Huygens's experiment with this "transformation": he exposed a gun barrel filled with water on his windowsill in winter, and it cracked. In 4. and 5. Du Châtelet cites other evidence, such as the fact that ice does not always melt when the thermometer rises above 36°, and Amonton's observations of the relatively minor variation in temperature between winter and summer because of the absence of the Sun.]

6. . . . If freezing cannot be attributed to the absence of fire only, some other cause must be sought in nature. The accompanying circumstances afford the most help in discovering this cause, thus, they must be carefully examined.

THE HETEROGENEOUS PARTICLES PRESENT IN WATER ARE THE CAUSE OF ITS FREEZING.

We see that the particles of ice are in great motion; so heterogeneous particles must mix with water when it freezes and cause this constant agitation. For no fluid effervesces or becomes agitated if some heterogeneous body is not combined with it, with which it ferments.

The existence of these particles that mix with water and produce its freezing seems proved by a multitude of experiments. . . .

[Du Châtelet describes four experiments, some done for the Academy of Florence, which prove to her that a heterogeneous substance added to water makes it freeze.[71] She notes that the water from melted ice heats with much more difficulty than other water, and explains that it is not suitable for making either coffee or tea. "Those with a delicate palate" can tell from its taste that heterogeneous particles must have been mixed with this water, since its flavor and its quality changed. She concludes that these heterogeneous particles must be stiff and calls them *"frigerating* particles."][72]

But what are these *frigerating* particles? This remains for us to examine.

---

71. By Academy of Florence, Du Châtelet probably means the Accademia del Cimento [Academy of Experiment] mentioned before in this *Dissertation*.

72. Note that this conclusion is similar to the idea of *pabulum ignis*, a basic substance in all matter that causes inflammability. Similarly, Du Châtelet hypothesizes a substance that in itself

THE ICED WATER THAT WE MAKE REVEALS TO US THE FRIGERATING
PARTICLES THAT CAUSE ICE.

Men invented an art that can equally promote their instruction and their pleasures; the way in which *iced waters*, so called, are made can be a clue for discovering the manner in which natural freezing operates.

Everyone knows that the water contained in a vase surrounded with salt and snow will freeze, however hot the atmosphere, as soon as the salt begins to melt the snow; but if, in place of salt, anhydrous nitric acid [*esprit de nitre*] is mixed with the snow, the cold produced then makes the thermometer lower to 72° below freezing. Fahrenheit first did this experiment, and it incontrovertibly proves that there is still much fire in natural ice, since a kind of cold can be produced that surpasses 72° and that makes water freeze on Earth. And who will dare to put limits on this power to cause the cold! Thus, this experiment demonstrates that we know no more of the limits of freezing than of heating.

THESE PARTICLES ARE SALTS AND NITRATES IN THE AIR.

It is very likely that natural freezing operates in the same manner as our artificial freezing, and that particles of salt and nitrates that the Sun raises in the air, and which then fall back on the Earth, insinuate themselves into the water, stop up its pores, and plug, like so many nails, its interstices, expel the particles of fire, and finally make the water pass from the state of a fluid to that of a solid. Thus, the absence of fire is one of the causes of freezing, but it is not the sole cause. For, however true it may be that in all freezing the particles of fire fly away from between the pores of water, yet, without the rigid particles that insinuate themselves, the absence of fire alone would not suffice to reduce it to ice: this also appears in spiritous liquids, like concentrated nitric acid [*eau forte*], ethyl alcohol, etc. which do not freeze, although, in the cold many particles of fire leave their pores. . . .

[Du Châtelet believes that this is because the *frigerating* particles do not combine with these liquids as they do with water. She then describes a number of observations reported by Frenchmen in Persia and Armenia, and in the grottos of Besançon in France, which seem to prove her hypothesis about salt and nitrates. She writes that all countries have soil containing saline and nitrous particles that the wind can carry into the air. This is the only

---

has the ability to cause freezing, the opposite of heating and becoming inflamed. Her term "frigérifiques" is a French-like coinage, and probably is adapted from Boyle who used the term *frigerating*, meaning *cooling*.

part wind plays in the process of freezing, because she knows that "blowing with a bellows on a thermometer never makes it lower."]

WHY IT RARELY FREEZES IN SUMMER IN OUR CLIMATES.
It rarely freezes in summer in climates that do not have an abundance of these *frigerating* particles, because, the particles of salt and of nitrate being more divided and smaller, as a result of the agitation that the Sun causes in all of nature, remain suspended in the atmosphere when the Sun raises them from the Earth and do not fall back on the Earth as in winter. Moreover, the water particles being in great motion, the few that do fall on Earth are not enough to freeze it. . . .

I do not believe, after all these reasons, that one could refrain from acknowledging that these particles (of which all the phenomena of nature, and all our operations on ice demonstrate the existence) are absolutely necessary for the freezing of water and that without them no cause could be assigned for it.

XIV

Of the nature of the Sun

Most people have only a vague idea of the nature of the Sun. They see that its rays warm us and that they shine, and they conclude from this that the Sun must be an immense globe of fire, which constantly sends us the luminous material of which it is composed.

THE SUN CANNOT BE A GLOBE OF FIRE.
But what is meant by a globe of fire? If one means a globe consisting entirely of igneous particles, of elementary fire, I dare say that this idea is untenable. . . .

[1–4. From the fact that the Sun's rays do not dissipate in a variety of circumstances, Du Châtelet concludes that the Sun must be a solid.]

IF THE SUN WERE A GLOBE OF FIRE IT COULD NOT BE THE CENTER OF OUR PLANETARY WORLD.
5. The Sun is at the center of our planetary system; all philosophers agree about this. However, if it is a globe of fire, it appears that it cannot occupy this place; for either fire is heavy and tends toward a center, or it has no weight and does not tend toward any point more than toward another.

Now in the former case, all the corpuscles of fire that compose the body of the Sun would tend toward the center of this star, and then the propagation of light would be impossible. For how could the Sun, by its rotation on its axis, impart to the particles of fire that compose it, a centrifugal force great enough to compel them to fly off with so much force from the center of gravity toward which they tend, and for them to traverse only by this centrifugal force, 33 millions *lieues* in 7 or 8 minutes?

IF FIRE HAD WEIGHT IT COULD NOT EMANATE FROM THE SUN.[73]

If, on the contrary, fire has no weight, if it does not tend toward any point, what power will retain it at the center of the universe, and oppose the effort of the centrifugal force that the particles of fire that compose it must acquire by the rotation of the Sun? What, in a word, will stop it from dissipating? So the Sun must be a solid body, since it does not dissipate, and it is the center of our planetary world: and fire must be weightless, since it emanates from the Sun.

THE SUN CANNOT BUT BE A SOLID BODY IN M. NEWTON'S SYSTEM.

Allow me for a moment to consider Newtonian attraction. In this system the Sun is at the center of our planetary world, and this place it was assigned by the laws of gravity, for having more mass than the other globes, it forces them to turn around it. Now if fire has no weight (as I believe I have proved) how can the Sun be a body of fire, that is to say, a body without weight, and nonetheless attract all the celestial bodies toward it, by reason of its greater mass? So, it is necessary in the system of attraction either for the Sun to be a solid body or for fire to have weight and to tend toward its center; but if this fire of the Sun tends toward its center, by what force will it constantly move away from this center. Thus, M. Newton believes that the Sun is a solid body. . . .

[Du Châtelet speculates on what the Sun could be made of and suggests that it is the very dense fiery substance that releases igneous particles from great volcanoes.]

The light of the Sun appears to be close to yellow, this means that the Sun must, by its nature, project more yellow rays than rays of other colors,

---

73. This section is a continuation of the argument with Voltaire about whether fire has weight. His loyalty to weighty fire came from his absolute adherence to Newton's idea of attraction as the ultimate force in nature. Du Châtelet, perhaps with irony, turns his supposition into a contradiction of Newton's system of the universe in this section.

for M. Newton proved in his *Opticks*, p. 216, that the Sun abounds in this type of rays.[74]

It is very possible that in other systems there may be Suns which, projecting more red, green rays, etc. than the basic colors of the Sun that we do see, might be different from ours, and that, lastly there might be in nature other colors than those we know in our world.

XV

Of the central Fire

ALL FIRE DOES NOT COME FROM THE SUN.

All fire does not come from the Sun; two pebbles struck against each other suffice to convince us of this truth. Each body and each point of space has received from the Creator a portion of fire in proportion to its volume. This fire enclosed in the bosom of all bodies gives them life, animates them, impregnates them, maintains the motion among their particles, and prevents them from condensing entirely. . . .

[Du Châtelet describes experiments by Messrs. Mariotte and Dortous de Mairan proving that the temperature at the center of the Earth increases as one approaches its center.][75]

Since fire is spread equally all over, this stock of heat must have been placed in the center of the Earth by the Creator, from whence it spread itself equally to the same distance in all the bodies that compose it. So that if there were no Sun, all the climates of the Earth would be equally hot, or rather equally cold on its surface; but the heat would increase, as it increases in reality, as one would approach the center of the Earth.

Thus, the central fire appears proved by the phenomena of nature. And in order to explain it, it is not necessary to have recourse, as does a philosopher of our day, to a tendency of fire to go down, a tendency contradicted by the most common experiments as by the most subtle.[76] For the existence of this fire, the will of the Creator suffices; and for its conservation, the law that has fire withdraw more slowly from bodies the denser they are. For fire

---

74. In his *Opticks* (1704) Newton described his observations and experiments with lenses and other materials, such as soap bubbles, to determine the nature and properties of light.

75. The descriptions would have come from *Memoirs* presented to the Royal Academy of Sciences.

76. Du Châtelet is certainly referring to Voltaire.

at the center of the Earth must be retained by a weight, the resistance of which it cannot overcome.

When this fire finds some outlet, it bursts out of this subterranean furnace, and it is this that makes volcanoes, sulfurous vents, etc. but no more than a very small part of the fire held in the entrails of the Earth can ever escape. . . .

IT IS AN EFFECT OF THE PROVIDENCE OF THE CREATOR THAT FIRE BURNS WITH MORE DIFFICULTY THAN IT ILLUMINATES.

Fire illuminates us as soon as it can be transmitted in a straight line to our eyes, but it only heats us in proportion to the resistance that bodies offer it, and this is one of the greatest marks of the providence of the Creator. For if fire burned as easily as it illuminates, we would constantly be exposed to being consumed, and if it needed the resistance of bodies to illuminate, we would often be in darkness. But as soon as it strikes our eyes, it gives us a very bright light, and it never heats us enough to incommode us unless we fuel its power, the greatest heat of the summer being about three times less than that of boiling water.

IT IS VERY LIKELY THAT THE QUANTITY OF FIRE IN CELESTIAL BODIES IS PROPORTIONAL TO THEIR DISTANCE FROM THE SUN.

The fire that is in all bodies, independently of the Sun, and this central fire that one can, with much probability, suppose to be in all the celestial bodies, can lead to the belief that the quantity of fire in these planets is proportional to their distance from the Sun. Thus, Venus, which is the closest, has least, Saturn and the comets, which are very distant from it, will have more of it, each according to its distance. . . .

[In contrast to the material of Saturn] the material of comets must be very dense, since they go so close to the Sun, without dissolving in its heat. So God must have made up, by the quantity of central fire or by the fire he spread in the bodies that compose these celestial bodies, for their distance from the Sun, and perhaps also compensated for this distance by increasing the proportion in which the fire acts there, just as he provided for the illumination of Saturn and Jupiter by the quantity of their moons.[77] Thus it is useless to suppose heterogeneity of matter in the celestial bodies placed at different distances from the Sun, but only a more or less great quantity of fire; or an increase in the proportion according to which the rays act on bodies.

77. Du Châtelet assumes that all moons reflect the Sun's light onto their respective planets.

THE CENTRAL FIRE CONSERVES ALL THE PROPERTIES THAT WE KNOW OF
FIRE, BUT IT CANNOT USE THEM.

Fire conserves all its properties in the center of the Earth. It tends to equilibrium there, its particles seek to spread to all sides, etc., but it only actively exercises these properties in part, because it cannot entirely overcome the force that opposes its action. . . .

[Du Châtelet notes that there are other phenomena associated with this central fire, but space dictates an end to her dissertation.]

But I have already imposed too much on the patience of the respectable body to which I dare present this weak essay. I hope that my love for the truth will stand in place of talent,[78] and that the sincere desire I have to contribute to its knowledge will obtain forgiveness for my errors.

CONCLUSION OF THE SECOND PART

I conclude from all that has been said in this second part.

1. That fire is equally distributed in all inanimate bodies.

2. That animate creatures contain more fire in their substance than the others.

3. That friction is the most powerful means of exciting the fire contained between the particles of bodies.

4. That the mass of bodies, their elasticity, and the rapidity of motion that one imparts to them, infinitely increase the activity of the fire that they contain and that friction excites.

5. That fire rarefies all bodies, and extends them in all their dimensions.

6. That bodies inflame more or less quickly depending on their color, all things being equal, and that the most reflective are those that inflame last.[79]

7. That liquids do not acquire heat past boiling even from the greatest fire.

8. That the basic causative element of fire is not of fire, that fire only takes away the finest parts of bodies, and that they do not change into fire.

9. That, far from being a cause of elasticity, fire destroys it.

10. That fire appears to be the cause of electricity.

---

78. In the original 1738 version, she wrote "in place of eloquence." Note that this apologetic ending was typical not only of women's but also men's writing for such prize competitions.

79. Today, bodies would be described as "heating up" rather than "inflaming."

11. That fire does not act on bodies only in proportion to its quantity.

12. That the rays of fire acquire an activity from their proximity that infinitely increases the effects of fire.

13. That the time in which different bodies cool is almost the same as that in which they heat.

14. That the absence of fire is not the only cause of freezing, but that *frigerating* particles are involved.

15. That these *frigerating* particles are particles of salt and nitrate.

16. That the Sun is a solid body.

17. That all fire here on Earth does not come from the Sun, but that each body contains a certain quantity.

18. That there is in the Earth a central fire, which is the cause of the vegetation made in its bosom.

END.

III

FOUNDATIONS OF PHYSICS

VOLUME EDITOR'S INTRODUCTION

This set of selections comes from Du Châtelet's major work of natural philosophy, the *Foundations of Physics*, published in Paris in 1740 when she was just thirty-four. "Natural philosophy" is a phrase coined in seventeenth-century England. Like Du Châtelet, the "natural philosopher" sought to describe not only the mechanics of the natural world but also the first causes of phenomena such as motion and gravity, and the role of God. The French used three words to describe those who excelled in this kind of knowledge: *philosophe* [philosopher], *géomètre* [mathematician], and *physicien* [physicist]. Du Châtelet's letter to her bookseller and publisher, Laurent François Prault, included in this collection, gives some idea of the scope of her reading as she worked to become proficient in each of these categories.

Du Châtelet had three purposes in writing the *Foundations*. First, as she explained in her preface, she decried the lack of an adequate text in "natural philosophy," a problem she encountered when she took over the mathematical and scientific education of her son. Second, she wished to give Newtonian mechanics the metaphysical foundations the Englishman had failed to sort out. Voltaire's recently published *Elements of the Philosophy of Newton* on which she had collaborated, she described as too "narrow" in its scope. It had not solved what she and many Continental critics of the Englishman saw as his theory's principal weakness: it offered no metaphysical, causal explanations for the phenomenon of attraction, how and why it worked as it did. Voltaire, like so many Newtonians, fell back on the inconclusive phrases of Newton's "General Scholium," references to a God, and to an ethereal medium filling the universe.[1]

---

1. In the "General Scholium" at the end of Book III of the *Principia*, after a general description of the attributes of God to whom he ascribes design and governance of "this most elegant sys-

Third, Du Châtelet had found that she disagreed with certain key ideas presented by Newton and his followers and, in her desire to arrive at "the truth," she could not leave these uncontested and uncorrected. For example, in her chapter IV, she defended the use of hypothesis in science, a method, she explained, that made possible all of the advances in seventeenth- and eighteenth-century knowledge of the universe, and, it must be noted, subsequent discoveries in the nineteenth and twentieth centuries as well.[2] It was, as she described it, the scaffolding of a building, useless when completed, but essential to its construction. In her chapter XXI, Du Châtelet returned to the subject that had occasioned her corrections to her *Dissertation on the Nature of Fire*, the nature of force in collisions. Now, in the final chapter of the *Foundations*, she wrote at length on this subject, disputing with Jean-Jacques Dortous de Mairan by name, a leading member of the Royal Academy of Sciences, and defending the Leibnizian formula that she had first read about in the 1724 and 1726 Academy memoirs by the Swiss mathematician, Maupertuis' mentor, Johann Bernoulli.

Du Châtelet had a version of the *Foundations* ready for the printer by the fall of 1738.[3] Then, abruptly, she instructed her publisher, Prault, to stop. She had discovered in the translations of Christian Wolff's writings sent to her and Voltaire by Frederick of Prussia the metaphysical underpinning she sought. Wolff described Leibniz's "law of sufficient reason," that, in her words "shows that one thing is preferable to the other." In combination with the law forbidding contradictions and those mandating indiscernible differences and continuity in nature, it was "the thread which can guide us in the labyrinths of error which the human spirit has built to have the pleasure of being led astray." Thus, this law formed the basis of all certain knowledge in physics, "a compass capable of leading us in the moving sands

---

tem" of the cosmos, he speaks of attraction but gives as its cause an unnamed "force." In the concluding paragraph he avoids the issue altogether: "And it is enough that gravity really exists and acts according to the laws that we have set forth and is sufficient to explain all the motions of the heavenly bodies and of our sea." Isaac Newton, *Principia: Mathematical Principles of Natural Philosophy*, trans. I. Bernard Cohen and Anne Whitman, assisted by Julia Budenz (Berkeley: University of California Press, 1999), 940–43.

2. Newton's followers took a negative view of *hypothesis* based on one sentence of his "General Scholium." For discussion of this sentence and its subsequent translation and interpretation, see Cohen, "A Guide to Newton's *Principia*," in Newton, *Principia*, trans. Cohen and Whitman, 275–77.

3. Du Châtelet intended a second volume for the *Foundations*. Chapters "On Liberty" (or free will)," "Essay on Optics," and "Reasoned Grammar" exist in manuscript in the Voltaire Collection of the St. Petersburg National Library. They, in turn, refer to others: on the soul, on light and color, on elasticity in matter. In all probability, the "description of the planetary world" that she mentions in the *Foundations* became chapter 1 of the commentary that accompanied her translation of Newton's *Principia*.

of this science." These four premises became the strict rules of reasoning that she uses, for example, in chapter VII to deduce the nature of matter. Du Châtelet also expected the law of sufficient reason to govern human behavior. She could not imagine a person making a decision without a sufficient reason for, say, standing, sitting, or lying down. She even assumed that the Creator must choose to follow these logical imperatives. Otherwise all in the universe would be accident, unstable, similar to the world of dreams. No science, in the sense of certain knowledge, would be possible, for a balance might tilt in either direction regardless of the number of weights placed on one side or the other. The letters to Maupertuis chosen to accompany this selection show some of the issues she puzzled over and her method of resolving them.

The *Foundations* demonstrates Du Châtelet's mastery of mathematics, in particular Descartes' analytic geometry. Like many European physicists, she believed mathematics to be a more certain source of explanation and analogy than experiment. It was "the key to all discoveries" in nature. Given the irregularities in scientific apparatus, the variable capabilities of the person doing the procedure, and the unpredictable responses of those who observed it, Du Châtelet was not alone in her skepticism about experiment as more than a means of demonstrating a known truth. In her day, the difficulties encountered in replicating Newton's optical experiments were the classic example of these problems, as were the disputes over what could and could not be deduced about the nature of collisions from the varied efforts made to verify one explanation over another experimentally.[4]

To further her mathematical knowledge, Du Châtelet had engaged a tutor for herself, a Swiss mathematician and former student of Wolff recommended to her by Bernoulli's eldest son and by Maupertuis. Samuel König caused the first of the two controversies surrounding the publication of the *Foundations*. As early as the fall of 1739, Du Châtelet found that she must answer König's charges that she had merely taken his dictation in the composition of the final book. This she does in the letter to Johann Bernoulli II that precedes the selections from the *Foundations*. The second attack in the spring of 1741 came from Dortous de Mairan, newly appointed the secretary of the Academy of Sciences. Initially, he had thanked Du Châtelet for having sent him a copy of her book. Reading her final chapter, the explicit attack on his

---

4. For an idea of the problems associated with experiment, see Steven Shapin, "The House of Experiment in Seventeenth-Century England," in *The Scientific Enterprise in Early Modern Europe,* ed. Peter Dear (Chicago: University of Chicago Press, 1997), 273–304; Simon Schaffer, "Glassworks: Newton's Prisms and the Uses of Experiment," in *The Uses of Experiment: Studies in the Natural Sciences,* ed. David Gooding, Trevor Pinch, and Simon Schaffer (New York: Cambridge University Press, 1989), 67–104.

ideas, must have been an unwelcome surprise. Angered by the veiled sarcasm in which she had presented his views, he saw himself as the champion, not only of the Cartesian formula he had defended in 1728 against Bernoulli but also of the Academy of Sciences, now challenged by a young noblewoman. He quickly formulated a pamphlet that he distributed throughout the Republic of Letters that portrayed Du Châtelet as a silly coquette, seduced by Leibnizian ideas when she should have accepted those of men wiser and better than she. Du Châtelet surprised him again, however, and answered his pamphlet point by point in a style contemporaries acknowledged was his equal in wit and daring. Probably at the insistence of his friends, Dortous de Mairan let the matter drop.[5]

The controversy, and the approval and acclaim for the *Foundations* from mathematicians and physicists in England, France and Germany—Wolff himself wrote a congratulatory letter to her—secured Du Châtelet's reputation as a member of the elite, intellectual world of the Republic of Letters. For this translation, the first 1740 edition was chosen both for its spontaneity and for the insights it offers about the evolution of her thoughts and skills. It is also the most readily accessible, as it is included in the Landmarks of Science microprints collection. Du Châtelet did complete a revised edition, noteworthy for its clearer focus and phrasing, published in 1742. It is this version that was immediately translated into German and Italian, both in 1743.

All of the letters are taken from the Besterman collection.

ॐ

RELATED LETTERS

To Pierre-Louis Moreau de Maupertuis[6]

At Cirey, this 30th April [1738]

I would be inclined to believe, monsieur, that my last letter bored you so much, or that you found it so ridiculous, that you judge it not worthy of a response, if I did not receive on all sides assurances of the pleasure of having

5. The subtlety of the contested point of physics and the method of argumentation explain why this amazing interchange is not among the selections translated for this collection. Du Châtelet was so proud of her efforts that she included both of their pamphlets with the revised version of her *Dissertation on the Nature and Propagation of Fire* published in 1744.

6. No. 122 in the Besterman collection, vol. 1, 220–21; D1486 in the *Oeuvres complètes*.

you here soon. Now I cannot believe that you would wish to come and see someone to whom you would not write.

Meanwhile I awaited the insights I need from you. You undoubtedly will have found my question quite ridiculous when I asked you how it followed that the same quantity of motion could subsist in the universe, supposing that the force of bodies in motion is the product of their mass by the square of their speed. But you are the master in Israel and I am ignorant and seek to instruct myself trembling before you.

Since I wrote to you, I read what M. Leibniz gave in the *Acta Eruditorum* on *forces vives*, and I saw that he distinguished between the quantity of motion and the quantity of force; and then I found what I needed and I saw that I was only a stupid one, and that I ought not to have confused two very distinct things by making the forces the product of the mass by the square of the speeds. But the only thing that puzzles me at present is liberty, for in the end I believe myself free and I do not know if this quantity of force, which is always the same in the universe, does not destroy liberty. Initiating motion, is that not to produce in nature a force that did not exist? Now, if we have not the power to begin motion, we are not free.[7] I beg you to enlighten me on this point.

I need your letter to relieve my impatience waiting for you. M. de V. asks me to assure you that he shares in this. We have made ourselves philosophers to be worthy of you. I count on the pleasure of seeing you; then I will ask your pardon for all of my questions, and all of my importunities. You cannot imagine the pleasure that I will have to tell you myself the esteem and the affection I will hold for you all my life.

To Pierre-Louis Moreau de Maupertuis[8]

19 November 1738

When I addressed myself to you, monsieur, about the changes that I wished to make to my memoir on fire, I counted on addressing my friend, not a prize commissioner.[9] I know full well that what I ask of you is not within the

---

7. Du Châtelet's quandary over the relationship between her initiated motion and the Leibnizian concept referred to here—that of a fixed amount of force in the universe—shows the interconnections for her between science and metaphysics. It was Newton's position that force dissipated and was lost as a result of these collisions. He hypothesized that God would periodically renew it.

8. No. 151 in the Besterman collection, vol. 1, 270–71; D1661 in the *Oeuvres complètes*.

9. Submissions were judged by members of the Academy who in theory remained anonymous, as did those who entered the prize competition. The winners were then published in *Recueils*, collections of prize essays, but often a number of years after the contest.

letter of the law, but I also know that the letter kills and that the spirit vivifies. I am not asking for any change in the body of the work, for I sense very well that the Academy, having given its judgment on the memoir as it was presented, wants the public to see it in the same state, and that is only too just. But I only ask for the suppression of a footnote concerning *forces vives*, a subject entirely irrelevant to that of fire, and surely this note had no influence on the judgment of the Academy.[10] That is why I addressed myself to you who know how I currently think on this matter, and who have enough justice to feel that what I ask has nothing to do with the honor of the Academy nor its commissioners. Besides, if you want to do me this service, I will be very much obliged to you; but I beg you not to speak of this letter either to the Academy or to M. de Réaumur.[11]

As you know nothing of my memoir, and you have promised to give me the pleasure of reading it, it is very easy for you to ask for it. Printing has not yet begun, and you will be able to judge for yourself the unimportance of what I have the honor to ask of you, but this bagatelle would be disagreeable for me, since it is very sad to see, in the only work by me that will perhaps ever be printed, a sentiment so opposed to my current ideas. If you are willing then, monsieur, to give me the pleasure of erasing this footnote (for it is not in the body of the work) you would do me a service that I will never forget. You will find it in the first part.

I would be delighted if you should want to tell me your sentiment on this work, but I imagine that you scarcely have the time, as your letters have become so rare since you arrived in Paris. I am perceptibly very distressed, for you well know the extreme pleasure they give me, but what annoys me most is that you no longer speak of coming here. However, if you keep your word to me, this happy time must be near. The impatience that I have for it equals the extreme affection you have inspired in me and the importance I attach to the advantage of being able to consult you myself and instructing myself in your company. M. de V., who shares all these sentiments with me, begs me to assure you of it.

I have been told that M. Du Fay took four rays from the crown of Newton.[12] I am very curious to know the experiments that have led him to advance a proposition that must very much surprise all the people who know

10. In the errata that she was allowed to submit, she changed a reference to Dortous de Mairan's formula for *force* to that of Bernoulli. She had been reading Bernoulli and Leibniz on the nature of collisions and had changed her mind.

11. Du Châtelet was also negotiating indirectly with Réaumur, a leading academician.

12. Charles François de Cisternay Du Fay (1698–1739), a chemist and astronomer. Du Châtelet described this experiment in a subsequent letter to Maupertuis and sent her own thoughts on "the formation of colors" to Du Fay.

Newton's *Opticks* and its accuracy. I am, however, convinced that he did not advance it without having strong proofs to establish it. I must confess that I am very curious about it and, above all, to know what you think of it. I have also been told that a man named M. de Gamaches, an abbé, wrote a book of physics in which he reconciled the fullness of Descartes' universe and the void of Newton. This seems to me an odd enterprise. Do you know the author? As for the work, I do not know if he will dare speak to you about it.[13]

I implore you to be extremely discreet about this little review of *Newton* that I sent you, and about this footnote.[14]

Adieu, monsieur, be certain that when you want to see the people who love you and esteem you most, you must come to Cirey.

To Laurent François Prault [15]

The 16th [?February 1739]

I am delighted, monsieur, that you should be a little exact; I rely on you so much for my library of physics, and I will shortly send you the books I want to dispose of. I rely on your probity, as you will be the master of prices. I ask you to let me know if the Academy of Sciences, that you found for sale, is all bound and how much this will amount to and if it would be necessary to pay right away.[16] I also want the *Philosophical Transactions*, the *République des lettres* [Republic of Letters] up to the death of Bayle, and all the books of physics that you come across.[17] As I remember them, I will put them on a card and send the list to you. I have the *Opticks* of Newton, Rohault commented on by Clarke, Whiston, *La Figure de la terre* [The Shape of the Earth], *Figure des astres* [Shape of the Celestial Bodies], Musschenbroek *Physique* [Physics],

13. Etienne Simon de Gamaches was a confirmed Cartesian, as demonstrated in his *Astronomie physique* [Astronomy of Physics] (1740).

14. Du Châtelet wrote a not altogether positive letter about Voltaire's book on Newton, published in the *Journal des savants*, in September 1738. Both this review and the footnote she insisted be added to her *Dissertation on Fire* showed that she did not accept Newton's ideas as completely as Voltaire did.

15. No. 186 in the Besterman collection, vol. 1, 328–31; D1885 in the *Oeuvres complètes*. Note that Prault was Du Châtelet's publisher for her *Foundations*.

16. Du Châtelet is probably thinking of the *Memoirs* of the Academy published periodically and containing the essays that won the yearly prizes. In this era most books appeared as pages bound between plain cardboard covers. The buyer instructed the book dealer to bind whatever had been bought in a particular way, and at a specific price depending on the covers used.

17. The London Royal Society published its *Philosophical Transactions*, or papers presented to the Society; Pierre Bayle (1647–1706) was most famous for his *Dictionnaire historique et critique* [Historical and Critical Dictionary] and was briefly editor of the *République des lettres*, a learned journal of the seventeenth century.

'sGravesande *Physique* [Physics], Clarke, the *Entretiens physiques* [Conversations on Physics] of Father Renaut for what they are worth, Euclid, Pardies, Malezieu, *Application de l'algèbre à la géométrie* [Application of Algebra to Geometry] by Guisnée, the *Sections coniques* [Conical Sections] of M. de L'Hôpital, the *Mathématiques universelles* [Universal Mathematics] and the *Oeuvres* [Works] of Descartes.[18] That is about all. Please look for me for the *Principia mathematica* [Mathematical Principles] of M. Newton, a nice edition, and have them bound for me in red leather, gilt-edged, at your earliest convenience; I will be very obliged to you.

I assure you that I am very grateful for your two cartons, I know that it has been a burden, but think a little of the time one has had; it is impossible that something might not be missed, above all in a book where each word must be carefully weighed.[19] M. de V. is very pleased with you and loves you very much. If all book dealers were like you he would be happier. In the end there is always more to be gained from being an honest man, above all in a business where trust and reputation attract a thousand deals that one would not entrust to another. Rest assured that on all occasions when I can please you, I will not fail to do so. *I am your servant with a great desire to have my library entirely fill'd, that is, entirely finish'd of good books of your magazine.*[20]

I reread your letter and I find there that the collection of the Academy goes for six hundred *livres*.[21] The only question is the time of payment. I ask

18. This is the most explicit description known of Du Châtelet's library at this key time in her self-education. Samuel Clarke (1675–1729) annotated Rohault's work so as to make it a textbook of Newtonian not Cartesian physics, as it had been originally. William Whiston (1667–1752) was a follower of Newton who wrote on the predictability of comets. *La Figure de la terre* and the *Figure des astres* are both by Maupertuis. The work by Musschenbroek had been translated into French, the 'sGravesande would be the Latin edition. The "Clarke" may be a reference to the Leibniz-Clarke debate published in French and English in 1717. Father Noël Regnault wrote the *Entretiens physiques d'Ariste et d'Eudoxe* (1729) as a popular introduction to physics. The rest of the list consists of books of mathematics by the leading mathematicians of seventeenth- and early eighteenth-century France. The *Application de l'algèbre* by Guisnée was Du Châtelet's first textbook, recommended by Maupertuis. Guillaume François Antoine de L'Hôpital, the Marquis de L'Hôpital was one of the first French mathematicians to study Leibnizian calculus.

19. Du Châtelet is probably referring to the *Foundations*.

20. Although Du Châtelet could translate English into French and spoke it well, writing it still required practice. Note the common mistakes easily made in going from one to another language: of sound, "mi" for "my"; of presumed agreement between subjects and their adjectives as in French for "goods books"; and her assumption that "magazine" in English would mean the same as "magasin" in French. In French these last mistakes are known as "faux amis," "false friends," who mislead by their similarity to the other language.

21. The *livre* was the basic unit of French currency; half a *livre* was an average daily wage for a laborer in Paris.

you to send them to M. Robert; he tells me that he saw no one coming on your behalf. Leave a bill, for I believe him to have a poor memory.

I await with impatience the collection by Saurin and thank you for the box from the stage coach. M. de V. wanted to send you back *Cassandre*.[22] It is true that it is in my new books, but I wanted to keep it because I had asked for it. Please make a separate list of the books you send me.

Also send the Keill as soon as you have it.[23]

To Johann Bernoulli JJ[24]

In Paris, at the hôtel de Richelieu, this 15th of September [1739]

I am convinced monsieur, that you do not believe me capable of being in Paris for a month and of waiting all that time before answering you. But I beg you to have a good enough opinion of me to be convinced that there is no pleasure in this turbulent city more agreeable than correspondence with you.

I saw M. de Mairan, and I confess I very much feared this first meeting; however, it went marvelously. We dined together and he had no questions about any kind of force. I also dined with M. de Fontenelle and M. de Réaumur. I believe that they are a little ashamed of the foreign associate they have just chosen, but I dare not speak to them of that, for fear of offending their pride.[25]

M. de Maupertuis has at last returned to us. He will make in Paris the observations he ought to have made at Malvoisine, where they did not find a single stone at the time of M. Picard.[26] We have talked much about you, and we would like you to be here. It is a wish I make everywhere I am.

As for M. König, he is becoming more unreasonable every day about

---

22. Jacques Saurin's *Discours historiques, critiques . . . sur les événemens les plus mémorables du vieux et du nouveau testament* [Historical, Critical . . . Discourses on the Most Memorable Events of the Old and New Testament] (1720–39) was probably part of their readings on the Bible. *Cassandre* was a novel by Gautier de Costes La Calprenède (1609?–1667).

23. John Keill used his lectures as the basis of one of the earliest and best-known textbooks on Newton (published in 1701).

24. No. 221 in the Besterman collection, vol. 1, 376–78; D2073 in the *Oeuvres complètes*.

25. Du Châtelet had made an issue about changing her reference to Dortous de Mairan's views on the nature of collisions in her *Dissertation on Fire* when the Academy agreed to publish it. Clairaut had nominated König as a foreign associate of the Academy of Sciences. Fontenelle and Réaumur were the principal leaders if not always presiding officers of the Academy.

26. Jean Picard (1620–1682) was a famous French astronomer, but the reference to him in this bit of mockery of Maupertuis is not clear.

his departure, and I no longer hope to keep him here. I even believe that he will not go back to Brussels with me. I cannot guess what his purpose was in coming into my household, for even if I had been an angel, it would have been impossible in three months of travel and business affairs to learn what I wished to know. I took in his brother, who is being raised at Cirey as my son, and I have nothing to reproach myself with about the attentions I have given them; and I confess that I am very much offended by conduct that I am very far from having elicited. Moreover, you sense what trouble and what chaos this has put in my head, for I have combined my studies of geometry, of logic and of metaphysics, and nothing of all that being in place, you may judge what a potpourri this has produced. It would require none other than you to sort it out. I confess that the behavior of M. de König would make me hate all mathematicians and all Swiss if I did not know you, and it seems to me that, as a mathematician and as a Swiss, it would be a good deed if you could put it right.

I spoke to M. Clairaut about what he said about *forces vives*. He maintains that M. de Mairan and all the English agree on the conservation of live force, but they reject their assessment.[27] I do not know how they settle all that, but I believe they would be all too happy to find an honorable means to extricate themselves from the matter.

M. de König still maintains that the translation of M. de Maupertuis is done, but I do not want to be more involved in his affairs any longer. M. de Voltaire makes a thousand compliments to you. I intend to go back to Brussels very soon, to leave only after the end of my court case.[28] I will, however, have time to receive your news again, and flatter myself to have it.

I am very eager to set you a good example and to finish my letter without compliments. I hope you will imitate me; my feelings for you, monsieur, are above compliments.[29] Remember me to MM. Bernoulli.[30]

27. Du Châtelet means that they deny the formula, mv^2.

28. For the years 1739–43 Du Châtelet spent most of her time in Brussels defending her husband's claims to lands in Flanders. The court case ended with a monetary settlement. See Zinsser, *Emilie Du Châtelet*, 200–202.

29. Customarily, a letter of this sort would end with prescribed compliments.

30. Du Châtelet probably means his father and his brother, Daniel.

FOUNDATIONS OF PHYSICS

[*by Madame la Marquise Du Châtelet addressed to her son*]

COMPLETE TABLE OF CONTENTS OF THIS VOLUME

[*chapters translated are noted with an* *]

PREFACE

I

I have always thought that the most sacred duty of men was to give their children an education that prevented them at a more advanced age from regretting their youth, the only time when one can truly gain instruction. You are, my dear son, in this happy age when the mind begins to think, and when the heart has passions not yet lively enough to disturb it.

Now is perhaps the only time of your life that you will devote to the study of nature. Soon the passions and pleasures of your age will occupy all your moments; and when this youthful enthusiasm has passed, and you have paid to the intoxication of the world the tribute of your age and rank, ambition will take possession of your soul; and even if in this more advanced age, which often is not any more mature, you wanted to apply yourself to the study of the true Sciences, your mind then no longer having the flexibility characteristic of its best years, it would be necessary for you to purchase with painful study what you can learn today with extreme facility. So, I want you to make the most of the dawn of your reason; I want to try to protect you from the ignorance that is still only too common among those of your rank, and which is one more fault, and one less merit.

You must early on accustom your mind to think, and to be self-sufficient. You will perceive at all the times in your life what resources and what consolations one finds in study, and you will see that it can even furnish pleasure and delight.

II

The study of physics seems made for man, it turns upon the things that constantly surround us, and on which our pleasures and our needs depend. In this work, I will try to place this science within your reach, and to disengage it from this admirable art, called algebra, which separating things from images, eludes the senses, and speaks only to the understanding. You are not yet able to understand this language, which seems rather that of the mind than of the whole of man. It is reserved to be the study of the years of your life that will follow those of today; but the truth can take different forms, and I will try to give to it here that which suits your age, and only to speak to you of things that can be understood by resorting only to the standard geometry which you have studied.[31]

---

31. Du Châtelet is referring to Euclid's geometry (from the fourth century BCE), which she probably studied with her brother when he was just a bit younger than her son. She also stud-

UTILITY OF GEOMETRY.

Never cease, my son, to cultivate this science that you have learned from your very tender years. With no resort to it, one would hope in vain to make great progress in the study of nature. It is the key to all discoveries; and if there are still several inexplicable things in physics, that is because geometry has been insufficiently used to explain them, and one has perhaps not yet gone far enough in this science.

III

I am often surprised that so many clever people as France possesses have not preceded me in this work that I embark upon for you today. For, it must be admitted that, although we have several excellent books of physics in French, we have no complete book of physics, except the short treatise of Rohault, written eighty years ago.[32] But this treatise, although very good for the time when it was composed, has become very insufficient because of the quantity of discoveries that have been made since it was written; and a man who had studied physics only in this book, would still have many things to learn.

As for me, who in deploring this scarcity, am very far from believing myself capable of supplying it, I only propose in this work to gather together before your eyes the discoveries scattered in so many good Latin, Italian, and English books. Most of the truths they contain are known in France by only a few readers, and I want to spare you the trouble of drawing them from sources, the depth of which would frighten and might discourage you.

IV

Although the work I undertake requires much time and effort, I will not regret the trouble it will cost me, and I will believe it well spent if it can instill in you love of the sciences, and the desire to cultivate your reason. What trouble and what cares does one not give oneself every day in the uncertain hope of procuring honors and augmenting the fortune of one's children! Are the knowledge of the truth and the habit of looking for it and following it objects less worthy of my pains—especially in a century when a taste for

---

ied with Alexis-Claude Clairaut (1713–1765), a gifted young mathematician, whose *Elémens de géométrie* (1749) was probably based on their lessons. According to the publisher of Du Châtelet's translation of and commentary on Newton's *Principia*, she also used them in lessons for her son.

32. Du Châtelet is referring to the standard French physics text of her day, the *Traité de physique* [Treatise of Physics] by Jacques Rohault (1618–1672), published in 1671. His was the leading explanation of the universe according to Descartes.

physics has reached all ranks, and is beginning to become a part of the science of the world?

V

I will not write for you here the history of the revolutions experienced by physics, a thick book would be needed to report them all. I propose to make you acquainted *less with what has been thought than with what must be known.*

Up to the last century, the sciences were an impenetrable secret, to which only the so-called learned were initiated; it was a kind of cabal,[33] the cipher of which consisted of barbarous words that seemed to have been invented to confuse the mind and to discourage it.

HOW MUCH WE OWE TO DESCARTES.[34]

Descartes appeared in that profound night like a star come to illuminate the universe. The revolution that this great man caused in the sciences is surely more useful, and perhaps even more memorable, than that of the greatest empires, one, it can be said, that human reason owes most to Descartes. For it is very much easier to find the truth, when once one is on the track of it, than to leave those of error. The geometry of this great man, his dioptrics, his method, are masterpieces of sagacity that will make his name immortal, and if he was wrong on some points of physics, that was because he was a man, and it is not given to a single man, nor to a single century, to know all.

We rise to the knowledge of the truth, like those giants who climbed up to the skies by standing on the shoulders of one another.[35] The Huygenses,

33. The word *Cabal* in both French and English suggests *cabbala*, secret arts known only to the learned.

34. Du Châtelet is referring to many aspects of René Descartes' (1596–1650) thinking and writings: his assertions about the authority of one's own reasoning from his *Discours de la methode* [Discourse on Method] (1637); and two of the three essays published with it: the *Géométrie* [Geometry], his development of analytic geometry; the *Dioptrique* [Dioptrics], his study of refraction in optics. Du Châtelet also read his *Principes de la philosophie* [Principles of Philosophy] (1723 ed.). His greatest error in physics from Du Châtelet's perspective would be his system of the universe, a universe of constantly moving particles that formed vortices that carried the planets in their orbits through impulsion, the impact of the particles on one another. By the end of the seventeenth century, Descartes' description of the universe had been accepted by Continental physicists and astronomers. Sir Isaac Newton wrote opposing his ideas in his *Philosophiae naturalis principia mathematica* [Mathematical Principles of Natural Philosophy]. Du Châtelet studied two of the three editions of the *Principia*, those from 1713 and 1726. She favored Newton's planetary system of universal attraction, and wrote the *Foundations* in part to give his ideas a metaphysical basis.

35. An often-quoted statement by Newton, originally from his 1676 letter to Robert Hooke (1635–1703), an English experimental philosopher.

and the Leibnizes learned from Descartes and Galileo, these great men who, so far, are known to you only by name, and with whose works I hope soon to make you acquainted.[36] It is by making the most of the works of Kepler, and using the theorems of Huygens, that M. Newton discovered this universal force spread throughout nature, which makes the planets circle around the Sun, and that operates as gravity on Earth.

VI

Today the systems of Descartes and Newton divide the thinking world, so you should know the one and the other; but so many learned men have taken care to expound and to correct Descartes' system that it will be easy for you to learn from their works. One of my aims in the first part of this work is to put before your eyes the other part of this great process, to make you acquainted with the system of M. Newton, to show you how far making connections and determining probability are pushed, and how the phenomena are explained by the hypothesis of attraction.

You can draw much instruction on this subject from the *Elémens de la philosophie de Newton* [Elements of the Philosophy of Newton], which appeared last year.[37] And I would omit what I have to tell you about that—Newton's system—if the illustrious author had embraced a vaster terrain; but he confined himself within such narrow boundaries that he made it impossible for me to dispense with my own exposition of this matter.

VII

Guard yourself, my son, whichever side you take in this dispute among the philosophers, against the inevitable obstinacy to which the spirit of partisanship carries one: this frame of mind is dangerous on all occasions of life; but it is ridiculous in physics. The search for truth is the only thing in which the

36. Christiaan Huygens (1625–1695), Dutch physicist, was most famous for his discoveries about the rings of Saturn, and his experiments with the pendulum to establish general principles of motion. Gottfried Wilhelm Leibniz (1646–1716) wrote numerous works on a wide range of philosophical and scientific subjects. Du Châtelet made his ideas about knowledge, the role of God in the universe, and the laws of motion the basis for chapters of her *Foundations*. Galileo Galilei (1564–1642), the Italian astronomer, physicist, and mathematician, was known in Du Châtelet's time for his experiments with falling bodies and his affirmation of the Copernican system of the universe. Johannes Kepler (1571–1630), the German mathematician and astronomer, is most famous as the discoverer of the elliptical path of the planets.

37. This was Voltaire's *Elements of the Philosophy of Newton*, a product of collaboration with Du Châtelet while at Cirey, published first in 1738 in an unauthorized Dutch edition. In the revised editions, Voltaire added sections that explicitly argued against Leibnizian philosophy and physics, the metaphysical ideas and the laws of motion that Du Châtelet presented in her *Foundations*.

love of your country must not prevail, and it is surely very unfortunate that the opinions of Newton and of Descartes have become a sort of national affair. About a book of physics one must ask if it is good, not if the author is English, German, or French.

DISCUSSIONS OF ATTRACTION.

It seems to me, moreover, that it would be just as unfair on the part of the Cartesians to refuse to admit attraction as a hypothesis as it is unreasonable of a few Newtonians to want to make it an inherent property of matter.[38] It must be admitted that a few among them have gone too far in this, and it is with some reason that they are reproached for resembling a man at the opera whose bad eyesight prevents him from seeing the ropes that make flights possible, and who, for example, on seeing Bellerophon suspended in the air, said: *Bellerophon is suspended in the air because he is pulled equally on all sides from the wings.* For, in order to decide that the effects the Newtonians attribute to attraction are not produced by impulsion, it would be necessary to know all the ways in which impulsion can be used, but we are still very far from knowing that.[39]

We are still in physics, like this man blind from birth whose sight Chiselden restored.[40] At first this man saw nothing but a blur; it was only by feeling his way and at the end of a considerable time that he began to see well. This time has not quite come for us, and perhaps will never come entirely; there are probably some truths not made to be perceived by the eyes of our mind, just as there are objects, that those of our body will never perceive. But he who refused to learn because of this limitation would resemble a lame person who, having a fever, would not take the remedies which can cure it, because these remedies would not stop him from limping.

VIII

38. Two of Newton's followers, Henry Pemberton (1694–1771) in his commentary on the *Principia, A View of Sir Isaac Newton's Philosophy*, and Roger Cotes (1682–1716) in his preface to the second edition of Newton's work (1713), took this position. Du Châtelet certainly had access to both of these books.

39. *Impulsion* was the term used to describe the collision of particles in Descartes' universe. It was possible to believe in attraction and to think that impulsion was somehow its cause. Du Châtelet's contemporaries, the mathematicians, Johann Bernoulli (1667–1748) and Leonhard Euler (1707–1783) took this position and rejected gravity as a property of matter. See chapter XI of the *Foundations* where Du Châtelet discusses their work, if indirectly. She corresponded with both of them.

40. William Chiselden (1688–1752) was an anatomist and the surgeon for Chelsea College in England.

HYPOTHESES ARE NECESSARY IN PHYSICS.

One of the mistakes of some philosophers of our time is to want to banish hypotheses from physics;[41] they are as necessary as the scaffolding in a house being built; it is true that, when the building is completed, the scaffolding becomes useless, but it could not have been erected without it. All of astronomy, for example, is founded only on hypotheses, and if they had always been avoided in physics, it seems that fewer discoveries would have been made. So nothing is more likely to delay the progress of the sciences than to want to banish them, and to persuade oneself that one has found the great mainspring that moves all of nature, for one does not search for a cause that one believes one knows. This is why the application of the geometric principles of mechanics to physical effects, which is very difficult and very necessary, remains imperfect, and why we find ourselves deprived of the work and the research of several fine geniuses who would perhaps have been able to discover the true cause of phenomena.

WHEN THEY CAN BECOME DANGEROUS.

It is true that hypotheses become the poison of philosophy when they are made to pass for the truth, and perhaps they are then even more dangerous than was the unintelligible jargon of the Schoolmen;[42] for this jargon being absolutely meaningless, it only required a little attention from a clear-thinking mind to perceive how ridiculous it was, and to seek the truth elsewhere. But an ingenious and bold hypothesis, which has some initial probability, leads human pride to believe it, the mind applauds itself for having found these subtle principles, and next uses all its sagacity to defend them. Most great men who have made systems provide us with examples of this failing. These are great ships carried by the currents; they make the most beautiful maneuvers in the world, but the current carries them away.

IX

USEFULNESS OF EXPERIMENTS.

In all your studies, remember, my son, that experiment is the cane that nature gave to us blind ones, to guide us in our research; with its help we will make good progress, but, if we cease to use it, we cannot help falling. It is

---

41. Du Châtelet discusses this controversy at great length in chapter 4 of the *Foundations*.

42. By *Schoolmen* Du Châtelet means the learned of the universities such as the Sorbonne in Paris, who, through the works of St. Thomas Aquinas, used his system of reasoning and taught the ideas of Aristotle about philosophy in its broadest sense. It was their authority that Descartes so successfully challenged.

experiment that teaches us about the physical characteristics of things and it is for our reason to use it and to deduce from it new knowledge and new enlightenment.

X

TO WHAT EXTENT RESPECT IS OWED TO GREAT MEN.

If I thought it incumbent upon me to caution you against the spirit of partisanship, I believe it even more necessary to advise you not to carry respect for the greatest men to the point of idolatry, as the majority of their disciples do. Each philosopher has seen something, and none has seen all; no book is so bad that nothing can be learned from it, and no book is so good that one might not improve it. When I read Aristotle, this philosopher who has suffered fortunes so diverse and so unjust, I am astonished sometimes to find ideas so sound on several points of general physics, beside the greatest absurdities; but when I read some of the questions that M. Newton put at the end of his *Opticks*, I am struck with a very different astonishment.[43] This example of the two greatest men of their century can but make you see that he who is endowed with reason must take nobody at his word alone, but must always make his own examination, setting aside the consideration always allotted to a famous name.

XI

This is one of the reasons why I have not filled this book with citations, I did not want to seduce you with authorities; and more, there would have been too many. I am very far from believing myself capable of writing a book of physics without consulting any book, and I even doubt that without this help one might be able to write a good one. The greatest philosopher may well add new discoveries to those of others, but once a truth has been found, he has to follow it; for example, M. Newton had to begin by establishing Kepler's two analogies when he wanted to explain the course of the planets, without which he would never have arrived at the beautiful discovery of the gravitation of the celestial bodies.[44]

Physics is an immense building that surpasses the powers of a single

---

43. Du Châtelet probably knew of Aristotle, the fourth century BCE Greek philosopher, through his critics. Newton's *Opticks*, in contrast, she once wrote to her mentor, the mathematician and philosopher, Pierre-Louis Moreau de Maupertuis (1698–1759), she had studied so thoroughly that she knew it almost by heart.

44. Du Châtelet uses the term *analogy* instead of *law* in a Newtonian sense, a mathematical model from which, by analogy, laws of nature could be hypothesized and then verified by observation and experiment.

man. Some lay a stone there, while others build whole wings, but all must work on the solid foundations that have been laid for this edifice in the last century, by means of geometry and observations; still others survey the plan of the building, and I, among them.

In this work, I have not aimed at flaunting my intelligence, but at being right; and I have nurtured your reason enough to believe that you are capable of seeking the truth independently of all the alien adornments with which it is being overwhelmed in our day. I merely removed the thorns that might have wounded your delicate hands, but I did not think that I must replace them with alien flowers, and I am certain that a good mind, however weak it might still be, finds more pleasure, and a more satisfying pleasure, in clear, precise reasoning that it grasps easily, than in an ill-timed joke.

XII

In the first chapters I explain to you the principal opinions of M. Leibniz on metaphysics; I have drawn them from the works of the celebrated Wolff, of whom you have heard me speak so much with one of his disciples, who was for some time in my household, and who sometimes made abstracts for me.*[45]

M. Leibniz's ideas on metaphysics are still little known in France, but they certainly deserve to be. Despite the discoveries of this great man, there are no doubt still many obscure things in metaphysics; but it seems to me that with the principle of sufficient reason, he has provided a compass capable of leading us in the moving sands of this science.

The obscurities in which some parts of metaphysics are still shrouded serve as pretext for the laziness of the majority of men not to study it. They persuade themselves that because not everything is known, nothing can be. Yet, there certainly are points of metaphysics susceptible to demonstrations as rigorous as geometric demonstrations, although they are of another type. We lack a system of calculation for metaphysics similar to that which has been found for mathematics, by means of which, with the aid of certain *givens*, one arrives at knowledge of *unknowns*. Perhaps some genius will one day

---

45. Du Châtelet offered her own note here: "*See the *Ontology of Wolff*, and principally the following chapters: *De Principio Contradictionis, de Principio Rationis Sufficientis, de Possibili, and Impossibili, de Necessario and Contingente, de Extensione, Continuitate, Spatio, Tempore* [Of the Principle of Contradiction, of the Principle of Sufficient Reason, of Possibility, and Impossibility, of Necessity and Contingency, of Extension, Continuity, Space, Time]. Christian Wolff (1679–1754) was a German philosopher, student and explicator of Leibniz. Frederick of Prussia (1712–1786) first sent French translations of Wolff to Voltaire and Du Châtelet in 1736. She is here referring to Samuel König (1712–1757), the Swiss mathematician, who studied under Wolff and later became her tutor in advanced algebra and in Wolff's version of Leibnizian ideas.

find this system. M. Leibniz gave this much thought; he had ideas on this, which he unfortunately never communicated to anyone, but even if it could be invented, it seems that there are some unknowns for which no *equation* could ever be found. Metaphysics contains two types of things: the first, that which all people who make good use of their mind, can know; and the second, which is the most extensive, that which they will never know.[46]

Several truths of physics, metaphysics, and geometry are obviously interconnected. Metaphysics is the summit of the edifice; this summit is so elevated that our image of it often is a little blurred. This is why I thought I should begin by bringing it closer to you, so that, no cloud obscuring your mind, you might be able to have a clear and unassailable view of the truths in which I want to instruct you.[47]

CHAPTER ONE: *OF THE PRINCIPLES OF OUR KNOWLEDGE*

I

ON WHAT OUR KNOWLEDGE IS FOUNDED.

All aspects of our knowledge are born from each other and are founded on certain principles whose truth is known without even reflecting on it, because they are self-evident.

Some truths immediately depend on these first principles, and are derived from them as a result of a small number of conclusions only. In that case the mind easily perceives the sequence that has led to them; but it is easy to lose sight of this sequence in the search for truths that can only be reached by a great number of conclusions drawn one from another. There are a thousand examples of this in geometry; it is very easy, for example, to see that the diameter of a circle divides it into two equal parts, because only one conclusion is needed to pass from the nature of the circle to this property. But it is not so easily seen that the square of the ordinate BM is equal to the rectangle of line AB by line BC, although this property results from the

---

46. This sentence reflects an interchange late in the 1730s between Voltaire and Frederick of Prussia, in which Voltaire made this distinction. See D1376, Voltaire to Frederick of Prussia (15 October 1737) *Oeuvres complètes*, v. 88, 381.

47. Natural philosophers commonly offered a visual representation of the constituent parts of "Knowledge." Du Châtelet certainly knew of Descartes' Tree of Knowledge from his *Principles*, in which metaphysics forms the roots, physics the trunk, and the other sciences (mechanics, medicine, morals), the branches.

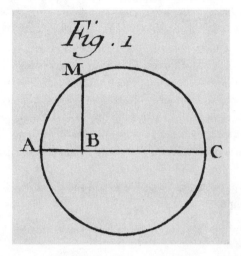

nature of the circle just as in the former case;[48] because there must be several intermediary conclusions before arriving at this last property of a circle. So, it is very important to be attentive to principles, and the manner in which truths result from them, if one does not want to go astray.

II

WHAT A PRINCIPLE IS.

The word *principle* has been much abused; the Scholastics who could demonstrate nothing chose unintelligible words for their principles.[49] Descartes, who sensed how much this manner of reasoning kept men away from the truth, began by establishing that one must only reason from clear ideas; but he pushed this principle too far: for he allowed a lively, internal sense of clarity and evidence to serve as the basis of our reasonings.

ABUSE OF THIS WORD BY M. DESCARTES.

In following this principle, this philosopher made a mistake about the essence of bodies that, according to him, consisted only of extension.[50] He be-

48. *Ordinate* means one of the points of a coordinate, in this case connecting the diameter AC to the exterior of the circle. See figure 1, $BM^2 = AB \times BC$.

49. The *Scholastics*, or *Schoolmen*, for Du Châtelet were catchall words for the thirteenth-century theologians such as St. Thomas Aquinas (1225–1274), who endeavored to reconcile reason and faith for the Catholic Church. In philosophy and physics, they used Aristotle's method of logic, syllogistics, and made his writings part of church dogma.

50. By *extension* she means the size and shape of a body in space.

lieved that in extension, he had a clear and distinct idea of a body, without troubling to prove the possibility of this idea that we will soon see to be very incomplete, since to it must be added the concepts of the force of inertia, and of the *force vive* (active force). This method, moreover, would only serve to perpetuate disputes, for among those with opposing views, each has this lively and internal sense of what it is they put forward. Thus, no one has to yield, since the evidence is equal on the two sides. So, one must substitute demonstrations for the illusions of our imagination, and not admit anything as truth, except what results incontestably from first principles that no one can call into question, and reject as false all that is contrary to these principles, or to the truths that one has established with them, whatever the imagination might say.

ONE MUST DISTRUST ONE'S IMAGINATION AND ONLY YIELD TO EVIDENCE.

§.3. A little attention to the manner in which one proceeds in science, where certainty is carried to its highest point, will suffice to make one aware of the utility of this method of reasoning. For instance, there is scarcely a clearer idea than that of the possibility of an equilateral triangle, and that the two sides of a triangle, taken together, are much longer than the third. Yet Euclid, this strict reasoner, was not content to appeal just to a lively and internal sense that we have of these truths, but he demonstrated them rigorously, showing what must be done in order to construct an equilateral triangle, and that it implies contradiction to say that two sides of a triangle, taken together, are not greater than the third.

ON THE PRINCIPLE OF CONTRADICTION.

§.4. *Contradiction* is that which simultaneously affirms and denies the same thing; this principle is the first axiom, on which all truths are founded. Everyone readily agrees on this, and it would even be impossible to deny it without lying to one's conscience; for we sense that we cannot force our minds to admit that a thing simultaneously is and is not, and that we cannot *not* have an idea while having it, nor see a white body as if it were black while we see it as white. Even the Pyrrhonists, who claimed to doubt everything, never denied this principle; they effectively denied that reality existed, but they never doubted that they had an idea while they had it in their minds.[51]

---

51. The Pyrrhonists were followers of the third century BCE Greek philosopher, Pyrrho. They became synonymous with the idea of complete skepticism.

IT IS THE FOUNDATION OF ALL CERTAINTY.

This axiom is the foundation of all certainty in human knowledge. For, if one once granted that something may exist and not exist at the same time, there would no longer be any truth, even in numbers, and every thing could be, or not be, according to the fantasy of each person, thus 2 and 2 could equally make 4 or 6, or both sums at the same time.

DEFINITION OF THE POSSIBLE AND THE IMPOSSIBLE.

§.5. It follows from this that the impossible is that which implies contradiction; and the possible does not imply it at all. Several philosophers give another definition of the possible and of the impossible, and regard as impossible that which does not give a clear and distinct idea, and as possible that which one can conceive, and which corresponds to a clear idea. This definition if well explained could be accepted, but it is necessary to be very careful that this definition does not induce us to take erroneous and deceptive notions for clear ones. For, it sometimes happens that we form deceptive ideas for ourselves that may appear evident for lack of attention, and because we have an idea of each term in particular, although it is impossible to have any idea of the sentence born from their combination.

EXAMPLES OF DECEPTIVE IDEAS.

Thus, at first one will believe that one understands what is meant by a triangle, if one defines it as *a figure enclosed between two straight lines*, and one thinks that one is speaking of a regular body, when speaking of a body with nine equal sides, because one understands all of the terms that enter into these propositions. Yet, it implies contradiction to say that two straight lines enclose a space and make a figure, and you have seen in geometry that it is impossible for a body to have nine sides, equal and alike.

There is yet another example of these deceptive ideas in the most rapid movement of a wheel, which M. Leibniz used to argue against the Cartesians; for it is easy to show that the most rapid movement is impossible to measure, since in extending any spoke this movement becomes more rapid to infinity. One sees, by these examples, that it is quite possible to believe that one has a clear idea of a thing of which we really have no idea.

So it is absolutely necessary, in order to preserve oneself from error, to verify one's ideas, to demonstrate their reality and not to admit any as incontestable, unless confirmed by experiment or by demonstration, which includes nothing false, or chimerical.

§.6. A very important rule results from the definition of the impossible that I have just given you; it is that when we advance that a thing is impossible, we are required to show that the same thing is simultaneously asserted and denied, or that it is contrary to a truth already demonstrated. This rule would avoid a great many disputes, if it were followed, for it would at once remove doubt from propositions, and expose the inadequacy of the proofs of those who treat as impossible all that does not conform to their opinions.

One should be just as cautious when maintaining that a thing is possible; for one must be in a position to show that the idea is free of contradiction. Without this condition our ideas are only more or less probable opinions, in which there is no certainty.

§.7. The principle of contradiction has always been used in philosophy. Aristotle, and after him all philosophers used it, and Descartes used it in his philosophy to prove that we exist. For it is certain that this one who doubted that he existed would have in the fact of his very doubt a proof of his existence, since it implies contradiction that one might have an idea whatever it be, and consequently a doubt, while at the same time not being in existence.

THE PRINCIPLE OF CONTRADICTION IS THE FOUNDATION OF ALL NECESSARY TRUTHS.

This principle suffices for all necessary truths, that is to say, for the truths which can only be determined in a single way, for this is what is meant by the term *necessary*. But when contingent truths are concerned, that is to say, when a thing can exist in various ways, none of its determinations is more necessary than another, then another principle becomes necessary, because that of contradiction no longer applies. Thus, the Ancients, who did not know this second principle of our knowledge, were wrong on the most important points of philosophy.

OF THE PRINCIPLE OF SUFFICIENT REASON.

§.8. The principle on which all contingent truths depend, and which is neither less fundamental nor less universal than that of contradiction, is *the principle of sufficient reason*. All men naturally follow it; for no one decides to do one thing rather than another without a sufficient reason that shows that this thing is preferable to the other.

IT IS FUNDAMENTAL TO ALL THE CONTINGENT TRUTHS.

When asking someone to account for his actions, we persist with our own question until we obtain a reason that satisfies us, and in all cases we feel that we cannot force our mind to accept something without a sufficient reason, that is to say, without a reason that makes us understand why this thing is what it is, rather than something completely different.

ABSURDITIES THAT RESULT FROM THE NEGATION OF THIS PRINCIPLE.

If we tried to deny this great principle, we would fall into strange contradictions. For as soon as one accepts that something may happen without sufficient reason, one cannot be sure of anything, for example, that a thing is the same as it was a moment before, since this thing could change at any moment into another of a different kind; thus truths, for us, would only exist for an instant.

For example, I declare that all is still in my room in the state in which I left it, because I am certain that no one has entered since I left; but if the principle of sufficient reason does not apply, my certainty becomes a chimera, since everything could have been thrown into confusion in my room, without anyone having entered who was able to turn it upside down.

Without this principle there would not be identical things, for two things are identical when one can substitute one for the other without any change to the properties which are being considered. This definition is accepted by everyone. Thus, for example, if I have a ball made out of stone, and a ball of lead, and I am able to put the one in the place of the other in the basin of a pair of scales without the balance changing, I say that the weight of these balls is *identical*, that it is the same, and that they are identical in terms of weight. Yet something could happen without a sufficient reason, and I would be unable to state that the weight of the balls is identical at the very instant when I find that it is identical; since a change could happen for no reason at all, happen in one and not the other; and, consequently, their weights would no longer be identical, which is contrary to the definition.

Without the principle of sufficient reason, one would no longer be able to say that this universe, whose parts are so interconnected, could only be produced by a supreme wisdom, for if there can be effects without sufficient reason, all might have been produced by accident, that is to say, by nothing.

THIS PRINCIPLE IS THE ONLY THING THAT CAUSES US TO DIFFERENTIATE WAKING FROM SLEEPING.

What sometimes happens in dreaming gives us the idea of a fabulous world, where all events could happen without sufficient reason.

I dream that I am in my room, busy writing; all of a sudden my chair changes into a winged horse, and I find myself in an instant a hundred leagues from the place where I was with people who have been dead for a long time, etc. All of this cannot happen in this world, since there would not be sufficient reason for all these effects; for when I leave my room, I can say how and why I leave it, and I do not go from one place to another without passing through intermediary places. Yet all these chimeras would be equally possible if effects could exist without sufficient reason; it is this principle that distinguishes dreaming from waking and the real world from the fabulous world that is depicted in fairy tales. Thus, those who deny the principle of sufficient reason are the inhabitants of a fabulous world that does not exist, but in the real world, all must happen according to this principle.[52]

In geometry where all truths are necessary, only the principle of contradiction is used. In a triangle, for example, the sum of the angles can only be determined in a single manner, and they absolutely must be equal to the sum of two right angles. But when it is possible for a thing to be in several states, I cannot be sure that it is in one state rather than another, unless I do give a reason for that which I affirm. Thus, for example, I can be sitting, lying down, or standing, all these determinations of my situation are equally possible, but when I am standing, there must be a sufficient reason why I am standing and not sitting or lying down.

ARCHIMEDES FIRST USED THIS PRINCIPLE IN MECHANICS.[53]
Archimedes, in passing from geometry to mechanics, recognized the need for sufficient reason; for, wanting to demonstrate that a pair of scales with arms of equal length loaded with equal weights would rest in equilibrium, he showed that in this equality of the arms and weights, the scales must stay at rest, because there was not sufficient reason why one of the arms should tilt rather than the other.

BUT IT IS M. LEIBNIZ WHO MADE EVIDENT ALL THE EXTENSION AND USEFULNESS OF IT.
M. Leibniz, who was very attentive to the sources of our reasoning, took this principle, developed it, and was the first who stated it clearly, and who introduced it into the sciences.

It must be acknowledged that one could not have rendered the sciences a greater service, for the source of the majority of false reasoning is forget-

---

52. This categorical statement is very provocative, as French and English natural philosophers rejected this Leibnizian principle. It is the principle that Voltaire later ridiculed in his tale of *Candide* (1759). Nineteenth- and early twentieth-century scientists used it as described here by Du Châtelet as a fundamental premise of their work, the presumption that there is a particular demonstrable cause of any given phenomenon.

53. Archimedes was the Greek mathematician of the third century BCE.

ting sufficient reason; and you will soon see that this principle is the only thread that could guide us in these labyrinths of error the human mind has built for itself in order to have the pleasure of going astray.

So we should accept nothing that violates this fundamental axiom; it keeps a tight rein on the imagination, which often falls into error as soon as it is not restrained by the rules of strict reasoning.

DIFFERENCE BETWEEN THE POSSIBLE AND THE ACTUAL.

§.9. It is necessary to distinguish between the possible and the actual. You have seen before that all that does not imply contradiction is possible, but is not actual. It is possible, for example, that this square table might become round, but this will perhaps never happen. Thus, all that exists being necessarily possible, one can conclude possibility from existence, but not existence from possibility.

So in order that a thing might be, it is not sufficient for it to be possible; this possibility must also be actualized, and this is called *existence*. Now a thing cannot come to exist without a sufficient reason, by which an intelligent being might understand why this thing becomes actual, having been possible before. Thus, a cause must contain not only the principle of the actuality of the thing of which it is the cause but also the sufficient reason for this thing, that is to say, what makes it possible for an intelligent being to understand why this thing exists. For any man who makes use of his reason must not be content with knowing that a thing is possible and that it exists, but he must also know the reason why it exists. If he does not see this reason, as often happens when things are too complicated, he must at least be certain that one could not demonstrate that the thing in question cannot have sufficient reason for its existence. Thus, in all that exists there must be something making it possible to understand why something that exists could exist; this is what is called *sufficient reason*.

THE PRINCIPLE OF SUFFICIENT REASON BANISHES FROM PHILOSOPHY ALL THE REASONING OF SCHOLASTICISM.

§.10. This principle banishes from philosophy all the reasonings of Scholasticism; for the Scholastics accepted that nothing happens without a cause, but they would allege as causes *plastic natures, vegetative souls,* and other meaningless words. But once it has been established that a cause is good only insofar as it satisfies the principle of sufficient reason, that is to say, insofar as it contains something making it possible to show how and why an effect can happen, then it becomes impossible to substitute these grand words for ideas.

For instance, when it is explained why plants appear, grow, and last, and that the cause advanced for these effects is a vegetative soul found in all plants, a cause of these effects is indeed advanced;[54] but it is a cause that is not admissible at all, because it contains nothing that helps us to understand how the vegetation of which I seek the cause operates. For assuming the existence of this vegetative soul does not promote understanding of why the plant that I am considering has a particular structure rather than any other, nor how this soul can give shape to a mechanism such as that of this plant.

IT IS THE FOUNDATION OF MORALS.

§.11. The principle of sufficient reason is also the foundation of the rules and customs founded only on what is called *propriety*. For the same men may follow different customs, they may determine their actions in many ways; and when one chooses to prefer those which are most reasonable over others, the action becomes good and could not be condemned; but the action is said to be unreasonable as soon as there are sufficient reasons for not committing it, and it is certainly on these same principles that one custom may be judged better than another, that is to say, when it has more reason on its side.

OF THE PRINCIPLE OF INDISCERNIBLES. HOW SUFFICIENT REASON FOLLOWS FROM THIS

§.12. From this great axiom of sufficient reason is born another that M. Leibniz calls *the principle of indiscernibles*. This principle banishes from the universe all similar matter, for if there could be two pieces of matter absolutely similar and identical, so that one might be put in the place of the other without it causing the slightest change (this is what is meant by entirely identical) there would be no sufficient reason why, for instance, one of these particles was placed on the Moon and the other on the Earth, since changing them and placing the one which is on the Moon on the Earth, and the one which is on the Earth on the Moon, all things would remain the same.

IT BANISHES ALL SIMILAR MATERIAL FROM THE UNIVERSE.

So one is obliged to recognize that the least particles of matter are discernible, or that each is infinitely different from all others, and that it could not be used in a place other than the one it occupies without disturbing the whole universe. Thus, each particle of matter is meant to have the effect that it produces, and from this, diversity is born, which is found between two grains of sand just as between our globe and that of Saturn, this diver-

---

54. This is an idea from Aristotle.

sity reveals to us that the wisdom of the Creator is no less admirable in the tiniest being than in the biggest.

The infinite diversity that reigns in nature is evident to us as far as our organs can sense. M. Leibniz, who advanced this truth first, had the pleasure of seeing it confirmed by the very eyes of those who denied it, on a walk with Madame the Electress of Hanover, in the garden of the Heurenausen.[55] For this philosopher, having stated that two entirely similar leaves could never be found in the almost innumerable quantity of those which surrounded them, several of the courtiers fruitlessly spent part of the day in this search, and could never find two leaves that did not have perceivable differences, even to the naked eye.

There are other objects that their smallness makes us see as alike, because we see them confusedly, but microscopes discover their differences for us. Thus experiments, which are not necessary for the truth of this principle, confirm it again.

OF THE LAW OF CONTINUITY.

§.13. From the axiom of sufficient reason there follows yet another principle, called *the law of continuity*, it is again to M. Leibniz that we are indebted for this principle, which is one of the most fruitful in physics. It is he who teaches us that nothing happens at one jump in nature, and a being does not pass from one state to another without passing through all the different states that one can conceive of between them.

The principle of sufficient reason is easily found in that truth, for each state in which a being finds itself must have its sufficient reason why this being is in this state rather than in any other, and this reason can only be found in the preceding state. Therefore this antecedent state contained something which gave birth to the current state that followed it, so that these two states are so completely interconnected it is impossible to put another state between the two. For if there was a state possible between the current state and that which immediately preceded it, the nature of the being would have left the first state without yet being determined by the second to abandon the first. Thus, there would be no sufficient reason why it should pass to this state rather than to any other possible state. Thus no being passes from one state to another without passing through the intermediate states, in the same way as one does not go from one city to another without traveling along the road between the two.

55. Leibniz told this story often, about Sophie, Electress of Hanover (1630–1714), and their walk in her garden at Herrenhausen (Du Châtelet spelled it incorrectly), probably around 1685.

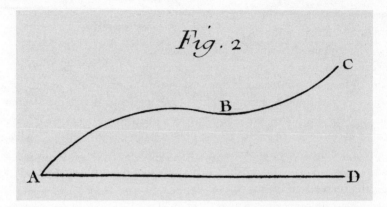

EXAMPLES OF THIS LAW IN GEOMETRY.

In geometry where everything happens in the greatest order, it can be seen that this rule is observed with an extreme exactitude, for all the changes which occur in lines that are one, that is to say in a line that is the same, or in those which together make up one and one whole only, all these changes, I say, exist after the figure has passed through all the possible changes that lead to the state it acquires. Thus, a line that is concave toward an axis, as line AB toward axis AD, does not become convex without passing through all the states between concavity and convexity, and through all the degrees that can lead from one to the other; thus concavity begins to diminish by infinitely small degrees up to point B, where the line is neither concave, nor convex, a point that is called the point of inflection. At this point the concavity ends and the convexity begins, and at this point B an infinitely small line parallel to axis AD forms; beyond this point B, the convexity begins and increases by infinitely small degrees, as mathematicians know.

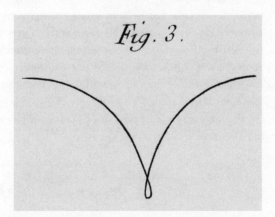

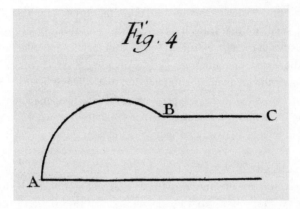

The points of retrogression found in many curves and that might appear to violate this law of continuity—because the line appears to end at this point and retrogress quickly in a contrary direction—do not, however, violate it at all; for it can be shown that at these points of retrogression nodes are formed as in figure 3,[56] in which it is clearly seen that the law of continuity is followed, for these nodes being diminished to infinity, in the end take the form of a perceivable point.

The law of continuity is not found in mixed figures, of which one cannot say that they form a true whole, because they have not been produced by the same law but are composed of several pieces, as if one added to the arc of a circle AB a straight line BC in order to make a single figure ABC. These figures violate the law of continuity, because the law by which one describes a circle AB ends at B and contains nothing in it that might give birth to line BC, but at point B another law begins, according to which line BC is described, and this second law bears no relationship to the first, which described circle AB.

The same thing happens in nature as in geometry, and it was not without reason that Plato called the Creator, *the eternal Geometrician*.[57] Thus, there are no angles properly speaking in nature, no inflexion nor abrupt retrogressions; but there are gradations in everything, and all prepares well in advance for changes that must be experienced, and goes by slight changes to the state it must be in. Thus a ray of light that is reflected on a mirror does not suddenly retrogress, and does not make an acute angle at the point of

56. *Nodes* had more than its specific astronomical meaning in the eighteenth century. It was a general term that could also be synonymous with a small loop, or knot.

57. Du Châtelet studied the dialogues of Plato (429–347 BCE), the Greek philosopher, during her time at Cirey.

reflection; but it passes to the new direction that it takes on being reflected through a small arc that leads it imperceptibly and through all the possible degrees between the two extreme points of incidence and reflection.

It is the same with refraction. The ray of light does not break at the point that separates the medium it penetrates and that which it leaves behind, but it begins to inflect before having penetrated the new medium; and the beginning of its refraction is a small curve that separates the two straight lines it describes in traversing two heterogeneous and contiguous mediums.

THIS PRINCIPLE SERVES TO DEMONSTRATE THE LAWS OF MOTION.

§.14. By this law of continuity the true laws of motion can be found and demonstrated, for a body that moves in any direction whatever could not move in an opposite direction without passing from its first movement to rest through all of the intermediate degrees of retardation, in order to pass again, by imperceptible degrees of acceleration, from rest to a new movement that it must experience.

THE PRINCIPLE OF CONTINUITY PROVES THAT THERE ARE NO PERFECTLY HARD BODIES IN THE UNIVERSE.[58]

§.15. This law shows that there is not a perfectly hard body in nature, for in the collision of perfectly hard bodies this gradation could not take place because the hard bodies would pass all at once from rest to movement, and from movement in one direction to movement in an opposite direction. Thus, all bodies have a degree of elasticity that renders them capable of satisfying this law of continuity which nature never violates.

§.16. It follows from what I have just said that when the conditions that give birth to a property come to change to other conditions from which another property must be born, so that finally these conditions become the same or identical, the property which resulted from the initial conditions must change by the same gradation into the property that is a continuation of the later conditions into which the first happened to change.

Geometry furnishes an infinity of examples that confirm and clarify this rule. The ellipse and the parabola, for example, describe very different lines, but when one makes the determinations of an ellipse vary (which are the conditions that render an ellipse possible) in order to make them approach those of the parabola, the properties of the ellipse also vary continually and

---

58. In the eighteenth century and in subsequent science, it had to be presumed that there could be no completely hard bodies in nature. Bodies were presumed to be *elastic*, meaning that their shape would be affected by an impact with another body, but that this shape would be resumed after the impact.

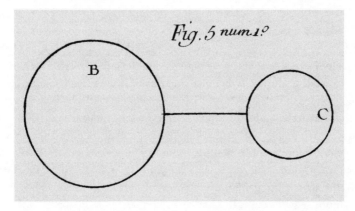

approach those of the parabola up to the point where finally the lines become the same. Thus, one of the foci of the ellipse remaining immobile, if the other moves away continually, the new ellipses that will be produced will continually become more like the parabola, and they finally will coincide with it, when the distance between the foci has become infinite. Thus, all the properties of the parabola will agree with those of an ellipse the foci of which will be infinitely distant, and the parabola can be considered as an ellipse whose foci are infinitely distant. By this same principle a decreasing movement finally becomes rest, and that ever-diminishing inequality turns into equality, so that rest may be considered as a very small movement, and equality as an infinitely small inequality. So, whenever this continuity of event does not obtain, it must be concluded that there are mistakes in the reasoning one has used.

DESCARTES' MISTAKE IN NOT HAVING PAID ATTENTION TO THIS LAW.

§.17. Descartes, for example, would have reformed his laws of motion had he paid attention to this law. He began by establishing as a first law that two equal bodies colliding with equal speeds must rebound with the same speed, and this is very true, for there being no reason why one of the two should continue in its path rather than the other, and these bodies being unable to penetrate each other or stay in repose, because the force of their equal speeds would be lost, which cannot happen, they must necessarily both rebound with the same speed with which they collided.

But M. Descartes' second law of motion and almost all the others are false, because they violate the principle of continuity. The second, for example, states that if two bodies B and C collide with equal speeds, but that body B is bigger than body C, then only body C will rebound and body B will continue on its path, both with the same speed that they had before the collision. This rule is denied by experience, and it is false because it is not in accord with the first rule of motion, or with the principle of continuity, for in always diminishing the inequality of the bodies, the effect that is a result of the inequality must always approach that which is a result of their inequality (§.16.), so that always diminishing the size of the largest body, its speed toward C must also diminish and finally become null when a certain proportion between B and C has been reached, beyond which point the inequality having completely vanished, the effect produced by the inequality of the two bodies will begin. That is to say, that then the movement of the greater body B will begin in an opposite direction, and the bodies will rebound with the same speed, according to the first law of M. Descartes. Thus, the second law cannot obtain since, according to this second law, although one may diminish the size of B and make it approach C so that the difference might be almost unassignable, the results will nonetheless remain very different and not be at all similar, which is totally contrary to the law of continuity. For when the inequality disappears, the effect creates a great jump, since the movement of body B changes direction all at once, passing through all the intermediary stages at one jump, while only an imperceptible change happens in the size of this body, which is nonetheless the cause of the great change that happens in the direction of its movement: thus the effect is greater than the cause. It can be seen by this example how important it is to pay attention to this law of continuity and in this way to imitate nature, which never transgresses this law in any of its operations.

CHAPTER TWO

Of the existence of God

THE STUDY OF PHYSICS LEADS US TO KNOWLEDGE OF A GOD.

§.18. The study of nature raises us to the knowledge of a supreme Being; this great truth is, if possible, even more necessary for good physics than for ethics, and it must be the foundation and the conclusion of all the research we make in this science.

PRÉCIS OF THE PROOFS OF THIS GREAT TRUTH.[59]

So, I believe that it is indispensable to begin by placing before you a précis of the proofs of this important truth, by which you will be able to judge its self-evidence for yourself.

59. Du Châtelet proceeds to give a combination of Cartesian and Leibnizian proofs of the existence of God.

§.19.1. Something exists, since I exist.

2. Since something exists, something must have existed from all eternity, otherwise nothingness, which is but a negation, would have produced all that exists, which is a contradiction in terms, for that is saying a thing has been produced, and yet not acknowledging any cause for its existence.

3. The Being that has existed from all eternity must exist necessarily and not owe its existence to any cause. For if it had received its existence from another Being, that other Being would have caused its own existence, and then, it is he of whom I am speaking, and it is God, or else he would owe his existence to another. It is easily seen, when ascending thus to infinity, that it is necessary to arrive at a necessary Being that exists by its own volition, or else admit that there is an infinite chain of Beings which, all together, will not have any external cause for their existence (since all Beings belong in this infinite chain) and that, each in particular, will have no internal cause, since none exists by its own volition, and they owe their existence to one another in an infinite gradation. Thus, it is supposing a chain of Beings that separately have been produced by one cause, and which all together have been produced by nothing, which is a contradiction in terms. So there is a Being that necessarily exists, since it implies contradiction that such a Being does not exist.

4. All that is around us is born and dies successively; nothing enjoys a necessary state, everything is successive, and we succeed one another. So there is only contingency in all the beings that surround us, this is to say, that the contrary is equally possible and does not imply contradiction (for this is what distinguishes a contingent being from a necessary being).[60]

5. All that exists has a sufficient reason for its existence. The sufficient reason for the existence of a being must be within it, or outside it. Now the reason for the existence of a contingent being cannot be within it, for if it carried the sufficient reason for its own existence, it would be impossible for it not to exist, which is contradictory to the definition of a contingent being. So the sufficient reason for the existence of a contingent being must necessarily be outside of it, since it cannot have it within itself.

6. This sufficient reason cannot be found in another contingent being, nor in a succession of such beings, since the same question will always arise at the end of this chain, however it may be extended. So it must come to a necessary Being that contains the sufficient reason for the existence of all contingent beings, and of its own, and this Being is God.

60. Du Châtelet explains the distinction between *necessary* and *contingent* in chapter 1, §.7.

The attributes of God

HE IS ETERNAL.

§.20. The attributes of this supreme Being follow from the necessity for its existence.

Thus, it is eternal, that is to say, it had no beginning and it will never have an end, for if the necessary Being had begun, it would have had to act before being, in order to produce itself, which is absurd, or something must have produced it, which is contrary to the definition of the necessary Being.

It cannot have an end, because the sufficient reason for its existence residing in itself, it can never abandon it; furthermore, what is contrary to a necessary thing implies contradiction and is consequently impossible. So it is impossible for the necessary Being to cease existing, just as it is impossible for 3 times 3 to make 8.

IMMUTABLE.

It is immutable, for if it changed, it would no longer be what it was, and consequently it could not have existed necessarily. Moreover, each successive state must have its sufficient reason in a preceding state, that one in another, and so on. Now, as in the necessary Being one would never reach the last state, since that Being never began, any successive state would be without sufficient reason, if it were susceptible to succession; thus, there cannot be change or succession in a necessary Being.

SIMPLE.

It follows clearly from what has just been said that the necessary Being cannot be a compound Being, which only exists as far as its parts are linked, and which can be destroyed by the dissociation of these same parts, and consequently the Being existing by its own volition is a simple Being.

NEITHER THE WORLD NOR OUR SOUL CAN BE A NECESSARY BEING.[61]

§.21. The world we see cannot be the necessary Being, for it is composed of parts and there is a continual succession in it, which is absolutely contradictory to the attributes I have just shown belong to a necessary Being.

By the same reasoning, neither matter nor the elements of matter can be the necessary Being.

---

61. Du Châtelet is using *soul* in the Cartesian sense of a thing separate from the body, having the properties of mind.

Nor can our soul be this necessary Being, for its perceptions, changing continually, it is in perpetual variation, but the necessary Being cannot vary. So our soul is not the necessary Being.

So the Being existing of its own volition is a Being different from the world we see, from the matter of which this world consists, from the elements which make up this matter, and from our soul; and it contains in itself the sufficient reason for its existence and of all the beings who exist.

THE NECESSARY BEING, THIS IS TO SAY, GOD, MUST BE UNIQUE.

§.22. It is easy to see by all that has been said that there can be only one necessary Being, for if there were two Beings that existed necessarily and independently of each other, each could exist alone, and consequently neither the one nor the other would exist necessarily.

§.23. It is evident that all that is possible does not exist, and that an infinity of things that could happen do not happen at all. Alexander, for example, instead of destroying the Empire of the Persians, could turn his armies against the peoples of the occident [west], or live peaceably in his kingdom. In a word, he could take an infinity of courses of action different from the one he took, which would have given rise to an infinity of combinations that were then possible and that would have produced events all different from those that occurred. This applies to the events contained in novels. They could happen if another succession of things took place; these are the stories of a possible world that lacks actuality, for each succession of things constitutes a world differing from all others by the events specific to it. Thus one can conceive of a succession of causes leading to the events in *Zaïde*, or those in the Queen of Navarre, for these events are possible, and they only lack actuality.[62] Similarly, one can conceive of possible universes, with other stars and other planets; and, as the different relationships of these universes can be combined in an infinity of ways, there are an infinity of possible worlds, only one of which actually exists.

When nothing had yet been produced, and none of these possible worlds existed, they all equally had the potential to come into existence; and they waited, so to speak, until an external force chose them, and made them actual. For what does not exist can only contribute to its existence ideally; that is to say, insofar as it contains certain determinations that the rest

---

62. *Zaïde* (1671) was a novel by Marie-Madeleine Pioche de la Vergne (1634–93), Countess de La Fayette. "Queen of Navarre" suggests *The Heptaméron* by Marguerite de Navarre (1492–1549), first published in 1558.

do not contain, and that can lead an Intelligent Being to choose it in order to give it existence.

There must be a sufficient reason for the actuality of the world we see, since an infinity of other worlds were possible. This reason can only be found in the differences that distinguish this world from all other worlds. This means, then, that the necessary Being must have envisaged all the possible worlds, considered their diverse arrangements and their differences, so as to be able to determine afterward to give actuality to the one that pleased him most.

GOD IS AN INTELLIGENT BEING.
The distinct representation of things constitutes understanding. Now the necessary Being who must have envisaged all the possible worlds before creating this one is therefore an intelligent Being, whose understanding is infinite, for all the possible worlds contain all the possible arrangements of all things possible. Thus, this Being we call God is an intelligent Being who sees not only all that actually happens but all that could happen in any possible combination of possible things; for all that is possible enters into the worlds that he contemplates never-endingly, and that are acted out, so to speak, before him.

AND HIS INTELLIGENCE IS INFINITELY BEYOND OURS.
§.24. As succession is an imperfection attached to the finite, there is no succession in the perceptions of God, who envisages at once all the worlds possible with all their possible changes; and as there are in our ideas an infinity of confused things, which we do not distinguish because of their multiplicity, the ideas that God has of things being infinitely distinct, they are infinitely different from ours, as would be, more or less, the idea that we have of the Moon in relation to the one a man who had lived a long time on that planet would have of it. The way in which God sees and envisages all possible things is, thus, incomprehensible to us. Thus we cannot form for ourselves a distinct idea of Divine understanding; it is like creation, among the things impossible for us to comprehend and deny. Let us always remember, when we wish to comprehend God's understanding, this child St. Augustine saw on the seashore who tried to put the ocean into a hazelnut shell; and this will give us some faint idea of the presumption of a being whose understanding is finite, and who wants to form a clear idea of the understanding of the Creator.[63]

63. The story of St. Augustine of Hippo (354–430), the child, and the nutshell have no basis in fact. However, it was the most popular representation of this Doctor of the Christian Church

HE IS FREE.

§.25. The choice that God made among all possible worlds of the world we see is a proof of his liberty, for having given actuality to a succession of things that contributed nothing of its own power to its existence, there is no reason preventing him from giving existence to other possible successions in the same category with regard to the possibility of actualization. So he chose the succession of things that constitute this universe to make it actual, because this succession pleased him most: He was the absolute master of his choice. The necessary Being is thus a free Being; for to act following the choice of one's own will is to be free.[64]

INFINITELY WISE.

§.26. But the choice he made of this world he did not make for no reason, for supreme intelligence will not behave without intelligence. Now since we judge here on Earth that a being is more or less intelligent according to whether he decides by reasons more or less sufficient, God, being the most perfect of all beings, none of his actions can be without a sufficient reason. So he had his own reason for determining to create a world, and this reason is the satisfaction he found in imparting a portion of his perfections, and the reason that determined him to give actuality to this world rather than to any other was the greater perfection he found in this one. But this reason is not outside of God, nor antecedent to him; he finds it in himself, it is part of his intelligence. For all the possible worlds being sequences of coexisting and successive things, these successions possess different degrees of perfection, according to whether they are more or less well linked and whether they tend more or less harmoniously to a general end. Now the contemplation of perfection is the source of pleasure in intelligent beings, for what has the most perfection pleases more, and a reasonable being only desires things in proportion to the perfections he notes in them. But as our understanding is limited, and we are liable to be wrong in the judgments we make, we often mistake an apparent perfection for a real perfection. In contrast, God, seeing things with an infinite understanding, cannot be deceived

---

in medieval European manuscripts. Augustine, puzzling over the possibility of the Trinity while walking at the seashore, questioned a child who was patiently filling a hole in the sand with water from the sea, nutshell by nutshell. Told by the saint that he would never complete the task, the child announced that he would complete his task before the man would ever understand the nature of three persons in one God. Augustine interpreted this answer as a sign from God. See James J. O'Donnell, Augustine: A New Biography (New York: HarperPerennial, 2006), 287–88. I am grateful to my colleague, the biblical scholar Edwin Yamauchi, for this reference.

64. Du Châtelet also made this her definition of human free will in her essay "On Liberty," which probably was originally intended for the Foundations but now exists only in manuscript.

by appearances, nor choose the bad because he failed to recognize the best; thus, he distinguishes among all possible worlds the best and the most perfect, and this greater perfection is the sufficient reason for the preference he gave to this world over all the other possible worlds. Thus the necessary Being is infinitely wise, for only a Being whose wisdom is infinite is able to choose what is most perfect.

§.27. It is in this infinite wisdom of the Creator that final causes, a principle so fruitful in physics, which some philosophers have tried to banish from it very inappropriately, originate; all indicates a design and it is to be blind, or to want to be, not to perceive that the Creator has intended, in the least of his works, purposes that he always achieves and that Nature unceasingly works to carry out. Thus, this universe is not in chaos, it is not a disordered mass without harmony and without connection, as some ranters would persuade us; but all its parts are arranged with infinite wisdom, and none could be transplanted or removed without harming the perfection of the whole.

In studying nature, one discovers some part of the intentions and the art of the Creator in the construction of this universe. Thus Virgil was right to say *Felix qui potuit, rerum cognoscere causas*,[65] since the knowledge of causes raises us to the level of the Creator and allows us to enter into the mystery of his designs by showing us the admirable order that prevails in the universe and the relationships of its different parts, which are not just necessary relationships of position, such as being above and below; but relationships of a design, of which everything carries the imprint. And the more the world ages, the further men take their discoveries, and the more one finds a design marked in the fabric of the world and of the least of its parts.

THIS WORLD IS THE BEST OF THE POSSIBLE WORLDS.

§.28. So this world is the best of the possible worlds, the one where the greatest variety exists with the greatest order, and where the largest number of effects is produced by the simplest laws. It is the universe that occupies the top of the pyramid,*[66] and that has nothing above it, but a real infinity below it which decreases in perfection and that consequently did not deserve to be chosen by an infinitely wise Being.

---

65. "Happy is he who understands the cause of things," from the Roman poet Virgil's *Georgics* (29 BCE), 2.490.

66. Du Châtelet contributed the following note: "*M. Leibniz continuing in his Théodicée the Dialogue between Boethius and Valla, introduces the Priest of Apollo, who wants to know the origin of the misfortunes of Sextus Tarquinus, and who seeks this origin in the Palace of the destinies, a pyramid consisting of all the possible worlds, in which this one, in which Tarquinus committed the crimes that led to Roman liberty, occupied the top."

THE IMPERFECTIONS OF THE PARTS CONTRIBUTE TO THE PERFECTION
OF ALL IN THIS UNIVERSE.

All the objections drawn from the evils prevalent in this world vanish by
this principle.[67] God allows them in the universe insofar as they enter into
the best succession of possible things and from which they could not be removed without removing some perfections from the whole; for all the universe is tied together, the least event is caused by an infinity of others that
preceded it, and an infinity of others are caused by it, and will arise from it.
Therefore, an event should not be judged apart from and outside of the relationship and succession of things; but it must be judged in relation to the
entire universe, and by the effects it produces in all places and at all times.
For to want to judge by an apparent evil the perfection of the universe is to
judge an entire painting by a single stroke of the brush, and it is a chimera to
imagine that all imperfections may be removed and everything stay the same
or become more perfect. The imperfection in the part often contributes to
the perfection of the whole; for when many rules must be obeyed at once to
reach a general perfection, the rules often contradict and generate unavoid-
able exceptions from which arise imperfections in the part, which nonetheless contribute to the most perfect whole that may be brought about.[68] The
human eye, for instance, could not see the least parts of an object without
losing sight of the whole; we would see a few points very distinctly if our
eyes were microscopes, but in so doing we would lose the whole. Therefore, our sight must be less distinct to be proportionate to our needs, since
distinguishing the least parts and a total view of the whole cannot be combined; for it is more useful to us to see the entire object than to distinguish
all its points one after the other. Thus it is a chimera to believe that the eye
of man would have been more perfect if it had distinguished the least parts
of things, since, on the contrary, such an ability would have been almost
useless to us.

The general will of God undoubtedly goes to the good and to the perfection of each thing in particular; but his consequent will, which is the result of all his previous acts of will, and which alone can be made actual, goes
to the good and the greatest perfection of the whole, to which the perfection of the parts must yield.

It is true that we cannot see all of this grand tableau of the universe, nor
show in detail how the perfection of the whole results from the apparent
imperfections we believe we see in some parts, for this would require en-

---

67. Here Du Châtelet deals with the ultimate metaphysical and theological contradiction: the
problem of a perfect Creator creating the imperfect, the ultimate Good creating evil.

68. Voltaire mocked this aspect of Leibniz's metaphysics in his tale of *Candide*.

visaging the entire universe and being able to compare it with all the other possible universes, which is an attribute of the Divinity (§.23). But our powerlessness in this respect cannot make us doubt that the supreme Intelligence has chosen the best world to which to give existence, for the necessary Being who is self-sufficient and who has no need for anything outside himself cannot have had other ends in the creation of this universe than to impart a portion of his perfections to his creatures, and to make a work worthy of himself, since he would have done something derogatory to himself and to his perfections if he had produced a world unworthy of his wisdom.

A consequence of the linkage between the parts and the whole is that all imperfection cannot be removed from man; man is a finite being, bounded and limited in all by his essence. How many evils happen to us because our understanding is limited, because we cannot know everything, understand everything, or be wherever our presence is necessary? But these are faculties the creature could not have without becoming a God; thus, the imperfections in the creature, a succession of his limitations, are necessary imperfections.[69]

THE SUPREME BEING IS INFINITELY GOOD.

§.29. It follows from all I have just said that the supreme Being is infinitely good; for having determined to create a world to which to impart a portion of his infinite perfections, he determined to grant actuality to the best succession of possible things. He granted to each thing in particular as much essential perfection as it could receive; and by his wisdom he directed the evils that would be inevitable in this succession of things to the greater good.

AND INFINITELY POWERFUL. HIS UNDERSTANDING IS THE PRINCIPLE OF THE POSSIBILITY, AND HIS WILL, THE SOURCE OF THE ACTUALITY OF THINGS.

§.30. He is infinitely powerful; for God having, for all eternity, envisaged all that is possible, his understanding is the source of all possibility, and as nothing can ever become possible other than what God conceived of as such, and nothing being actual except what he was pleased to grant existence to, he is the principle of the possibility and the actuality of all that is actual and possible.

§.31. God is the absolute Master of this succession of things to which

69. Du Châtelet continued to think about the distinctions between God and man; she explored it fully in her *Examinations of the Bible*, and mentioned it again in the *Discourse on Happiness*, both of which she was working on in this same period, 1735–39.

he has granted existence. He can change it and annihilate it, but (as we have seen) a contingent being cannot give itself existence, nor can it conserve it for a moment by its own force. Thus the reason for continuous existence cannot lie in the creature, who can neither begin nor continue to be but by the will of the Creator, which it needs at all times to sustain itself in the actuality that he has given it.

CHAPTER FOUR

Of hypotheses

THE USEFULNESS OF PROBABILITIES IN PHYSICS.

§.53. The true causes of natural effects and of the phenomena we observe are often so far from the principles on which we can rely and the experiments we can make that one is obliged to be content with probable reasons to explain them. Thus, probabilities are not to be rejected in the sciences, not only because they are often of great practical use, but also because they clear the path that leads to the truth.

USE OF HYPOTHESIS.

§.54. There must be a beginning in all researches, and this beginning must almost always be a very imperfect, often unsuccessful attempt. There are unknown truths just as there are unknown countries to which one can only find the good route after having tried all the others. Thus, some must run the risk of losing their way in order to mark the good path for others; so it would be doing the sciences great injury, infinitely delaying their progress, to banish hypotheses as some modern philosophers have.

MISUSES OF HYPOTHESES BY THE DISCIPLES OF DESCARTES.

§.55. Descartes, who had established much of his philosophy on hypotheses, because it was almost impossible to do otherwise in his time, gave the whole learned world a taste for hypotheses; and it was not long before one fell into a taste for fictions. Thus, the books of philosophy, which should have been collections of truths, were filled with fables and reveries.

THE DISCIPLES OF M. NEWTON HAVE FALLEN INTO THE
OPPOSITE EXCESS.

M. Newton, and above all his disciples, have fallen into the opposite excess: disgusted with suppositions and errors that they found filled books of philosophy, they rose up against hypotheses and tried to make them suspect and ridiculous, by calling them *the poison of reason and the plague of Philosophy*.[70] Moreover, he alone, who was able to assign and demonstrate the causes of all that we see, would be entitled to banish hypotheses from physics; but, as for us, who do not seem to be cut out for such knowledge, and who can only arrive at the truth by crawling from probability to probability, it is not for us to pronounce so boldly against hypotheses.

HOW HYPOTHESES ARE MADE.

§.56. When certain things are used to explain what has been observed, and though the truth of what has been supposed is impossible to demonstrate, one is making a hypothesis. Thus, philosophers frame hypotheses to explain the phenomena, the cause of which cannot be discovered either by experiment or by demonstration.[71]

HYPOTHESES ARE THE THREADS THAT LEAD US TO THE MOST SUBLIME DISCOVERIES.

§.57. If we take the trouble to study the way in which the most sublime discoveries were made, we will see that success only came after many unnecessary hypotheses had been made and yet the duration and unprofitableness of this work had not proved discouraging; for hypotheses are often the only available means to discover new truths. It is true that the means is slow and requires labor all the more onerous because for a long time one cannot know if it will be useful or fruitless. Similarly, when one takes an unknown route and finds several paths, it is only after walking a long time that one can be sure if one has taken the good route or if one has been mistaken. But if the uncertainty over which of these routes is the right one were a reason not to take any of them, it is certain that one would never arrive; whereas when one has the courage to set off, there is no doubt that of the three routes, of which two have misled us, the third will infallibly lead us to our goal.

---

70. In the "General Scholium," which he added to Book III of the *Principia*, Newton said that he refused to "feign" hypotheses in order to explain how gravity and attraction worked. His followers took this to mean that hypotheses in general had no place in science, and that only conclusions deduced from observation and experiment were useful and valid. As Du Châtelet explains, modern science would not have been possible without the use of hypotheses. See I. B. Cohen's "A Guide to Newton's *Principia*," 274–77 and his note to the *Principia*, n.oo, 943.

71. Du Châtelet and her contemporaries distinguished between *experiment*, seeking some new information about phenomena, and *demonstration*, the replication of someone else's experiment. Newton's experiments with prisms told him about the nature of light, but it was some time before others could replicate his results and thus, confirm his discoveries by demonstration.

WITHOUT HYPOTHESIS FEW DISCOVERIES WOULD HAVE BEEN MADE
IN ASTRONOMY.

It is in this manner that astronomy has brought us to the point where we admire it today; for, if, to calculate the path of the celestial bodies, astronomers had waited until the true theory of the planets had been found, there would be no astronomy now.

IT IS TO THESE THAT WE OWE THE TRUE SYSTEM OF THE WORLD.

The first idea of those who applied themselves to this science, just like the first idea of all other men, must have been that the Sun and all the celestial bodies turned around the Earth in twenty-four hours. Thus, they began to explain and to predict phenomena by this hypothesis, called *Ptolemy's hypothesis*, until the insurmountable difficulties of the consequences that derived from it when compared with observations, and the impossibility of constructing tables according to this hypothesis which were in accord with the phenomena of the sky, brought Copernicus to abandon it entirely and to test the opposite hypothesis, which is so much in agreement with the phenomena, that its certitude is at present not far from demonstration; and that no astronomer dares adopt that of Ptolemy.[72]

THEY OFTEN GIVE THE IDEA OF HOW TO DO NEW,
VERY USEFUL EXPERIMENTS.

§.58. Hypotheses must then find a place in the sciences, since they promote the discovery of truth and offer new perspectives; for when a hypothesis is once posed, experiments are often done to ascertain if it is a good one, experiments which would never have been thought of without it. If it is found that these experiments confirm it, and that it not only explains the phenomenon that one had proposed to explain with it, but also that all the consequences drawn from it agree with the observations, its probability grows to such a point that we cannot refuse our assent to it, and that is almost equivalent to a demonstration.

The example of astronomers can further serve marvelously well to clarify this matter; for the true orbits of the planets were ascertained by first supposing they made their revolutions in circles, of which the Sun occupied the center; but the variation in their speed and their apparent diameters being

72. Nicolaus Copernicus (1473–1543), the Polish astronomer, by his observations and new hypotheses about the universe began the shift in Europe from an Earth-centered to a Sun-centered solar system.

contradictory to this hypothesis, it was supposed they moved in eccentric circles, that is to say, in circles of which the Sun did not occupy the center. This supposition, which corresponded to the movements of the Earth well enough, deviated greatly from what is observed about the planet Mars; and to remedy this, attempts were made to make a new correction to the curve the planets describe in their annual revolution. This procedure succeeded so well that finally Kepler, going from supposition to supposition, found their true orbit, which admirably corresponded to all the appearances of the planets, and this orbit is an ellipse, of which the Sun occupies one of the foci.

By means of this hypothesis of the elliptic nature of orbits Kepler came to discover the proportionality of the areas and the times, and that of the times and the distances. And these two famous theorems, called the *Analogies of Kepler*, put M. Newton in reach of demonstrating that the supposition of the elliptic nature of the planets' orbits agrees with the laws of mechanics, and of assigning the proportion of forces that direct the movements of celestial bodies.[73]

Thus, it is evident that it is to hypotheses first made and then corrected that we are indebted for the beautiful and sublime knowledge of which astronomy and its subsidiary sciences are filled at present. It is impossible to see how men could have arrived there by other means.

IT IS BY MEANS OF HYPOTHESES THAT M. HUYGENS DISCOVERED THAT SATURN WAS SURROUNDED BY A RING.

By the same means we know today that Saturn is surrounded by a ring that reflects light, and is separated from the body of the planet, and inclined to the ecliptic. For M. Huygens, who discovered it first, did not observe what astronomers now describe; but he observed several phases of it, which sometimes resembled nothing less than a ring. Next, comparing the successive changes of these phases and all the observations that he had made of it, he sought a hypothesis that fit them, and explained these different appearances. That of a ring succeeds very well; it not only explains the appearances but also predicts the phases of this ring very accurately.

This correspondence between hypothesis and observation has finally converted this supposition of M. Huygens into a certainty; and no one doubts now that this ring is very real; thus, hypotheses brought us this beautiful discovery of the ring of Saturn.

73. Du Châtelet uses the term *analogies* in the Newtonian sense of laws based on mathematical models and experiment, though by her day Kepler's suppositions had been long accepted as proved by observations of the celestial bodies.

The same can be said of the ingenious explanation the same M. Huygens gave for halos, that is to say, of these sorts of colored crowns that sometimes appear around stars. No one before him had conceived of what might be the cause of these phenomena; but M. Huygens, after several fruitless suppositions, finally found that, by supposing the air to be filled with frozen grains of hail with a kernel of snow in their center, all the circumstances that accompany these phenomena could be explained, and no one has ventured to call into question M. Huygens's explanation.

DIVISION IS FOUNDED ON HYPOTHESES ONLY.

§.59. It is the same with numbers. Division, for example, is founded on hypotheses only. Without hypotheses you could not divide, for when you begin division, you suppose that the divisor is contained in the dividend as many times as the first number of the divisor is contained in the first number, or in the first two numbers of the dividend; and then you verify this supposition by multiplying the divisor by the quotient, and by subtracting from the dividend the product of this multiplication. If you find that this subtraction cannot be done, you conclude that the quotient is too big, and then you correct it. This whole operation is done by means of hypotheses.

HYPOTHESES ARE NOT ONLY VERY USEFUL, BUT EVEN SOMETIMES
VERY NECESSARY.

§.60. So making hypotheses is allowed, and it is even very useful in all cases when we cannot discover the true reason for a phenomenon and the attendant circumstances, neither a priori, by means of truths that we already know; nor a posteriori, with the help of experiments.

HOW ONE MUST PROCEED WHEN MAKING A HYPOTHESIS.

§.61. Without doubt there are rules to follow and pitfalls to avoid in hypotheses. The first is, that it not be in contradiction with the principle of sufficient reason, nor with any principles that are the foundations of our knowledge. The second rule is to have certain knowledge of the facts that are within our reach, and to know all the circumstances attendant upon the phenomena we want to explain. This care must precede any hypothesis invented to explain it; for he who would hazard a hypothesis without this precaution would run the risk of seeing his explanation overthrown by new facts that he had neglected to find out about. This is what happened to him who wanted to explain electricity after having only seen how Spanish wax, rubbed vigorously, attracts bits of paper. For it would have been easy to make other bodies do what happens with Spanish wax; rubbed in the same way, they would also have been electrified. Thus, the explanation of electricity by Spanish wax alone had been insufficient and precipitous.

When one can hope to know the greatest number of circumstances attendant upon a phenomenon, then one can seek the reason for it by means of hypotheses, taking the risk of having to correct it or having to be corrected; but the efforts made to find the truth are always glorious, even though they might be fruitless.

PITFALL TO AVOID IN HYPOTHESES.

§.62. Since hypotheses are only made in order to discover the truth, they must not be passed off as the truth itself, before one is able to give irrefutable proofs. So it is very important for the progress of the sciences not to delude oneself and others with the hypotheses one has invented, but one should estimate the degree of probability in a hypothesis, and never impose it on others by detours and a semblance of demonstration, which has much too often led people with a thirst for knowledge into error.

With this precaution one does not run the danger of taking for certain that which is not; and one inspires those who follow us to correct the faults in our hypotheses and to provide what they lack to make them certain.

§.63. Most of those who, since Descartes, have filled their writings with hypotheses to explain facts, which very often they only knew imperfectly, have sinned against this rule and have tried to pass off their suppositions as truths; this is partly the source of the disgust for hypotheses in this century. But the excessive resort to a useful thing does not detract from its utility and must not prevent us from making use of it when this can bear fruit.

A SINGLE CONTRARY EXPERIMENT SUFFICES TO REJECT A HYPOTHESIS.

§.64. One experiment is not enough for a hypothesis to be accepted, but a single one suffices to reject it when it is contrary to it. It follows, for example, from the hypothesis in which one supposes that the Sun moves around the Earth, which serves as the center of its orbit, that the diameters of the Sun must be equal at all times of the year; but experience shows that they appear unequal. From this observation, one can therefore conclude with certainty that the hypothesis, of which this equality is a consequence is false, and that the Earth does not at all occupy the center of the Sun's orbit.

A HYPOTHESIS CAN BE TRUE IN ONE OF ITS PARTS AND FALSE IN ANOTHER.

§.65. A hypothesis may be true in one of its parts and false in another; then the part that is found to be in contradiction with experiment must be corrected.

But great care must be taken to put in the conclusion only what must be there, and not to attribute to the entire hypothesis a fault which only applies to one of its parts. For example, M. Descartes attributed the fall of bodies toward the center of the Earth to a vortex of fluid matter impelling bodies to move toward the center by its rapid swirling around the Earth; but M. Huygens demonstrated by an irrefutable experiment that, according to this supposition, bodies should be directed in a fall perpendicular to the axis of the Earth and not toward its center. It can be concluded from that, that a vortex of fluid matter, such as M. Descartes conceived, would not cause bodies to fall toward the center of the Earth; but it would be too precipitous to conclude that no fluid material caused the phenomenon of the fall of bodies. It is the same with the other vortices, which, according to M. Descartes, carry the planets around the Sun; for M. Newton demonstrated that this supposition did not agree with the laws of Kepler. So it must be inferred that the movements of the planets are not the effect of the vortices of fluid material that M. Descartes supposed explained them. But it cannot be legitimately concluded that a vortex, or several vortices, conceived of in a different way, cannot be the cause of these movements.[74]

§.66. Thus, in making a hypothesis one must deduce all the consequences that can legitimately be deduced, and next compare them with experiment; for should all these consequences be confirmed by experiments, the probability would be greatest. But if there is a single one contrary to them, either the entire hypothesis must be rejected, if this consequence follows from the entire hypothesis, or that part of the hypothesis from which it necessarily follows.

Astronomers give us another example of this rule; indeed, a plethora of discoveries would not have come about in astronomy if no attempt had been made to verify by experiment what was deduced from hypotheses. It follows, for example, from Copernicus's hypothesis that if the distance of a star to the Earth has a relationship comparable to the diameter of its orbit, the height of the pole and the fixed stars must vary at different times of the year. The desire to verify this consequence led several astronomers to make observations on this annual parallax or height of the fixed stars; among others, M. Bradley, in whose hands this consequence was not only confirmed

---

74. Du Châtelet believes that attraction explains these phenomena but continues to give credence to Descartes' explanation to prove her point about the correct use of hypotheses. Leonhard Euler and Johann Bernoulli would fit this latter description, for they accepted attraction but continued to seek a material explanation for it. Note that Newton never successfully explained the cause of attraction, only its observed effects in the universe.

but then gave birth to this beautiful theory of the aberration of the fixed stars, which would never have been thought of before.[75]

DEFINITION OF HYPOTHESES. WHAT MAKES THEM PROBABLE.

§.67. Hypotheses, then, are only probable propositions that have a greater or lesser degree of certainty, depending on whether they satisfy a more or less great number of circumstances attendant upon the phenomenon that one wants to explain by their means. And, as a very great degree of probability gains our assent, and has on us almost the same effect as certainty, hypotheses finally become truths when their probability increases to such a point that one can morally present them as a certainty; this is what happened with Copernicus's system of the world, and with M. Huygens's on the ring of Saturn.

WHAT MAKES THEM INVALID.

By contrast, a hypothesis becomes all the more improbable as it fails to explain more of its attendant circumstances, as in Ptolemy's hypothesis.

§.68. When a hypothesis is made, one must have reasons for preferring the supposition on which it is founded to all other suppositions; otherwise one spews forth chimeras, and precarious principles that have no foundation.

§.69. So it is necessary not only that all one supposes be possible, but also that it be possible in the manner one uses it; and that the phenomena result necessarily, and without the obligation to make new suppositions. Otherwise, the supposition does not deserve the name of hypothesis; for a hypothesis is a supposition that explains a phenomenon. When the necessary consequences do not follow from it, and to explain the phenomenon, a new hypothesis must be created in order to use the first, this hypothesis is only a fiction unworthy of a philosopher.

§.70. If those who wanted to explain so many surprising effects by means of hooked, branchlike, and serrated particles had paid attention to what is required to make a truly philosophical hypothesis, they would not have slowed, as they did, the progress of the sciences by creating monsters that subsequently had to be fought against as realities.[76]

75. James Bradley (1693–1762) was professor of astronomy at Oxford, the successor to Edmund Halley (1656–1742) as astronomer royal. Both were known for their astronomical observations.

76. Nicolas Lémery (1645–1715), French chemist and public lecturer, developed a corpuscular theory of pointed particles that could unite with others that were serrated.

HYPOTHESES ARE ONE OF THE GREAT MEANS OF THE ART OF INVENTION.
GOOD HYPOTHESES HAVE ALWAYS BEEN MADE BY THE GREATEST MEN.

§.71. By distinguishing between the good and the bad use of hypotheses, both extremes are avoided, and without giving oneself up to fictions, a method very necessary to the art of invention is not denied to the sciences, a method that is the only means that can be used in difficult researches requiring correction over several centuries and the work of several men before attaining a certain perfection. And it must not be feared that by this method philosophy might become a heap of fables; for we have seen that a good hypothesis can only be made when a great number of facts and circumstances attendant upon the phenomenon one wants to explain have been observed (§.61), and that the hypothesis is only true and only deserves to be adopted when it explains all the circumstances (§.66). Therefore, the good hypotheses will always be the work of the greatest men. Copernicus, Kepler, Huygens, Descartes, Leibniz, M. Newton himself, have all imagined useful hypotheses to explain complicated and difficult phenomena; and the examples of these great men and their success must show how much those who want to banish hypotheses from philosophy misunderstand the interests of the sciences.

CHAPTER SIX

Of time

ANALOGY BETWEEN TIME AND SPACE.

§.94. The notions of time and space are very similar. In space, one simply considers the order of the coexistents insofar as they coexist; and in time, the order of successive things, insofar as they succeed each other, discounting any other internal quality than simple succession.

THE ORDINARY IDEA OF TIME IS FALSE. IT LEADS TO THE SAME
DIFFICULTIES AS THAT OF PURE SPACE.

§.95. Ordinarily our image of time, as of space, is produced by confused ideas: thus, one imagines it as a being composed of continuous, successive parts, which flow uniformly, which subsists independently of things existing in time, which has been in a continual flux from all eternity, and which will continue in the same way. But it is obvious that this notion of time as a being composed of continuous and successive parts that flows uniformly, being once accepted, leads to the same difficulties as those of absolute space; that is to say that, according to this notion time would be a necessary being, immutable, eternal, subsisting by itself, and consequently it would have all the attributes of God.

THE PRINCIPLE OF SUFFICIENT REASON PROVES THAT TIME IS NOT SEPARATE FROM THINGS.

§.96. From this idea of time M. Clarke put the famous question to M. Leibniz: *why God had not created the universe six thousand years earlier or later?*[77]

M. Leibniz had no trouble countering this objection of the English doctor and his opinion on the nature of time by the principle of sufficient reason; he only needed to use M. Clarke's own objection on the time of the creation. For if time is an absolute being consisting in a uniform flow, the question of why God did not create the world six thousand years earlier or later becomes a meaningless question, and forces one to acknowledge that something happened without sufficient reason. For the same succession of beings in the universe being kept, God could make the world begin earlier or later, without thereby causing any disturbance. Now since all instants are equal, when only succession is attended to there is nothing in them that could have led to a preference for one over another, to the extent that no diversity in the world would have been caused by this choice. Thus, one instant would have been chosen in preference to another to make this world actual without sufficient reason, which cannot be accepted (§.8).[78]

But furthermore, we are going to see, by the analysis of our ideas, that time is only an abstract being, which has no separate existence from things and which, consequently, cannot be endowed with the properties imagination attributes to it.

HOW ONE COMES TO FORM THE IDEA OF TIME AS AN ABSOLUTE BEING, EXISTING INDEPENDENTLY OF SUCCESSIVE BEINGS.

§.97. When we pay attention to the continuous succession of several beings, and we conceive of the existence of the first A as distinct from that of the second B, and this second B distinct from that of the third C, and so on, and we observe that two never exist together; but that A, having ceased

---

77. Like many of her contemporaries, Du Châtelet read Pierre Desmaizeux's French translation of the correspondence between Samuel Clarke (1675–1729) and Gottfried Wilhelm Leibniz in 1715–16 (1717 ed.). Leibniz had been the Princess of Wales, Caroline of Ansbach's tutor, Clarke was her chaplain. She asked the two men to discuss the nature of the universe and its relation to God, in particular the ideas of Sir Isaac Newton and those of Leibniz.

78. In chapter one of the *Foundations*, Du Châtelet established the basic rules governing all knowledge; in chapter two she explained that God would also choose to be subject to these rules; as a perfect being he would always have a sufficient reason for his actions.

to exist, B soon succeeds it; that B, having ceased, C succeeds it, and, we form a notion of a being we call *time*. And insofar as we construe the permanent existence of one being in terms of these successive beings, we say *that it lasted a certain time*, insofar as this being is conceived of as existing with several others that succeed each other.

So it is said that a being lasts since it coexists with several other successive beings in a continuous sequence; thus, the duration of a being becomes explicable and commensurable by the successive existence of several other beings. For one takes the existence of a single one of these successive beings for *one*, that of two for *two*, and so on with the others; and as the being that coexists with all, its existence becomes commensurable by the existence of all these successive beings.

A thousand examples can clarify what I have just said. It is said, for example, that a body uses time to traverse a space, because the existence of this body at a single point is distinguished from its existence at any other point; and it is observed that this body could not exist at the second point without having ceased to exist at the first, and that the existence at the second point follows immediately on the existence at the first. And, insofar as one gathers together these diverse existences, and considers them as making *one*, it is said that this body used time to create a line. Thus, the time is not actual in the things that last, but is a simple mode or exterior relationship, which depends only on the mind insofar as it compares the duration of beings with the movement of the Sun, and of the other exterior bodies, or with the succession of our ideas.

§.98. When we pay attention to the links between our ideas, we grasp that in the abstract notion of time the mind only considers beings in general; and that having discounted all the determinations these beings can have, one only adds to this general idea, that of their non-coexistence, that is to say, that the first and the second cannot exist together but the second immediately succeeds the first, and with no possibility of the existence of another being between the two, discounting here again internal relationships and causes that make them succeed each other. In this manner, one creates an ideal being, consisting in a uniform flow, which must be similar in all its parts, since to create it one uses the same abstract notion for each being without determining anything of its nature, and one considers in all these beings only their successive existence without caring about how the existence of one gives birth to the next.

§.99. This abstract being we thus created must appear to us independent of existing things and subsisting by itself. For since we can distinguish the successive manner for beings to exist, the manner of their internal determinations, and of the causes which gave birth to this succession, we must regard time as a being apart, separate from things and able to subsist without actual and successive things, since we can still think of this successive existence, after having destroyed with our minds all the other realities, that is to say, having discounted them.

§.100. But as we can also restore to these general determinations particular ones that make them beings of a certain type, in applying our attention simultaneously to their successive existence and to their particular determinations it must appear to us that we make something exist in this successive being that could not exist there before, and that we can remove it again without destroying this being.

§.101. Time must also necessarily be considered as continuous; for if two successive beings, A and B, are not conceived of as continuous in their succession, it will be possible to place one or several between two that will exist after A has existed, and before B exists. Now by this very reasoning one accepts time between the successive existence of A and B; thus, time must be considered as continuous.

Thus our imagination creates a notion of time in considering it as a being composed of successive, continuous parts without internal differences, with which all successive beings coexist, and which becomes their common measure. This notion can have its uses, when it only concerns the magnitude of the duration and comparisons of the duration of several beings together. As in geometry one is only concerned with these sorts of considerations, so the imaginary notion can easily be put in place of the actual one. But we should refrain from making the same substitution in metaphysics and physics; for then we would fall into these difficulties of making the duration an eternal being, and with all the attributes of God, discussed above.

TIME IS NO OTHER THING THAN THE ORDER OF COEXISTENCE.

§.102. So time is really nothing other than the order of successive beings; and one forms the idea of it inasmuch as one considers only the order of their succession. Thus there is no time without true, successive beings arranged in a continuous sequence; and there is time as soon as such beings exist.

IT IS DIFFERENT FROM SUCCESSIVE BEINGS, AS THE PLACE AND NUMBER OF NUMBERED AND COEXISTING THINGS DIFFER.

§.103. But this resemblance in the manner in which these beings succeed each other, and this order that results from their succession, are not the things themselves, as we saw above (§.87.), that the number is not the things numbered, and the place is not the things placed in this place.[79] For the number is only an aggregate of the same units, and each thing becomes a unit when all is considered simply as a being; thus the number is only a relationship of a being considered in regard to all, and however different it may be from the numbered things, nonetheless it only actually exists insofar as things exist that can be reduced to units of the same class. Once these things have been set, one can place a number, and remove it, and there are none left. In the same way time, which is nothing other than the order of continuous successions, could not exist if things in a continuous sequence did not exist. Thus, there is time when things are, it is removed when one removes these things; however, it is like the number, different from the things that succeed each other in a continuous sequence. This comparison between time and number can help to form the true notion of time; and to understand that time, like space, is nothing absolute outside of things.

GOD IS NOT IN TIME AND ALL SUCCESSION IS IMMUTABLE FOR HIM.

§.104. With regard to God, it cannot be said that he is in time, for there is no succession in him, nor can any change happen to him. Thus, he is always the same, and his nature does not vary; and as he is outside the world, that is to say, as he is not linked with the beings whose union constitutes the world, he does not coexist with successive beings as creatures do; thus he cannot be measured by that of successive beings. For though God continues to exist during time, as time is only the order of the succession of beings, and this succession is immutable in relation to God, to whom all things with all their changes are present at once, God does not exist in time. God is at once all that he can be, whereas creatures can only successively achieve the states of which they are capable.

§.105. One can accept as actual parts of time only those denoted by actually existing beings; for actual time being only a successive order in a continuous sequence, one can accept portions of time only insofar as real things have existed and ceased to exist. For successive existence makes time, and a being that coexists with the least actual change in nature has lasted the shortest actual time; and the least changes, as for example, the movements of the smallest animals, denote the smallest actual parts of time we can perceive.

79. §.87 comes from chapter five on space where Du Châtelet explains that "space" is not a "real being" but an abstraction of coexisting things, as numbers are an abstraction of things numbered. Without the things, there would be no space and no numbers, and neither space nor the numbers are the real things.

§.106. Time is usually represented by the uniform movement of a point that describes a straight line, because the point is there a successive being, present successively at different points, creating by its flow a continuous succession to which we attach the idea of time. We also measure time by the uniform movement of an object; for when the movement is uniform, the moving body will, for example, travel a *pied* in the same time in which it traveled a first *pied*.[80] Thus the duration of things that coexist with the movement of a moving body, while it travels a *pied*, being taken as *one*, the duration of those that will coexist with its movement while it travels two *pieds* will be *two*; and so forth. In this way time becomes commensurable, since a ratio of one duration in relation to another can be made, with the duration taken as *one*. Thus in clocks the hand moves uniformly in a circle, and the twenty-fourth part of the circumference of this circle makes *one*; and time is measured with this unit, by saying two hours, three hours, etc.; in the same way, one takes one year for *one*, because the revolutions of the Sun in the ecliptic are equal and are used to measure other durations in relation to this unit.[81]

§.107. Astronomers' efforts to find a uniform motion, which put them in reach of measuring time exactly, are well known, and it is what M. Huygens found with pendulums, of which he is the inventor, and which will be discussed further on.

THE SUCCESSION OF OUR IDEAS, AND NOT THE MOVEMENT OF BODIES, GIVES US THE IDEA OF TIME.

§.108. We have seen that the successive existence of beings gives rise to the notion of time; now, as it is our ideas that represent to us these beings, the notion of time is born from the succession of our ideas and not from the movement of exterior bodies; for we would have a notion of time even if nothing other than our soul existed, and insofar as things that exist outside of us are similar to the ideas our soul creates of them, they exist in time.[82]

Movement, by itself, is so far from giving us the idea of duration, as some philosophers have claimed, that we only acquire the mere idea of movement by our reflections on successive ideas, which the moving body causes in our mind by its successive existence with the different bodies that surround it.

80. For French measurements in Du Châtelet's day see chapter 1 (of this volume), note 29.

81. The *ecliptic* means the apparent path of the Sun, indicated by the plane of the Earth's orbit extended as if to meet the Sun, inclined to the Earth's equator by an angle of 23°27″.

82. Du Châtelet is again speaking of the *soul* in the Cartesian sense, as distinct from the body and having the properties of mind.

WE DO NOT PERCEIVE MOVEMENT, WHEN IT IS TOO FAST OR TOO SLOW. This is why we have no idea of movement when looking at the Moon or the hand of a watch, though both are moving, for this movement is so slow that the moving body appears at the same point, while we have a long succession of ideas; and because we cannot distinguish the parts of space the body traveled in this interval, we believe that the moving body is at rest. But when at the end of a certain length of time, the Moon and the hand of this watch have gone a considerable way, then our mind, joining the idea of the point where it left them, that is to say, their past coexistence with certain beings, to that of their current coexistence with other beings, acquires by this means the idea of this body's movement.

Similarly, when the moving body goes with such rapidity that we have had no succession of ideas while it was going from one point to another, we say that the moving body traveled the distance in an instant, that is to say, that it used no perceivable time. Almost for the same reason, when the impressions each of the seven colors makes on our retina are too fleeting, we do not distinguish each color in particular, but have a common sensation of all these colors, which we have called *whiteness*.[83]

§.109. Thus only mediocre movement can give us the notion of time, because there is some relationship with the succession of our ideas; but it only gives us this notion because the soul can then distinctly represent the different states of the moving body one after another, without amalgamating several. Time, which is an ideal being, is very different from movement, which is something real.

M. DE CROUSAZ'S MISTAKE ABOUT TIME. THERE WOULD BE A TIME EVEN IF THERE WERE NO MOVEMENT.[84]

§.110. Thus I cannot imagine how it could be said in a memoir that won the first prize of the Academy of Sciences (and which, moreover, has other excellent things in it), *that the existence of movement in a body is the existence of time in the body; that the time and the movement of a body are the same thing; and finally that it is*

---

83. Du Châtelet sent her brief "Essay on Optics" to other physicists. It bears the designation, "Chapter 4," and perhaps was intended for the *Foundations*. It certainly was part of her reading and thinking in her collaboration with Voltaire on his *Elements of the Philosophy of Newton*. See Ira O. Wade, *Studies on Voltaire, with Some Unpublished Papers of Mme. Du Châtelet* (Princeton NJ: Princeton University Press, 1947), 188–207.

84. Jean-Pierre de Crousaz (1663–1748) was a Swiss mathematician known for his adherence to Descartes and opposition to Leibniz and Wolff. In his 1721 prize essay, he wrote on the mechanics of collisions between bodies. He and Du Châtelet corresponded after the publication of the *Foundations*.

a child's notion to believe that time is the measure of rest, as it is of movement. Certainly, I could always remain in the same place and still have successive ideas; I would exist during a certain time, and I would have an idea of the duration of my being from the succession of my thoughts, even if I had never moved and never seen a body in motion, and consequently had no idea of movement. Thus, as long as there are beings with successive existence, there will necessarily be a time, whether the beings are in motion or at rest.

TIME MUST BE CAREFULLY DISTINGUISHED FROM THE MEASUREMENT OF IT.

§.111. The reason why motion and time have been confused is that time has not been carefully enough distinguished from the measurement of it.

§.112. The measurements of time taken from exterior bodies were necessary to us for putting order in facts past, present, and even to come; and to give to ourselves and others an idea of what we mean *by such a portion of time.* For the succession of our ideas cannot be used for any of these purposes; it cannot serve as a rule to us, for nothing can assure us that between two perceptions that appear to follow each other immediately, an infinite number did not happen that we have forgotten and that are separated by vast expanses of time.

Neither can this succession of our ideas be used as the means to make others understand what we mean *by such a portion of time,* because ideas succeed each other faster or slower in different heads.

WHY TIME IS MEASURED BY THE MOVEMENT OF EXTERIOR BODIES.

This is why we have been obliged to take the measurements of time outside of ourselves. Nearly all peoples agree on using the course of the Sun to measure time, and it is apparently because it seems to proceed overhead that men have confused time and movement, and for lack of a distinction between time and the measurements established for measuring its parts. For if the Sun, for example, were extinguished and relit at equal intervals, it would also serve us as a measure of time, even if the Earth and it were immobile.

THERE IS NO WAY TO MEASURE TIME VERY ACCURATELY, AND WHY.

§.113. There is not, and cannot be, a very accurate measurement of time; for one cannot apply a part of time to itself to measure it, as one measures extension by *pieds* and *toises*, which are themselves portions of extension. Each has his own measurement of time in the quickness or slowness with which his ideas succeed each other, and it is these different speeds in which ideas follow each other in different persons, and in the same person at different times, that have resulted in many ways of expressing oneself, as, for example in this phrase, *I found the time very long;* for the time seems long to us, when the ideas have succeeded each other slowly in our mind.

§.114. It is easy to grasp that measurements of time can be different for different peoples, the annual and daily course of the Sun, the vibrations of a pendulum (which are, of all measurements, the most accurate) have provided us with those *of minutes, hours, days, and years;* but it is quite possible that other things have been used as measurements by other peoples. The only one that might be universal is what is called *an instant;* for all men necessarily know this portion of time, which flows while a single idea stays in our mind.

§.115. All measurements of time are founded only on the duration of our being, and on that of the beings that coexist with us and whose existence we view in terms of our own. For, having acquired the idea of successions and of time, although we have successive ideas, we transfer this idea to time during which we did not have any ideas as, for example, when one has fainted, and thus do we acquire the idea of the duration of the world and the universe, by relating the idea that we have of the duration of our existence to the time that passed when we did not yet exist, and to that which will pass when we are no longer.

HOW WE ACQUIRE THE IDEA OF ETERNITY.

§.116. To us the duration of all finite beings has a beginning and an end; so, if we subtract from this idea that of the beginning, then the duration is *eternity a parte ante;* if we remove the end, the kind of duration is called *eternity a parte post,* and it is thus that the soul of man is eternal.[85] Finally, if we remove its beginning and its end from the idea that we have of the duration of finite beings, the duration will become *the eternity of God,* for only God can be eternal *a parte post,* and *a parte ante,* that is to say, have neither beginning nor end. Thus, men acquire the idea of infinite duration, as of all other ideas of the infinite, by additions and subtractions, the end of which we can never see.

CHAPTER SEVEN

Of the elements of matter

WHAT WERE, ACCORDING TO ANCIENT PHILOSOPHERS, THE PRINCIPLES OF THINGS.

§.117. Philosophers from all times have exerted themselves about the origin of matter, and the elements. The Ancients each had their own different sentiment on this subject, some thought water the basic element of all bodies;

---

85. Again Du Châtelet has used Descartes' idea of the nature of the soul.

others, air; others, fire; Aristotle bringing these diverse sentiments together, admitted four elements of things: water, air, earth and fire. He believed that the mixture of these four principles, which, according to him, were basic, because they were not resolvable into other combinations, resulted in all that surrounds us.[86]

DESCARTES' IDEA ON THE ELEMENTS OF MATTER.

§.118. Descartes, who despite the interval of time between Aristotle and him was Aristotle's successor, also made elements, but in his own way. For Aristotle's four principles Descartes substituted three kinds of small bodies of different sizes and shapes, these small bodies, or elements, resulted, according to him, from the original divisions of matter, and were formed by their combination: fire, water, earth, air, and all the bodies that surround us.[87]

NEW OPINION ON THE ELEMENTS THAT WAS FORMED FROM DESCARTES'.

Most philosophers today have abandoned Descartes' three elements and conceive of matter simply as a mass, uniform and similar, with no internal difference; but, the small particles have such diversified forms and sizes that the infinite variety existing in this universe can result from them. Thus, they suppose the only difference between the constituent particles of gold and paper, for example, to be that which comes from the shape and the arrangement of these particles.

THIS OPINION IS CLOSE TO THAT OF EPICURUS ON ATOMS.

This opinion, which, like that of Descartes, is very well known, is nearly that of Epicurus on atoms, as revived by Gassendi in our own day; these solid and indivisible particles of matter, distinguished from one another by their shape and size, only differ in name from Epicurus's atoms.[88]

THE PRINCIPLE OF SUFFICIENT REASON SHOWS THAT ATOMS ARE INADMISSIBLE.

§.119. M. Leibniz, who never lost sight of the principle of sufficient reason, found that these atoms did not explain extension in matter, and, seeking

---

86. Aristotle's ideas of the elements and the nature of matter prevailed into the seventeenth century.

87. In the course of the seventeenth century Descartes' ideas became the prevailing wisdom as described in his *Principles of Philosophy*.

88. Pierre Gassendi (1592–1655), the French experimentalist and philosopher, brought the ideas of Epicurus and the Stoics to his contemporaries. Du Châtelet is referring to his support for the Greek philosopher's concept, "atomism."

to discover this reason, he believed that it could only lie in a different idea of particles, those without extension, which he named *monads*.

EXPOSITION OF M. LEIBNIZ'S SYSTEM OF MONADS, OR ELEMENTS
OF MATTER.

Few people in France know anything of this opinion of M. Leibniz's but the word, *monads;* the books of the famous Wolff, in which he explains so clearly and eloquently M. Leibniz's system, which, in his hands, took a totally new form, have not yet been translated into our language.[89] So, I am going to try to explain the ideas of these two great philosophers on the origin of matter; an opinion, which half of learned Europe has embraced, which deserves serious attention.

§.120. All bodies are extended in length, width, and depth. Now, as nothing exists without a sufficient reason, it is necessary for this extension to have a sufficient reason that explains how and why it is possible; for, saying *that there is extension, because there are small extended particles*, comes to saying nothing, since the same question will be asked about these small extended particles as about extension itself, and the sufficient reason for *their* extension will be asked about in turn. Now, as sufficient reason obliges us to state that a thing is different from what one is asking about, since otherwise no sufficient reason is provided and the question always remains the same, if one wants to fulfill this principle about the origin of extension, it is necessary to come in the end to something that is without extension, that has no particles, to give a reason for that which is extended and has particles. Now, a being without extension and without particles is a simple being; so, compounds, extended beings, exist because there are simple beings.[90]

It must be confessed that this conclusion astonishes the imagination, simple beings are not within its province, they cannot be represented by images, and only the understanding can conceive of them. To have simple beings accepted less reluctantly, the Leibnizians use quite an accurate comparison; they say: if someone asked why there were watches, he certainly would not be content if he was answered, *it is because there are watches;* but in order to give sufficient reasons and ones that satisfy the questioner about the possibility of a watch, it would be necessary to come to things that were

89. Strictly speaking, there *were* translations which Du Châtelet and Voltaire received from Frederick of Prussia, done by one of his courtiers.

90. "Etre Simple," *simple being*, is the term Du Châtelet used to signify *monad*, perhaps to dissociate her explanation from the previous and current accounts by Leibniz's adversaries of this concept as ridiculous. By avoiding the word, she probably assumed that she could gain a better hearing for Leibniz's ideas.

not *watches*, this is to say, to springs, to cogwheels, to pinions, to the chain, etc. The same reasoning applies to extension; for, when one says there are bodies with extension because there are atoms, it is as if one said: *there is extension, because there is extension*; this is in effect saying nothing at all. Thus, the sufficient reason for an extended and composed being can only be found in simple beings without extension, just as the sufficient reason for a compound number can only be found in a noncompound number, that is to say, in a unit. So, it must be admitted, these philosophers conclude: there are simple beings, since there are compound beings.

ATOMS CANNOT BE THE SIMPLE BEINGS OF WHICH MATTER IS COMPOSED.

§.121. Atoms, or indivisible particles of matter, cannot be the simple beings; for, these particles, though physically indivisible, are extended and are consequently of the same case as the bodies they compose. Thus, the principle of sufficient reason equally denies this simplicity to very small bodies and very big ones, which is necessary if the reason for extension in matter is to be found in them.

As it is finally necessary to arrive at necessary things when explaining the origin of beings, one cannot say that it is enough only to suppose that atoms are necessarily extended and indivisible, and that then there is no longer a need to seek the reason for their extension, since all philosophers agree that what is necessary requires no demonstration of its existence. For only something the contrary of which implies a contradiction must be recognized as necessary (§.20.).[91] What is necessary needs a sufficient reason showing why it is necessary; and this reason cannot but be the contradiction to be found in what is opposed to it. Now, as there is no contradiction in the divisibility of extended beings, the indivisibility of atoms cannot be accepted as necessary; thus, one must come to simple beings.

The will of the Creator, to which atomists turn to explain the extension of the atom, cannot, according to M. Leibniz, resolve this issue, because the question is not why extension exists, but how and why it is possible. Now, we saw above that the will of God is the source of the actuality, but not of the possibility of things. So, it cannot be resorted to in order to explain the possibility of extension.

---

91. Du Châtelet is referring to chapter one of the *Foundations* in which she established the underlying principles governing all knowledge: the laws of contradiction, of sufficient reason, the principle of indiscernibles, and the difference between necessary and contingent. She is now applying these rules of reasoning to hypotheses about the nature of matter.

SIMPLE BEINGS, OR MONADS, HAVE NO PARTICLES.

§.122. M. Leibniz, after having established the necessity of simple beings, explains their nature and their properties.

Simple beings having no particles, none of the properties arising from composition could fit them; thus, simple beings having no extension, are indivisible, for, not having several particles, which make up into *one*, they could not be separated.

NOR SHAPE.

§.123. They have no shape, for shape is the limitation of extension; now, these simple beings, being without extension, cannot have shape. For the same reason, they have no height and fill no space, and have no internal motion; for all these properties suit compounds, and result from composition. Thus, simple beings are completely different from compound beings, and they cannot be seen, touched, or represented in the imagination by any perceivable image.

§.124. A simple being cannot be produced by a compound being, for all that can come from a compound is born either of a new association or of the dissociation of its particles. Now, association can only produce a compound being, and dissociation, when pushed to its ultimate degree, can only end in the simple beings, those which already existed in the compound. They have not been produced by this dissociation; so, a simple being cannot originate from a compound being.

THE SUFFICIENT REASON FOR SIMPLE BEINGS IS IN GOD.

Neither can a simple being originate in another simple being; for the simple being, being indivisible and not having particles that can be separated, nothing can be detached from it. Thus, a simple being cannot originate in a simple being; now, since simple beings can originate neither in compound beings nor in simple beings, it follows that the reason for these beings must be in necessary beings, this is to say, in God. And it cannot be said that the explanation for atoms or for the indivisible particles of matter could be in God as is that of simple beings; for God cannot have created extension without creating simple beings first. The particles of compounds must have existed before the compound, but these particles not being resolvable into others, their first cause must be found in the Creator.

SIMPLE BEINGS CONTAIN THE SUFFICIENT REASON FOR ALL THAT IS
FOUND IN COMPOUND BEINGS.

§.125. Simple beings, being the origin of compound beings, the sufficient reason for all that is found in compound beings must be found in simple beings; so, simple beings must have intrinsic determinations, enabling us to understand why the compounds that result from them are such as they are, rather than quite different; that is to say, why they have such and such attributes, such and such properties, etc. Now, as you have seen here above that there are no identical beings in nature, all simple beings must be dissimilar and contain within them the differences that prevent one from being put in the place of another in a compound, without changing its determination, since if these simple beings were not all dissimilar, the resulting compounds could not be so either.[92]

THE SIMPLE BEINGS HAVE A PRINCIPLE OF ACTION THAT ONE
CALLS FORCE.[93]

§.126. Perpetual change can be observed in compounds; nothing stays in the same state; all tends to change in nature. Now, since the primary cause for what happens in compounds must ultimately be found in simple beings, from which the compounds resulted, there must be found in simple beings a principle of action capable of producing these perpetual changes, and by which may be understood why the changes happen in such a time, rather than in any other, and in such a manner, rather than in any other.

The principle that contains the sufficient reason for the actuality of any action is called *force*; for the simple power or ability to act is only a possibility of action or of passion in beings, which requires a sufficient reason for its actuality. In this way it is said that an animal has the ability to walk; a bow, to drive an arrow; a watch, to mark the hours, because one can explain by the structure of the animal, the bow, and the watch, how and why these effects are possible, but it does not follow from this that these effects are actual; if this were so, the animal would always walk, and the watch would always mark the hours, but this does not happen. So, in addition to this possibility, a sufficient reason must be admitted for the actuality, that is to say, a force that sets to work this power that a being has to act. Now, the sufficient reason for all that happens to compounds lying ultimately in simple beings, it follows that simple beings have this force, which consists in a continual tendency to action, and this tendency always has its effect when there is no sufficient reason preventing it from acting, that is to say, when there is no

92. Here and in chapter three of the *Foundations*, Du Châtelet describes simple beings as similar to what we now call DNA. It is interesting to note how other aspects of her explanation in this chapter suggest our modern understanding of atoms.

93. *Force* in the eighteenth century had a general meaning of the impetus for motion. Note Du Châtelet's image of a universe where motion, not rest, is the natural state of all matter.

point of resistance; for one must call resistance that which contains the sufficient reason why an action does not become actual, though the reason for its actuality remains.

SIMPLE BEINGS ARE IN CONTINUAL MOTION.

So, simple beings are endowed with a force, whatever it may be, as a result of whose energy they tend to act, and act indeed as soon as there is no point of resistance.[94] Now, as experiment proves that the force of simple beings is deployed continuously, since it constantly produces perceivable changes in compounds, it follows that each simple being is, by virtue of its nature and by its internal force, in a motion that produces in it perpetual changes and a continuous succession; and that its internal state and the sequence of successions that it experiences are different from the internal state, and from the successions experienced by any other simple being in the entire universe.

SIMPLE BEINGS ARE THE ONLY REAL SUBSTANCES.

§.127. Compounds last in spite of changes they endure, matter remains the same while it takes different forms, neither our body nor that of the plants, nor the air, nor anything that surrounds us is annihilated; however, the state of these beings changes constantly. Thus, simple beings from which compound beings result cannot but endure, this is to say, that they cannot but have constant and invariable determinations, while at the same time they have others within them that vary continually; for if simple beings were not durable by nature, the compounds could not last. So, simple beings are the real substances, beings that are durable and susceptible to the modifications which their internal force produces (§.52).[95]

Nothing can stop this internal force of simple beings, nor change the effects that are a consequence of it, because no natural agent can either break or destroy simple beings.

§.128. This makes it clear that the real substances (that is to say), simple beings, are active, since they carry in them the principle of their changes, that is to say, this force which to them is essential, which never leaves them, and which cannot cease. And we can understand what M. Leibniz meant

94. In the eighteenth century the word *energy* simply connoted *force* and could be applied to all kinds of situations. However, Du Châtelet, following the ideas of Leibniz and Johann Bernoulli on force and motion, is close to our more modern understanding of *energy*.

95. In §.52, the concluding section of chapter three of the *Foundations* on the "Essence, Attributes, and Modes" of matter, Du Châtelet explained the Leibnizian concept of "essential determinations" of a substance from which its permanent attributes and variable, possible modes or modifications follow.

when he said that the true character of the substance is to act, that it is distinguished from accidents by action, and that it is impossible to conceive of it without force.

I said here above that, in M. Leibniz's manner of thinking,[96] each monad, or simple being (for it is the same thing), contains a sequence of changes that is different from the sequence of changes of every other simple being— a necessary consequence of the principal of indiscernibles. We have an example of this in our souls, for no one doubts that the sequence of ideas of one soul is different from the sequence of ideas of all other souls in existence.[97]

§.129. The different states of a simple being depend on one another; for, such a successive state being no more necessary than another, there must be a sufficient reason why such a state is actual, and why, rather in such a time than in any other. Now, this reason can only be found in the preceding state, and the reason for that will be in the state antecedent to it, and so on back to the first. This first state, which supposes no other antecedent to it, depended on God, but all subsequent states are linked, so that from the first follows the last which was contained in it, and which must be such, because the first was so and not otherwise. In the same way the actual state of a clock depends on the preceding state, and that on another, and so on back to the first, which depended on the way in which the craftsman arranged the wheels; thus does the forty-seventh proposition of Euclid follow from the first, in which it is contained.[98]

IN THE UNIVERSE EVERYTHING IS LINKED TO EVERYTHING ELSE.

§.130. All is linked in the world; each being has a relationship to all the beings that coexist with it, and to all those that preceded it, and that must follow it. We ourselves sense at every moment that we depend on the bodies that surround us; if the nourishment, the air, a certain degree of heat, are taken from us, we perish, we can live no longer. The whole Earth depends

---

96. Du Châtelet uses the French word *sentiment*, meaning a sensory perception on which the mind reflects, thus, indicating the operation of both the senses and reason, translated here as "manner of thinking." She probably derived this from her reading of John Locke's *Essay concerning Human Understanding* (1689), translated into French by Pierre Coste (1668–1747) in 1700. Coste also translated Newton's *Opticks* (1722).

97. Du Châtelet is again using the term *soul* in the Cartesian sense.

98. Euclid's forty-seventh proposition (of his Book 1) is the first of two proofs he gave of the Pythagorean Theorem for a right triangle: $a^2 + b^2 = c^2$. Proposition 1.1 was the construction of an equilateral triangle.

on the influence of the Sun, and the Earth could not conserve itself, or grow vegetation, without the Sun's aid. It is the same with all other bodies; for, though we do not always distinctly see their mutual connection, we cannot, by the principle of sufficient reason and by analogy, doubt that there is one, and that this universe does not make a whole, an entire and a single machine, all the parts of which relate to one another, and are so linked that they all tend to the same end.

§.131. The original reasons for all that happens in bodies lie necessarily in the elements of which they are composed. It follows that the original reason for the connection of bodies to each other, insofar as they coexist and succeed each other, lies in simple beings. So, the connection of the parts of the world depends on the connection of the elements, which is the foundation and the first origin. Thus, the state of each element contains a relation to the present state of the entire universe, to all the states that will be born from the present state, just as in a well-made machine, the least part has a relation to all the others. For the state of any element A being determined, harmony and order require that the state of its neighbors B, C, D, etc. should also be determined in a particular manner rather than in any other, to work in harmony with the state of the first; and as the same reason continues for all states of the elements, all future states of the elements will also have a relation to the present state that must coexist with them, to past states from which this present state results, and to the states that will follow it, and of which it is the cause. Thus, it can be said that in M. Leibniz's system, it is a metaphysical-geometric problem, *the state of an element being given, to determine the past state, present, and future of all the universe.* The solution of this problem is reserved to the Eternal Geometrician who solves it at every moment insofar as he sees distinctly the relation of the state of each simple being to all the states, past, present, and future of all the other beings of the universe: but it will always be impossible for finite beings to have a distinct idea of this infinite relationship, that all things that exist have between them, because then they would become God. . . .

[§.132.] . . . In truth we can only have a clear representation of the most marked changes, and those that affect our organs with a certain force; but all these representations exist, though our soul does not perceive them, because of their weakness and their infinite multiplicity, which make it impossible to distinguish them, and consequently means that they only excite in us dim representations. That an infinity of dim representations accompany our clearest ideas is something we cannot deny, if we pay a little attention to ourselves. For example, I have a very clear idea of this paper, on which I write, and of the pen I use; however, how many dim representations are enclosed and hidden, so to speak, in this clear idea. For there is an infinity of things in the texture of this paper, in the arrangement of the fibers that compose it, in the difference and the resemblance of these fibers that I do not distinguish, of which, however, I have a dim representation. For the fibers, their differences and their arrangement subsisting, there is no reason why they would not cause impressions on my organs, and consequently representations in my soul; but these impressions being too weak and too compounded, I cannot distinguish them at all, and they cause dim representations in my soul. Thus, the total representation that results from this paper as a whole is clear, but the representations of its parts are dim. It is easy to see from this why in our mother's womb we are in a state where all our ideas are dim; it is that our body, not having yet developed, our limbs and organs are weighed down and concentrated almost in a point; consequently, it is impossible for the animal not to be equally affected everywhere by the same impression. Thus, the least movement shakes the entire animal so strongly that it cannot distinguish one impression from another, nor consequently form distinct ideas. Whereas, when we have left the envelopment of the *uterus*, our body is so disposed that the movement of rays of light, for example, cannot shake our acoustic nerves, nor sounds the optic nerve, and thus embroil us in very different ideas, which must be conceived of and sensed separately for them to be distinct.

§.133. So, this connection between our soul and the entire universe comes from the union of the elements among themselves, and from the relationships they all have with one another, and these relationships spring from their dissimilarity; for this dissimilarity causes each element, by its essence and its intrinsic determinations, to require the coexistence of a particular element with it rather than any other, and an element could not be removed from its place and substituted with another while conserving the same sequence of things, such a change would change the universe, and a new universe would appear. Whence it can be seen that the dissimilarity of elements explains why this universe is such as it is rather than completely different. . . .

[Du Châtelet continues her explanation; similarly, extension is an assemblage of "several, diverse, coexisting things," an assemblage of simple beings. In §.134 Du Châtelet then explains that compound beings, also an assemblage of simple beings, are not therefore true substances, for their particles could be reassembled as another compound being, just as one might take a watch apart and reuse the parts.]

WHY SIMPLE BEINGS SO DISGUST THE IMAGINATION.

§.135. One's reluctance to conceive how simple beings without extension can by their assemblage make up beings with extension is no reason for rejecting them; this rebellion of the imagination against simple beings comes probably from our habit of representing our ideas with perceivable images, which cannot help us here.

In things for which we cannot make perceivable images and that one cannot represent by characters, we must endeavor to supply them, never losing sight of the irrefutable principles by drawing conclusions from the consequences linked together with them, and without ever jumping in our reasoning.[99]

Mathematical truths would be no different from simple beings; if signs had not yet been invented to represent them to the imagination, these truths would be no less certain. Perhaps some day a calculus for metaphysical truths will be found, by means of which, merely by the substitution of characters, one will arrive at truths as in algebra. M. Leibniz believed he had found it; but sadly, he died without imparting his ideas on this, which at least would have put us on the right path, even if they had not yielded all that the name of such a great man promised.

§.136. It is regrettable, no doubt, that thinking people are not in agreement on the first principles of things; it would seem that the claim the truth has on our assent should extend to all notions and for all times. However, many truths have been fought over for whole centuries before being accepted; such was, for example, the true system of the world, and in our day, *forces vives*. It does not rest with me to decide if the monads of M. Leibniz are of the same case; but whether they are accepted or rejected, our researches on the nature of things will be no less certain; for, in our experiments we never will arrive at these first elements of which bodies are composed and the physical atoms (§.172), though in their turn composed of simple beings, are more than sufficient to exercise our desire for knowledge.[100]

99. Similar reasoning governed hypotheses about subatomic particles, hypotheses subsequently proved with the development of new scientific technologies.

100. In §.172, in chapter nine of the *Foundations* on the "Divisibility and Subtlety of Matter," Du Châtelet, having repeated the theory about the divisible parts of matter into elements, atoms, and simple beings, then asserts the probability that there is a fixed number of such particles in the universe. In this she follows Leibniz and disagrees with Newton. Her explanation was a precursor to the proofs Antoine Laurent Lavoisier (1743–1794), the celebrated French chemist, gave at the end of the eighteenth century for the conservation of mass.

CHAPTER ELEVEN

Of motion, and of rest in general, and of simple motion

DEFINITION OF MOTION.

§.211. Motion is the passage of a body from the place it occupies to another place.

THREE KINDS OF MOTION.

§.212 One distinguishes three kinds of motion: absolute motion, common relative motion, and motion relative to itself.

OF ABSOLUTE MOTION.

§.213. Absolute motion is the successive relationship of a body to different bodies considered as stationary, and strictly speaking, this is real motion.

OF COMMON RELATIVE MOTION.

§.214. A body experiences common relative motion when, being at rest in relation to the bodies that surround it, it nonetheless develops with them successive relationships, in relation to other bodies that are considered stationary; and this is the case in which the absolute place of the bodies changes, though their relative place stays the same; this is what happens to a pilot who sleeps at the tiller while his ship moves, or a dead fish carried along by the current.

OF MOTION RELATIVE TO ITSELF.

§.215. Motion relative to itself is the one experienced when, being transported with other bodies in a common relative motion, one nonetheless changes one's relationship with them, as when I walk on a sailing ship; for I keep changing my relationship with the parts of this ship that transports me.

EXAMPLES OF DIFFERENT KINDS OF MOTION.

§.216. The parts of all moving objects are in relative common motion; but, if they came to separate and went on moving as before, they would acquire a motion relative to themselves.

§.217. If a ship went toward the orient [east], and a man walked in this ship from the prow to the poop deck, that is to say, from the orient toward the occident [west], with the same speed as the boat, this man would have, while he traverses the length of this boat, a motion relative to himself, but his absolute motion would only be apparent, since in constantly changing his position in relation to the parts of this ship, he would remain in correspondence to the same points outside the ship.

If, on the contrary, this man walked on this ship from the poop deck to the prow, that is to say, in the same direction as the ship carrying him, he would have at the same time a common relative motion with the ship, and a motion relative to himself; for he would constantly change his position with the parts of this ship and with the bodies outside of the ship. It is this kind of motion that all bodies walking on Earth experience, for the Earth moves ceaselessly.

§.218. If instead of this man, one imagines a stone thrown horizontally in this ship, in a direction contrary to that in which the ship goes, but with a speed equal to that at which it is carried, this stone will appear to those who are on the ship to have a motion relative to itself, in the direction in which it was thrown; but those who are on the shore will see it in absolute rest, in relationship to its horizontal direction, and this rest is its real state.

This stone is in absolute rest in relation to its horizontal motion, because, moving with this ship, it acquired in the direction in which this ship goes a force equal to that by which the ship was carried. Now, supposing that it is thrown in a contrary direction by a force equal to that which carries the ship; these two equal and opposite forces cancel each other, and the stone stays in absolute rest in relation to the horizontal motion; for the hand that threw it found in it a real force, and the one the hand imparted to it was consumed, canceling that force. It would be otherwise if this stone were thrown into the ship by a hand outside the ship; for then the stone would really have motion relative to itself from the orient toward the occident, and it would fall into the sea surrounding the ship.

§.219. In regard to the motion of this stone toward the center of the Earth, it never stops; for neither the horizontal motion which has been imparted to it, nor that of the ship is opposed to the motion that its gravity imparts to it toward the center of the Earth.[101]

---

101. The image of the ship comes from Newton and was commonly used to illustrate these relationships. Du Châtelet uses it in chapter 21 as part of her proof of *forces vives*, and in a famous description in chapter 12 "Of compound motion" (§.284), where she listed all the possible motions acting on the body thrown on the ship: gravity, the flow of the river, the turning of the Earth on its axis and around the Sun, and so on. She notes, however, that "it is only the first two which pertain to it in relation to those who are transported with the body in this ship; for all bodies which have a motion in common with us are as if at rest in relation to us . . ." (Emilie Du Châtelet, *Institutions de physique* [Paris: Chez Prault fils, 1740], 251–52).

WHY THE RIVERBANK SEEMS TO MOVE AWAY, WHEN ONE DISTANCES
ONESELF FROM IT.

He who is on the ship and who believes that the stone went from orient to
occident attributes to the stone the motion that only pertains to the ship;
and he is deceived by his senses just as we are when we believe that the shore
we are leaving behind is moving away, though it is the ship we are on that
is sailing away. For we judge objects to be at rest when their images always
occupy the same points on our retina. Thus, as we walk with this ship, its
parts always occupy the same place in our eyes, but the parts of the shore,
in contrast occupying now one part and now another, we judge them to be
in motion for this reason. Thus, true motion and apparent motion are sometimes very different.

OF REST IN GENERAL.

§.220. Rest is the continuous existence of a body in the same place.
One makes a distinction between relative rest and absolute rest.

OF RELATIVE REST.

§.221. Relative rest is the continuation of the same relationships of the
body being considered to the bodies which surround it, though these bodies move with it.

OF ABSOLUTE REST.

§.222. Absolute rest is the permanence of a body in the same absolute
place, this is to say, the continuation of the same relationships of the body
being considered to the bodies that surround it, considered as stationary.

§.223. When the active force or the cause of motion is not in the body
which can move, this body is at rest, and this is, strictly speaking, real rest.

EXAMPLES OF THESE TWO KINDS OF REST.

§.224. No body on Earth is in absolute rest, for the Earth constantly
changes its relationship to all the bodies around it.

The bodies attached to the Earth such as the trees, plants, etc. are in
relative rest. For these bodies do not change the relationship between them,
but the Earth to which they are attached, and the bodies which surround
them, being in constant motion, are in relative common motion. Thus, a
body can be in relative rest, though it moves in a relative common motion.

§.225. But, in order to avoid the complexities these distinctions would
introduce in discourse, it is ordinarily supposed, when speaking of motion
and rest, that this is absolute motion and absolute rest; for the only real mo-

tion is that which operates by a force residing in the body that moves, and the only real rest is the absence of this force.

In this sense, there is no rest in nature, for all the particles of matter are always in motion, though the bodies of which they are composed may be at rest; thus it can be said that there is no internal rest.[102]

§.226. There is no degree of rest, as there is of motion; for a body can move more or less fast, but when it is once at rest, it is neither more nor less so.

However, rest and motion are often only comparative for us, for bodies that we believe at rest and that we see as at rest are not always so.

§.227. A body at rest will never begin to move by itself; since all matter is endowed with passive force, by which it resists motion, it cannot move by itself. In order for motion to happen with sufficient reason, there must be a cause that sets this body in motion. Thus, any bodies at rest would forever stay at rest, if some cause did not set them in motion, as, for example, when I withdraw a plank on which a stone is placed, or when some moving body communicates its motion to another body, as when one billiard ball pushes another.

§.228. By the same principle of sufficient reason, a moving body would never cease to move if some cause did not stop its motion, consuming its force; for by its inertia matter resists motion and rest equally.

§.229. The active and passive force of bodies is modified by their impact, according to certain laws that can be reduced to three principles.

GENERAL LAWS OF MOTION.[103]

First Law

A body perseveres in the state it is in, be it rest or motion, unless some cause brings an end to its motion or to its rest.

Second Law

The change that happens in the motion of a body is always proportional to the motor force that acts on it; and no change can happen to the speed and the direction of the moving body except by an exterior force; for without that, this change would happen without sufficient reason.

---

102. Again, Du Châtelet, like Leibniz, is envisioning a universe of constant motion, and one in which all bodies have an inherent force, similar to the modern concept of *energy*. See also §§.227–28.

103. Du Châtelet is restating Newton's "Laws of Motion" from Book I of the *Principia*, but she is adding metaphysical causation with her references to "sufficient reason." Note that she does not dwell on the corollaries to the Third Law here, but has already presented a number of their essential ideas in earlier sections and will cover others in later sections.

Third Law

The reaction is always equal to the action; for a body could not act on another body if this other body did not resist it. Thus the action and the reaction are always equal and opposite.

WHAT MUST BE CONSIDERED IN MOTION.

§.230. Several things are considered in motion:

1. The force that imparts the motion to the body.

2. The time during which the body moves.

3. The space the body traverses.

4. The speed of motion, this is to say, the relationship between the space the body has traversed and the time used to traverse it.

5. The mass of the bodies, according to which they resist the force that wants to impart or to take away motion from them.

6. The quantity of motion.

7. The direction of motion, be it simple or compound.

8. The elasticity of the bodies to which the motion is imparted.

9. The effect of the force of the moving bodies, or the quantity of obstacles that they can disrupt in consuming their force.[104]

10. Finally, the way in which the motion is communicated.

§.231. There is no motion without a force that imparts it.

1. OF MOTOR FORCE.

The active cause that imparts the motion to the body, or which incites it to move, is called motor force.

The effect of this force, when it is not destroyed by an invincible resistance, is to make the body traverse a certain space, in a certain time, in a space that does not perceptibly resist; and in a space that resists, its effect is to make it overcome some of the obstacles it encounters.

This cause, which draws the stationary body from the state of rest it was in, and which makes it traverse a certain space and overcome a certain quantity of obstacles, communicates to this body a force that it did not have when at rest, since according to the First Law, the body by itself would never leave its place.

§.232. By the same Law, when a moving body ceases moving, some

---

104. Du Châtelet here enunciates Johann Bernoulli's and Leibniz's ideas, that a body's motion can be understood and calculated by "obstacles overcome." See also her section on "motor force," §§.234–35, and §.268.

force equal and opposite to its own must have stopped its motion and consumed its force.

§.233. Any efficient cause is equal to its fully completed effect; thus equal forces when used up will always produce equal effects.

§.234 One calls an *obstacle* all that opposes the motion of a body, and that consumes its force completely, or in part.

MOTION WOULD BE ETERNAL IN A VOID.

§.235. Since, according to the First Law of motion, a body of its own accord always perseveres in the state in which it is; and the force by which a body moves, can only be consumed completely or in part by overcoming obstacles, a body, once moving in the absolute void (if an absolute void were possible), would continue to move for all eternity in this void, and would forever traverse there equal spaces in equal times, since in the void no obstacle would consume the force of this body neither in total nor in part.

§.236. So all motion contains an infinity of time, since all motion could last forever in the void. But all motion does not contain an infinity of speed, for a body that moved forever in the void could only move with a more or less great speed.

§.237. The space traversed by a body is the line described by this body during its motion.

2. OF THE SPACE TRAVERSED.

If the body in motion were a point, the space traversed would only be a mathematical line; but as there is no body without extension, the space traversed always has some width. When the path of a body is measured, only its length is taken into account.

3. OF THE TIME DURING WHICH THE BODY MOVES.

§.238. If body A traverses the space CD a certain portion of time will elapse, during which it will go from C to D, however small the space CD might be; for the moment when the body is at C will not be that when it will be at D, it being impossible for a body to be in two places at once; thus any space traversed is traversed in some time.

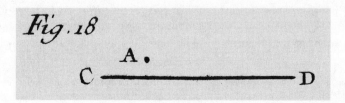

4. OF THE SPEED OF A MOVING BODY.

§.239 In addition to the space that a moving body traverses, the force that makes it traverse it, and the time that it uses to do it, one identifies yet another thing in motion that one calls *speed;* by this word is understood the property that a mobile body has of traversing a certain space in a certain time.

[*Fig. 18 also appeared here in the original.*]

The speed of a body is known by the space it traverses in a given time; thus the speed is all the greater when the mobile body traverses more space in less time. Consequently, if body A traverses the space CD in two minutes, and body B traverses the same space in one minute, the speed of body B will be twice that of body A.

THERE IS NO MOTION WITHOUT A DETERMINED SPEED.

There is no motion without some speed, for any space traversed is traversed in a certain time; but this time can be more or less long to infinity; for space CD, which I suppose to be a *pied*, can be traversed by body A in one hour, or in one minute, which is the 60th part of one hour, or in one second which is the 3600th part of it, etc.

§.240. Motion, that is to say, its speed, can be uniform or not uniform, accelerated or slowed, equally or unequally accelerated and slowed.

OF UNIFORM MOTION.

§.241. Uniform motion is that which makes the mobile body traverse equal spaces in equal times; thus, in uniform motion the spaces traversed are as the speeds of the mobile body, and as the time of its motion.

§.242. In an infinitely small time, one always considers the motion as being uniform, that is to say, that at each infinitely small instant the mobile body is supposed to traverse equal spaces, whether its motion in a finite time be accelerated or slowed, uniform or not uniform.

§.243. It is only in a space offering no resistance that a perfectly uniform motion could happen, just as it is only in such a space that perpetual motion would be possible; for in this space nothing would be encountered that could accelerate or slow the motion of bodies.

PROOF OF THE IMPOSSIBILITY OF PERPETUAL MECHANICAL MOTION.[105]

§.244 The inequality of all motions that we are familiar with is a demonstration against the mechanical perpetual motion that many people have

---

105. The concept of perpetual motion fascinated the natural philosophers and skilled mechanical artisans of eighteenth-century Europe. Basing their designs on complex clocklike mech-

sought: for this inequality only comes from continual losses of force that moving bodies experience from the resistance of the surrounding matter in which they move, the friction of their particles, etc. Thus for perpetual mechanical motion to happen, it would be necessary to find a body that was exempt from friction or had received an infinite force from the Creator, since it would be necessary to make this force surmount constantly repeated resistance, and without it ever running out, which is impossible.

PERFECTLY EQUAL MOTION IS UNKNOWN TO US.

§.245. Although, precisely speaking, there is no perfectly uniform motion, nevertheless, when a body moves in a space without perceptible resistance, and this body is subjected neither to acceleration nor perceptible slowing in its motion, this motion is considered as if it were perfectly uniform.

OF NONUNIFORM MOTION.

§.246. Nonuniform motion is that which is subjected to some increase or some decrease in its speed. . . .

[§§.247–51. Du Châtelet defines accelerated and slowed motion.]

MORE FORCE IS NECESSARY TO ACCELERATE MOTION THAN TO IMPART IT.

§.252. A greater quantity of force is necessary to increase the speed of a body by a degree than to impart to it the first degree of speed, when it is at rest.

§.253. If the motion is uniform, that is to say, if the speed of the body remains the same, the space traversed will increase in the same proportion as the time of the motion of this body (leaving aside obstacles), so that if one multiplies the speed of this body by the time of its motion the product will be the space traversed; if the space is divided by the time, the product will provide the speed, and this same space, divided by the speed, will give the time. Thus in uniform motion, when one has two of these things, space, time, and speed, one will necessarily have the third. . . .[106]

[§§.254–55. Du Châtelet describes a variety of ways in which these relationships change when one of the three factors changes, and when the motions of several bodies are compared.]

---

anisms, inventors such as Jacques Vaucanson (1709–1782) made a mechanical duck, a mechanical boy playing a flute. At Lunéville, King Stanislas of Poland, who reigned as Duke of Lorraine, 1737–66, had an entire life-size mechanical rural village designed and constructed in his garden.

106. Du Châtelet is describing the now familiar s [speed/rate] $\times t$ [time] $= d$ [distance], and its variations.

WHAT IS MEANT BY ABSOLUTE SPEED AND RESPECTIVE SPEED.

§.256. One distinguishes between *absolute* and *respective* speeds.

A body's own, or absolute speed, is the relationship between the space it traverses and the time during which it moves.

The respective speed is the speed with which two bodies move toward or away from each other in a certain space in a determined time, whatever their absolute speeds might be; thus, absolute speed is something positive, but respective speed is only a simple comparison the mind makes of two bodies, according to whether they move toward or away from each other.

5. OF THE MASS OF BODIES.

§.257. Bodies equally resist motion and rest. This resistance being a necessary consequence of their inertial force, it is proportional to the quantity of their own matter, since inertial force pertains to even a *minimum* of matter. So a body resists the motion that one wants to impart to it, all the more when it contains a greater quantity of its own matter in a similar volume. This is to say that the more mass, the more it resists, all other things being equal.

Thus, the more mass a body has, the less speed it acquires by the same pressure, and vice versa.

§.258. It is easier to impart a certain speed to a body than to impart to the same body a speed twice that of the first; thus, a double pressure is necessary to impart to the same body a doubled speed. And precisely the same pressure is necessary to give a body two degrees of speed, or to give one degree of speed to another body whose mass is double that of the first.

Thus the pressure that makes different bodies move with the same speed is always proportional to the mass of these bodies, all other things being equal.

The motion of a body is all the more difficult to stop, as this body has more mass. Thus the same force is necessary to stop the motion of a body moving with any speed whatever, and to communicate to this same body the same degree of speed it was forced to lose.

OF THE EQUALITY OF ACTION AND REACTION.

§.259. This resistance that all bodies present when one wants to change their current state is the foundation of the Third Law of motion, by which the reaction is always equal to the action.

The establishment of this law was necessary so that bodies might act on one another, and that motion, being once produced in the universe, might be communicated from one body to another with sufficient reason.

THERE CANNOT BE ACTION WITHOUT RESISTANCE.

In any action, the body that acts, and that against which it acts, fight each other, and without this kind of fight there can be no action; for I ask how a force can act against that which does not offer any resistance.

When I pull a body attached to a rope, however easily I may pull it, the rope is stretched taut equally on both sides, which indicates the equality of the reaction, and if this rope was not stretched taut, I could not pull the body.

OBJECTION TO THE EQUALITY OF THE ACTION AND REACTION.
ANSWER.

But someone will ask: can I make this body move forward if I am pulled by it with a force equal to that which I am using to pull it? Those who ask this question do not pay attention to the fact that when I pull this body and make it move forward, I do not use all my force to overcome the resistance with which it opposes me, but when I have overcome it, a part of the force still remains with me, which I use to move forward myself; and this body advances by the force I communicated to it and that I used to overcome its resistance. Thus, although the forces are unequal, the action and reaction are always equal.

The reason for this equality of action and reaction is that a body could not use a degree of force to overcome the resistance of another body without itself losing a quantity equal to what it used to achieve that; for this body cannot keep and use its force at the same time. This force that it uses to overcome this resistance is not lost, but the body that resists acquires it. . . .

[Du Châtelet explains that this Law applies to bodies of unequal masses.]

We have seen above that the communication of motion happens in relation to mass, which is again a proof that action is equal to the resistance; for bodies resist in direct proportion to their mass.

§.260. Bodies react from their force of inertia and, in reacting, tend to change the state of the body pushing them and that they resist, and in this reaction they acquire the force that the body acting on them consumes in acting on them, for these bodies resist in acquiring the motion; thus, the force bodies acquire in order to move they acquire in part by their force of inertia, which is the principle of their reaction. So precisely speaking, all the force of matter, be it at rest or moving, be it communicating motion or receiving it, all its action and its reaction, all its impulsion and its resistance, are nothing other than this *vis inertia* in different circumstances.

IT IS THE EQUALITY OF ACTION AND REACTION THAT MAKE A BOAT GO
BY ITS OARS.

§.261. A boat goes by means of oars because the oars push the water
in the opposite direction, and the water reacts against the oars and repulses
with them the boat to which they are attached with a force equal to that
with which the oars have cut it. Thus the vessel goes all the faster as there
are more oars, as the oars are bigger, and as they are moved faster and with
more strength.

It is by this method that one supports oneself swimming in the water;
for feet and hands then serve as oars.

It is the same with birds. When they fly, they move in the air with their
wings in the same way as men swimming in the water with their feet and
hands.

[6.] OF THE QUANTITY OF MOTION.

§.262. There is yet another thing to consider in motion, namely, the
quantity; for the quantity of motion in an infinitely small instant is proportional to the mass and the speed of the moving body, so that the same body
has more motion when it moves faster. And of two bodies moving with equal
speed, the one having the most mass has the most motion; for the motion
imparted to any body can be conceived of as divided into as many particles
as this body contains of its own matter, and the motor force belongs to each
of these particles that participate equally in the motion of this body, in direct proportion to their size. Thus the motion of the whole is the result of
the motion of all of the particles, and consequently the motion is doubled
in a body whose mass is double that of another, when these bodies move
with the same speed. . . .

7. OF THE DIRECTION OF MOTION. OF SIMPLE MOTION.

§.263. There is no motion without a particular direction; thus any mobile body that is moving tends toward some point.

When a moving body obeys only a single force directing it toward a
single point, this body moves in a simple motion.

OF COMPOUND MOTION.

§.264. Compound motion takes place when the mobile body obeys several forces that make it tend toward several points at the same time.

Simple motion is the only one I examine here. I will discuss compound
motion in the next chapter.

§.265. In simple motion the straight line drawn from the mobile body to the point toward which it tends represents the direction of motion of this body, and if this body moves, it will certainly follow this line.

Thus, all bodies that move in simple motion describe a straight line while they move.

Strictly speaking, we know no simple motion other than that of bodies falling perpendicularly toward the center of the Earth by the sole force of gravity, unless bodies move on a stationary plane; for gravity acting equally on all bodies at each indivisible instant, its action mixes with all motions, and if simple, it makes them become compound.

§.266. Gravity or weight is also one of the reasons why there could not be uniform motion in an absolute void or on a stationary plane; for this force causes bodies to traverse unequal spaces in equal times.

8. OF THE ELASTICITY OF BODIES.

§.267. Bodies receiving or communicating motion can be either completely hard, that is to say, incapable of compression, or completely soft, that is to say, incapable of reconstitution after the compression of their particles, or again elastic, that is to say, capable of regaining their original shape after the compression. We do not know any body that is completely hard, or completely soft, or perfectly elastic; for as M. de Fontenelle says, *nature does not allow any such precision.*[107]

But to make explanations more intelligible, the most exact precision is presumed; thus it is assumed that all bodies that recoil spring back perfectly.

Hard bodies are those whose shape is not altered perceptibly by impact, such are, for example, diamonds; and *soft*, the bodies an impact causes to take a new shape which they keep after the impact, like wax, clay, etc. Further on in this work I shall discuss elastic bodies and the way in which motion is communicated between them.

9. OF THE FORCE OF MOVING BODIES.

§.268. When a moving body encounters an obstacle, it strives to displace this obstacle; if this effort is destroyed by an invincible resistance, the

---

107. Bernard le Bovier de Fontenelle (1657–1757), a mathematician and physicist, was a leading member of the French Royal Academy of Sciences. He served as its perpetual secretary from 1697 until 1741. He presented numerous memoirs to the Academy on a variety of subjects and was the author of the best-selling *Entretiens sur la pluralité des mondes* [Conversations on the Plurality of Worlds]. He was a family friend of Du Châtelet's.

force of this body is a *force morte* [dead force], that is to say, it does not produce any effect, but it only tends to produce one.

If the resistance is not invincible, the force then is *force vive* [live force], for it produces a real effect, and this effect is called *the effect of the force of this body*.

The quantity of this *force vive* is known by the number and the size of the obstacles the moving body can displace by using up its force.

There are great disputes among philosophers about whether this *force vive* and the *force morte* must be estimated differently, and this will be discussed in chapter 21 of this work.[108]

10. OF THE COMMUNICATION OF MOTION.

§.269. Finally, the last thing that remains for me to examine about motion is the way it is communicated; for experiment teaches us that a moving body that encounters another at rest communicates to it part of the force that it had in order to move, and then the body with which it has collided passes from the state of rest in which it was, to that of motion, and it continues to move after the collision until some obstacle has consumed its force.

§.270. The reason why this body continues to move after the absence of a driving force is a consequence of the inertial force of matter, the force by which bodies stay in the same state, if some cause does not take them out of it. Now when my hand throws a stone, this stone and my hand begin moving together; I withdraw my hand, and there is a cause that stops its motion in that direction, but the stone I have not withdrawn continues to move until the resistance of the air has caused it to lose projectile motion, which I had communicated to it, or gravity makes it fall down toward the Earth. Thus the continuation of the motion of this stone, after the absence of my hand, is the effect of the force that I communicated to it.

It is for this reason that when a vessel is sailing very fast and is stopped suddenly, the things in this ship, tending to conserve the motion they ac-

---

108. In her description of force Du Châtelet is measuring it in the way in which Johann Bernoulli advocated in his two memoirs for the French Royal Academy of Sciences (1724 and 1726). In modern terms *force morte* is equivalent to potential energy and *force vive* to kinetic energy. There was confusion in Du Châtelet's day about these two concepts. Leibniz favored the concept of "force vive," which Du Châtelet discusses at length in her chapter twenty-one where she disputes the 1728 memoir written by Jean-Jacques Dortous de Mairan supporting Descartes' views on motion. Dortous de Mairan was close to Fontenelle and succeeded him as perpetual secretary of the Academy in 1741. I am grateful to Paul Veatch Moriarty for his clarifications of these concepts.

quired by being transported with it, might well be hurled forward, if they were not tied down.

WHY THE ROLLING OF A SHIP CAUSES VOMITING.

It is for the same reason that the sea causes the rolling of a ship, and even more the turmoil of a storm, makes men sick and makes them vomit, especially if they are not accustomed to the sea. For, the liquids in their bodies only gradually gain a movement in *harmony* with that of the ship, and until they have acquired it, there is disorder and commotion in the body, which takes the form of vomiting and other illnesses; and so, almost the same thing happens in the bodies of men as in a vase filled with water that is spun around; for the water only slowly acquires the motion of the vase, and it maintains it some time after this motion is stopped.

CHAPTER TWENTY-ONE

Of the force of bodies

[In §§.557–72 Du Châtelet describes the ways in which force acts on bodies; for example, force is successively acquired. Force acts even if just as a tendency when a body resists; for example, a body resists the force of gravity when placed on a table. This "harmless effect" of the force is *force morte* and is retained by the body as long as its motion is opposed by an invincible obstacle. The formula for determining *force morte* is mv (mass × velocity). She explains that all mathematicians agree on this definition and this formula for its determination. Leibniz was the first to distinguish between *force morte* and *force vive* in his memoirs for the *Acta Eruditorum* (1686 and after). As she describes it, *force vive* is the successive acquisition of force by a body. She uses gravity as an example and cites Galileo's formula measuring the force of gravity as the square of the speed of fall. Thus, she concludes that *force vive* is measured by the square of the speed of motion of the body multiplied by its mass, mv2 (expressed as ½ mv2 today). Although experiments confirm this conclusion, she notes that it is considered a "kind of heresy in physics." She then answers the principal objection to *force vive*, the argument of "time," as a determining factor in the measurement of force. Opponents, she explains, argue that a force increases as it takes longer to act on a body; for example, a spring closing. She responds that the force can only be measured by the obstacles it overcomes and by which it is consumed; the time is of no consequence, and making time a determining factor in the equation for force leads to "absurdities," such as perpetual mechanical motion, continual motion in infinite time, which all agree is impossible.]

SOME REFUSE TO ACCEPT *FORCES VIVES* WHILE ACKNOWLEDGING THE
EXPERIMENTS THAT ESTABLISH THEM.

§.573. *Forces vives*[109] may be the only point of physics which some still dispute while acknowledging the experiments that prove it; for if you ask those who reject them what would be the effects of two bodies equal in mass on two equal obstacles, but the speeds of which are 4 and 3, they will answer that one will be an effect, as 16 and the other as 9. Now, it is easy to see that, whatever distinction and whatever modification they next bring to this acknowledgment that the force of truth draws from them, it always remains certain that the effect being squared, there must have been a squared force to produce it.

§.574. It would be pointless to report to you here all the experiments that prove this truth, you will one day see them in the excellent memoir that M. Bernoulli presented to the Academy of Sciences in 1724 and in 1726, and found in the *Recueil des pièces* [Collection of Memoirs], which won, or merited the prizes it awards. And you have already seen a part of it in the memoir that M. de Mairan gave in 1728 to the Academy against *forces vives* that we read together, and in which the famous proceeding is explained with much clarity and eloquence.

EXAMINATION OF SOME PARTS OF M. DE MAIRAN'S MEMOIR AGAINST
FORCES VIVES.

As this work appears to me to be the most ingenious that has been produced against *forces vives*, I will pause to take the time to remind you here of some passages, and to refute them.

M. de Mairan says, numbers 38 and 40 of his memoir: "That the force of bodies should not be measured by the spaces traversed by the moving body in the slowed motion, nor by the obstacles overcome, springs closed, etc. but by the spaces not traversed, by the parts of matter not displaced, the springs not closed, or not flattened: now," he says, "these spaces, parts of matter, and springs are like simple speed. Thus, etc."

One of the examples he gives is that of a body that goes back up to the same height from which it fell with the force acquired in falling, and that in going back up overcomes the obstacles of gravity: "For a body fallen from

109. Du Châtelet, like her contemporaries, refers to *force vive* in the singular and the plural, *forces vives*, but it is the same phenomenon.

a height 4 and which acquired 2 of speed in falling, would in going back up by a uniform motion, and with this speed 2, travel a space 4 in the first second; but gravity which pulls it down, making it lose in this first second 1 of force and 1 of speed, it only traverses 3 in the first second, the same as in the second second where it still has 1 of speed and 1 of force; whereas it would traverse 2 in a uniform motion, it only traverses one, because gravity makes it lose *one*. What are the losses of this body, *one*, in the first second, and *one* in the second? This body that had 1 of speed, has lost 2 of force, so the forces were as its speeds," concluded M. de Mairan, "not as the square of its speeds."

But to see the fault in this reasoning, it suffices to consider (as in §.567)[110] the action of gravity as an infinite sequence of equal springs, which communicate their force to falling bodies, and which the body contains in rising again; for, then, it will be seen that the losses of a body that rises are as the number of closed springs, that is to say, as the spaces traversed, not as the spaces not traversed.

In the obstacles overcome, as with the displacements of matter, the closed springs, etc. even by way of hypothesis or supposition, it is impossible to reduce slowed motion to uniform motion, as M. de Mairan advances in his memoir, and whatever esteem I have for this philosopher, I dare insist that when he says in numbers 40, 41, and 42, *That a body, which by a slowed motion, closes three springs in the first second, and 1 in the second, would close 4 in this first second, and 2 in the second by a uniform motion and a constant force*, he is saying, I am not afraid to venture this, an entirely impossible thing. For it is as impossible for a body with the force necessary to close 4 springs to close 6 (whatever supposition is made), as it is impossible that 2 and 2 make 6. For if one supposes with M. de Mairan that the body has not consumed any part of its force to close 4 springs in the first second of a uniform motion, I say that these 4 springs would not be closed, or that they would be so by some other agency. If one supposes the contrary, that, having exhausted a part of its force to close these first three springs in the first second, and having only the necessary force to make it close a spring in the second second, the body would take back a part of its force to close two in this second second by a uniform motion (for one or the other of these suppositions must be made), one obviously supposes in this last case that the body has renewed its force, which is beside the question. Thus, it is not true that the total force of a body is represented by what it would have done if it had not been consumed; for it could

110. In §.567 Du Châtelet describes the accumulation of force as a body ceded to successive pressures exerted on it and thus acquired by it.

never make an effect greater than that which destroyed it, and it only po-
tentially contained what it deployed in the effect produced. Thus, this very
subtle reasoning, which initially might seem alluring, relies only on this false
principle, that the quantity of motion and the quantity of the force are the
same thing, and that the force can be supposed to be uniform like the motion, although it has overcome part of the obstacle that must consume it.
But that is entirely false, and cannot be accepted even as supposition. For to
suppose simultaneously that a force stays the same, and meanwhile produces
part of the effects which must consume it, that is to suppose contradictory
things. Thus the measure of the force in slowed motions *is not the parts of un-
displaced matter, the springs not pulled, the spaces not traversed in going up; but, the spaces
crossed in rising, the parts of displaced matter, and pulled springs.*

M. de Mairan goes on to say in number 33 that "just as a force is not
infinite, because the uniform motion it produced in an unresisting space
would never cease, it does not strictly follow either that the motor force of
this same body is bigger because it lasts longer." But it is easy to see that in
uniform motion supposed eternal, there is no destruction of force, whereas
when the motor force during a doubled time has disturbed squared obstacles, there has been a real expenditure of force, which cannot have happened without a base of force squared, and that, thus, the two cases cannot
be compared.

I flatter myself that M. de Mairan will consider the remarks I have just
made on his memoir as proof of the regard in which I hold this work. I confess that he has said all that could be said in favor of a bad cause; thus, the
more seductive his reasoning, the more I felt obliged to make you see that
the doctrine of *forces vives* is not undermined by it.

VERY OBVIOUS REASONING WHICH PROVES FORCES VIVES.

§.575. This doctrine can be confirmed by a very simple argument,
which everyone makes naturally when the occasion arises: if two travelers
walk equally fast, and one walks for one hour, and makes one *lieue*, and the
other two *lieues* in two hours, everyone acknowledges that the second made
double the distance of the first, and that the force he used to cover two *lieues*
is double that which the first used to walk one *lieue*. Now, supposing that a
third traveler covers these two *lieues* in one hour, that is to say, that he walks
at double the speed, it is evident again that the third traveler, who makes
two *lieues* in one hour, uses two times the force used by the one who walked
these two *lieues* in two hours. For we know that the faster a courier must walk
to cover the same distance in less time, the more force he needs, which all
couriers understand so well that they all want to be better paid the faster

they go. Now, since the third traveler uses two times more force than the second, and the second uses two times more than the first, it is obvious that the traveler who walks at double the speed during the same time, uses four times more; and consequently the forces that these travelers expended will be as the square of their speeds.

§.576. The enemies of *forces vives* manage to discount most of the experiments that prove them, because they cannot deny them. They reject, for example, all those done showing the impression bodies make in soft materials, and it is true that there is inevitably always confusion in the results of these experiments, and in the examples one deduces from animate creatures, strange circumstances which prolong the disputes.[111]

ACADEMY OF PETERSBURG, FIRST VOLUME. DECISIVE EXPERIMENT OF
M. HERMANN IN FAVOR OF FORCES VIVES.

§.577. But M. Hermann reports a case that leaves no place for any subterfuge, and in which it cannot be disputed that the force of a body was squared by virtue of a doubled speed.[112] This is the case in which, for example, a ball A which has 1 of mass, 2 of speed, successively hits on a horizontal plane, supposed to be perfectly smooth, a ball B at rest, which has 3 of mass, and a ball C that has 1 of mass; for this body A will give a degree of speed to ball B whose mass is 3, and it will give the remaining degree of speed to ball C, which it

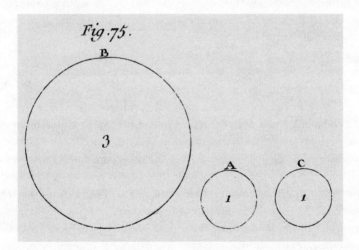

Fig.75.

111. This is a description of 'sGravesande's experiments from his *Physices elementa mathematica* [Mathematical Elements of Physics] (1720).

112. Jacob Hermann (1678–1733), a mathematician, was known for his work in mechanics. Du Châtelet is referring to a memoir he wrote while at the Academy of St. Petersburg.

next encounters, and whose mass is 1, that is to say, equal to its own; and this body A, having then lost all its speed, will stay at rest.

Now let us examine what the force will be of bodies B and C to which body A communicated all its force and all its speed; certainly the mass of body B being 3 and its speed 1, its force will be 3 even in the opinion of those who refuse to accept *forces vives*; body C, whose speed is 1 and mass 1 will also have 1 of force: thus body A will have communicated the force of 3 to body B and the force of 1 to body C. Thus body A with 2 of speed gave 4 of force. This means that it had this force; for, if it had not had it, it could not have given it; thus, the force of body A, which had 2 of speed and 1 of mass, was 4, that is to say, as the square of this speed multiplied by its mass.

[*Fig. 75 appeared here in the original.*]

§.578. There is an admirable correspondence between the way body A loses its force by the impact in this experiment, and the way a body that rises up again by the force acquired falling, loses its own because of the redoubled pull of gravity. For a body that, with a speed of 2, will rise up to a height 4, loses 1 of speed when it has risen up again to a height 3, just as ball A loses 1 of speed in setting ball B in motion, whose mass is 3; and the body that rises up again loses the second degree of speed that remains to it, in rising from a height of 3 to a height of 4, that is to say, in traversing a space one-third of the first, just as body A loses the degree of speed left to it in hitting body C, one-third of body B. Thus the same thing happens, either because the force of bodies is communicated to them by impulsion or as an effect of their gravity.

HOWEVER, THE DIFFICULTY WITH TIME ALWAYS REMAINS IN THIS EXPERIMENT.

§.579. Although in this experiment of M. Hermann's, a body with 2 of speed communicated 4 degrees of force to bodies equal to it, which can then exert this force and communicate it to other bodies, which leaves no place for pretexts that one alleged against most of the other experiments which prove *forces vives*. However, the difficulty with time (if it is one) always remains in this experiment, since the ball A only communicated its force to balls B and C successively. Thus all the adversaries of *forces vives*—M. Papin who rejected them and M. Leibniz, their inventor,[113] and M. Jurin, who recently declared against this opinion—have always challenged M. Leibniz and the partisans of *forces vives*, to demonstrate to them a case in which a doubled speed produced

113. *Inventor* in the sense of first conceptualizing them as distinct.

a squared effect in the same time, in which a simple speed produces a simple effect, going so far as to promise to accept *forces vives*, if such a case could be found in nature. This is how M. Jurin puts it: *Id si facere dignati fuerint me ipsis discipulum, parum id quidem est, at multos egregios viros ausim promittere.*[114]

§.580. As the laws of motion do not permit, when a body hits a single other one, for it to transmit all its force to another with four times the mass by a single hit, M. Leibniz, in order to meet this kind of challenge, resorted to a lever, by means of which he succeeded in transmitting by a single hit all the force of a body to another with four times the mass, to which it communicated half of its speed. But the fact of the lever gave rise to exceptions that made M. Leibniz's experiment unproductive for the purpose of converting his adversaries. Thus the objection based on the difficulty with time always remained.

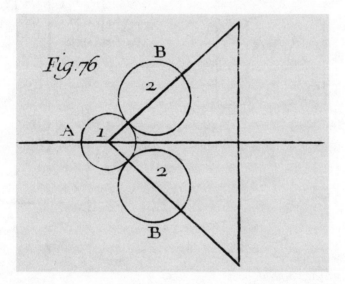

EXPERIMENT THAT ENTIRELY DESTROYS THE OBJECTION BASED ON TIME.

§.581. But this objection was completely overturned by finding the case the adversaries of *forces vives* believed could not be found. This is the case in which a body A freely suspended in the air whose speed is 2, and the mass supposed as 1, at the same time hits at an angle of 60 degrees two bodies B

114. Du Châtelet here makes her own note to give the translation of Jurin's quotation: "*And if they can find such an effect in nature, I promise them, not only to be their disciple, which

and B, the mass of each of which is 2; for, in this case the body striking A stays at rest after the hit, and the bodies B and B divide its speed between them, each moves by a degree of speed. Now these bodies B and B, whose mass is 2 and who have each received a degree of speed, have each acquired 2 of force, whichever way one looks at it. Thus body A with a speed of 2 communicated a force of 4 at one and the same time. This is precisely the case required by the adversaries of *forces vives*; thus, this experiment makes the objection based on the difficulty of time, about which up to the present the enemies of *forces vives* have made such a fuss, collapse entirely.

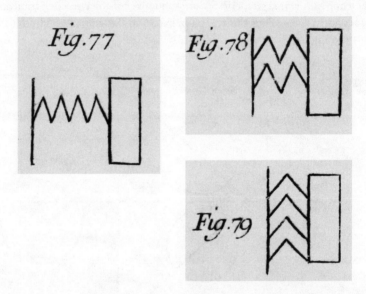

OTHER PROOF DRAWN FROM THE TIME IN WHICH SPRINGS COMMUNICATE THEIR FORCE.

§.582. In addition, the force is always the same, whether it has been communicated in a short time or a long time. The time in which springs communicate their force, for example, depends on the circumstances in which they are deployed; for there are circumstances in which the force of

---

would mean much; but to find more distinguished ones for them." Denis Papin (1647–1712?), though French, worked with Huygens and became a professor of mathematics at the University of Marburg. James Jurin (1684?–1750), the English mathematician and well-known supporter of Newton, later corresponded with Du Châtelet on this particular point in her *Foundations*.

a spring can be transmitted in the same body faster than in other circumstances. Yet the force that this spring communicates is always the same.
Thus, four equal springs will communicate the same force to the same body,
whether they communicate it in one, two, or three minutes, as in Figs. 77,
78, and 79, and this time could be infinitely varied, depending on whether
these springs were more or less at liberty to act, though the force communicated was always the same; thus, the time is immaterial to the communication of motion.

ANOTHER OBJECTION TO FORCES VIVES.

§.583. There is yet another objection to *forces vives*, which at first appears fairly strong; it arises from what happens when two bodies hit each
other with speeds that are in inverse proportion to their mass, for if these
bodies are without perceptible spring they will stay at rest after the collision. At first it would seem as if the body, which has the most speed having
the most force, according to the doctrine of *forces vives*, must push the other
body before it.

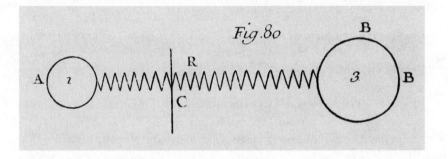

RESPONSE.

MACLAURIN PRIZE ACADEMY PIECES. BERNOULLI PRIZE PIECES.

DISCOURSE ON MOTION.[115]

But to understand how two bodies with unequal force can, nonetheless, stay
at rest after the collision, let us consider a spring R, which releases its tension
at the same time on both ends, and which pushes at either end bodies of unequal mass, the inertia of these bodies being the only obstacle they oppose

115. Colin Maclaurin (1698–1746), a Scots mathematician and professor at Edinburgh University, was active in the prize competitions of Europe's Academies.

to the release of the tension of the spring; and this inertia being proportional to their mass, the speeds the spring will communicate to these bodies will be in inverse proportion to their mass; and consequently they will have equal quantities of motion, but their forces will not be equal, as M. Jurin and some others would infer. These forces will be to each other as the length CB to length CA, that is to say, as the number of springs that acted on them; thus, their forces will be unequal and will be to each other as the square of the speed of these bodies multiplied by their mass.

Now when the spring R is released up to a certain point, if these bodies returned toward it with the speeds it communicated to them by releasing, it is easy to see that each of these bodies would have precisely the necessary force to return the parts of the spring that acted against it in their first state of compression, and that they would use unequal force to close this spring, since in releasing, it had communicated to them unequal forces, which they consumed in closing it; and if the spring was stopped in its state of compression when these bodies have just closed it, the two bodies, all of whose forces had been used to close it, would then remain at rest.

Now, when two bodies that are not elastic collide with speeds that are in inverse proportion to their masses, they have on each other the same effect as one has just seen, the effect that body A and body B had on the parts of spring R in order to close it, and it is easy to see by this example how bodies can consume unequal forces in the giving way of their parts and stay at rest after the collision.

EXPERIMENT THAT CONFIRMS THIS ANSWER.

§.584. M. 'sGravesande created an experiment that wonderfully confirms this theory. He took a firm ball of clay and, using Mariotte's Machine,[116] he made it collide successively with a copper ball, whose mass was 3 and speed 1, and with another ball of the same metal whose speed was 3 and mass 1, and it happened that the impression made by ball one, whose speed was 3, was always much greater than that made by ball three with the speed of 1, which indicates the inequality of the forces. But when these two balls with the same speeds as before collided at the same time with the clay ball freely suspended from a thread, then the clay ball was not set in motion

---

116. This is a second experiment by 'sGravesande to support the concept of *forces vives*. Mariotte's Machine was a simple structure with balls of different materials hung so that collisions could be enacted and observed. As modified by Musschenbroek, the Dutch experimentalist and instrument maker, the collisions on the machine happened against a board with markings that could be used to measure the recoil of the balls.

and the two copper balls stayed at rest and equally depressed the clay; and these equal impressions having been measured, they were found to be much greater than the impression that ball three with the speed of 1 had made when it only hit the firmed clay ball, and less than that which had been made by ball one with the speed 3. For ball three had used its force to make an impression on the clay ball, and its impression having been augmented by the effort of ball one that pressed the clay ball against ball three, diminished the impression of this ball one. Thus soft bodies that collide with speeds in inverse proportion to their masses, stay at rest after the collision, because they use all their force to mutually impress their parts. For it is not simple rest that holds these parts together, but a real force, and in order to flatten a body and drive into its parts, this force, named *coherence*, must be overcome, and in the collision the force used to drive into and impress these parts is consumed.

M. JURIN'S REASONING AGAINST *FORCES VIVES*.

§.585. The most specious reasoning made against *forces vives* is that of M. Jurin, reported in the *Philosophical Transactions*. . . . [117]

[Du Châtelet describes his assertion that the force of a body on a moving plane "will be its simple speed multiplied by its mass, and not as the square of this speed," and then proceeds to answer it.]

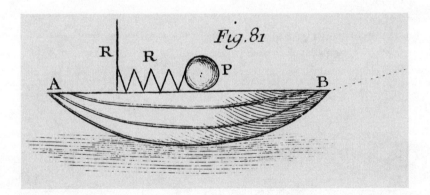

WHERE THE FLAW IN THIS REASONING LIES.

Here is where the flaw in this reasoning lies. Let us suppose, instead of the moving plane of M. Jurin, a boat, AB moving on a river in the direction BC

---

117. The *Philosophical Transactions* was the publication of the Royal Society of London, the English equivalent to the Royal Academy of Sciences in Paris.

with the speed 1, and the body P carried on the boat. This body acquires the same speed as the boat; thus its speed is 1. If a spring is attached in this boat, capable of giving to body P a degree of speed, this spring, which communicated to body P the speed of 1 off the boat, will not communicate it any more when it is carried on the boat; for the rest, against which the spring presses in the boat, not being immovable rest, and the boat yielding to the effort that the spring makes toward A, this spring releases at the same time from both ends, and this reaction must be taken into account. Thus, the spring will not communicate to body P the speed 1 in the boat, but it will communicate this speed less something, and this difference will be more or less great, according to the proportion that exists between the mass of boat AB and that of body P and the same quantity of *force vive*, which was in the boat AB in the spring R and in the body P before spring R was released, will exist after its release in the boat and in the body taken together. Thus, this case that M. Jurin defies all philosophers to reconcile with the doctrine of *forces vives* is only founded on this false supposition that the spring R will communicate to body P, carried on a moving plane or in a boat, the same force that it would communicate to it if the spring were pressing against an immoveable obstacle and at rest, but this is not the case, and cannot be, except in the case when the mass of the boat is infinite in relation to that of the body.

M. NEWTON MADE THE FORCE OF BODIES PROPORTIONAL TO THE QUANTITY OF THEIR MOTION.

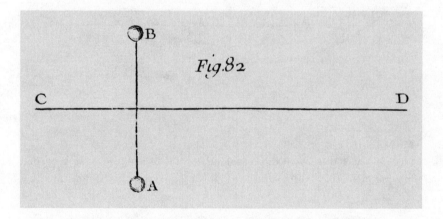

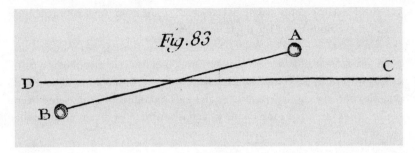

INEXPLICABLE PHENOMENON WITHOUT THE DOCTRINE OF *FORCES VIVES*, WHICH LED M. NEWTON TO CONCLUDE THAT THE TOTAL FORCE IN THE UNIVERSE WAS VARIABLE.

§.586. Although authority must be counted when truth is at issue, I feel obliged to tell you that M. Newton did not acknowledge *forces vives*, for the name of M. Newton is in itself nearly an objection. In the last question of his *Opticks* this philosopher examines the movement of an inflexible stick AB, at both ends of which have been attached bodies A and B, and he supposes that the center of gravity of this stick AB that he only considers as a line, moves the length of the straight line CD, while the bodies A and B turn continuously around this center, when the line AB is perpendicular to CD (as in figure 82) the speed of body A is zero, and that of body B is 2. Thus the motion of these bodies is then 2; but when this line AB is coincident or almost coincident with line CD (as in figure 83) then the sum of the motions of bodies A and B becomes 4. M. Newton concludes from this consideration and that of the inertia of matter that motion is constantly diminishing in the universe; and lastly that our system will some day need to be formed anew by its Author, and this conclusion was a necessary consequence of the inertia of matter, and the opinion held by M. Newton that the quantity of force was equal to the quantity of motion.[118] But when the product of the mass by the square of the speed is taken as force, it is easy to prove that the *forces vives* always remain the same, although the quantity of motion varies perhaps at each instant in the universe, and in all the cases, and especially in that which I have just cited from M. Newton, the *forces vives* stay invariable; whatever the position of the line AB in relation to line CD described by its center of

118. Du Châtelet is referring to Query 31 of the *Opticks* in which Newton considers the nature and behavior of particles. From the description of the motion of *globes* on the balance, he concludes that motion, or force, is lost. The consequence of this would be a need for divine intervention to replenish the force in the universe. See Isaac Newton, *Opticks* (New York, 1979) [1730 ed.], 397–98.

gravity. Thus, the continual miracles, which result from the position of this line AB have no place in the doctrine of *forces vives*.

[In §§.587–90 Du Châtelet concludes by presenting metaphysical arguments in favor of *forces vives* drawn from Descartes and given mathematical expression by Leibniz. She explains that this "conservation of an equal quantity of force" in the universe in all kinds of collisions negates the need for "miracles," for divine intervention, and is worthier "of the grandeur of the wisdom of the Author of nature."]

IV

EXAMINATIONS OF THE BIBLE

VOLUME EDITOR'S INTRODUCTION

With the publication in 1740 of the *Foundations of Physics*, and then her successful defense of her position in her dispute with Dortous de Mairan over the nature of collisions, Du Châtelet interspersed revisions to the *Foundations* and her *Dissertation* on fire with a return to the project she and Voltaire had embarked upon long before at Cirey. In the late 1730s they had begun a systematic study of the Old and New Testaments and of one of the most authoritative commentaries on the Bible, the twenty-three-volume exegesis by the learned Augustinian scholar, Dom Augustin Calmet (1672–1757).[1] His multivolume work presented them with all of the past exegetical arguments about each chapter and verse of the Bible. Coupled with writings by seventeenth-century English Deists, in particular, Thomas Woolston, whose six essays on Jesus Du Châtelet summarized in French, her manuscript grew to over seven hundred pages.

The style is sometimes satirical and argumentative, sometimes scientific and didactic, with no effort at creating a smooth, continuous narrative. Rather, Du Châtelet methodically goes from book to book, matching each remark to a particular chapter or verse. In this manner, she presents damning criticisms of the Old Testament God, declares the prophecies of a redeemer to be ambiguous at best, and disproves the divinity of Jesus, both his miracles and his resurrection.[2] She uses her knowledge of science to point

---

1. Du Châtelet knew Dom Calmet, and she had commissioned him to do a genealogical study of the marquis's lineage that was published at about the same time as the *Foundations*.

2. No part of this translation and accompanying remarks could have been done without the generous cooperation and tutelage of Bertram Eugene Schwarzbach, the editor of the critical edition of the *Examens de la Bible*. For example, he told us of Thomas Woolston, *Six discours sur les miracles de Notre Sauveur: Deux traductions manuscrites du XVIIIe siècle dont une de Mme Du Châtelet*, ed. William Trapnell (Paris: Honoré Champion, 2001).

out inaccuracies: snakes do not eat dirt; there is not enough water on the Earth to create Noah's flood; why have Joshua stop the sun when it does not rotate around the earth? Her astonishment at her first close reading of the Bible is evident, as is her disapproval of the apparent irrationality of it all and the numerous contradictions among different verses and accepted Christian dogma: a deity who gambles with Satan over his obedient worshiper, Job; a redeemer who speaks in incomprehensible parables and hides his miracles; a resurrection in one and a half, not three, days.

When she had finished, Du Châtelet had accomplished what must have been an underlying purpose, to destroy biblical authority and in particular the Old Testament God, a vengeful, cruel worker of miracles, with no respect for the natural laws he supposedly had designed. Her new adherence to Leibniz had led her in the *Foundations* to present a different kind of deity, one governed by the same strict laws of reasoning as those directing scientific discovery. This Creator had no need to intervene. The laws of nature had been set when the cosmos became "actual," realized out of all the possible contingent universes that her God could conceive of. Thus, she challenged another aspect of the argument over force and matter. Newton believed that force was dissipated in collisions, lost to the universe, and would on occasion need divine replenishment. Du Châtelet saw this stratagem as a negation of any concept of natural law. With Leibniz's understanding of force, the problem did not arise. With her condemnation of the miraculous, it could not be possible.

In this way, Du Châtelet presented herself as a heretic in two senses. First, unlike Voltaire, she could not accept all that the great Newton hypothesized. Second, her major criticisms of the Bible would have drawn condemnation from the French Church and the royal censors. Interestingly, throughout her life, she adhered to the outward conventional forms of worship. The priest at Cirey performed mass regularly. She expected to observe the specified number of weeks after her confinement when she became pregnant in 1748. And, she did not seek publication of the mammoth tome. Instead, it circulated in manuscript, part of the clandestine literature of the eighteenth century, writings passed from reader to reader, smuggled across borders to new audiences. The three versions of the *Examinations* that exist illustrate this phenomenon: one is in the library of the Academy of Sciences in Brussels; another in a private collection in Paris; the third in the municipal library of Troyes, a small town south of the capital. Bertram Eugene Schwarzbach's critical edition will be the first publication of this remarkable document.

Given the length of the *Examinations*, chapters have been selected from the Old and New Testaments that represent Du Châtelet's views and her method of biblical exegesis. She customarily gave the Latin Vulgate version of a passage she cited; for purposes of brevity these have sometimes been cut. Spellings of biblical proper names and places have been standardized according to the revised edition of *The New Oxford Annotated Bible*.[3]

The accompanying letters show all of the other concerns and activities that took her time during this period. Du Châtelet wrote to the Duke de Richelieu of her frustrations with Voltaire's continual provocation of the royal ministers and the king. By courting Frederick of Prussia, he caused anxiety for his safety and aroused her jealousy. Du Châtelet feared losing him, either to forced exile in England or the Netherlands or to an honored position at the Prussian court. In the end, Voltaire did not leave her, but he did go to Berlin in 1750 after her death. The letters also reveal her successes in the Republic of Letters, culminating in her election to the Bologna Academy.

Those to Richelieu and Bernoulli are from the Besterman collection. The letter to F. M. Zanotti, the secretary in Bologna, is to be found as an appendix to the article by Mauro De Zan cited previously.

⸱ↄ⸱

RELATED LETTERS

To the Duke de Richelieu[4]

In Paris, this 23rd November [1740]

I have been cruelly repaid for all I did at Fontainebleau. I brought to a satisfactory ending the most difficult affair in the world: I procured for M. de Voltaire an honorable return to his country; I win for him the benevolence of the minister, I reopen the way for him to the academies: In a word, I win back for him in three weeks all that he had undertaken to lose over six years.[5]

Do you know how he repays so much zeal and so much fondness? By

---

3. *The New Oxford Annotated Bible with the Apocrypha*, ed. Herbert G. May and Bruce M. Metzger (New York: Oxford University Press, 1977).

4. No. 253 in the Besterman collection, vol. 2, 33–34; D2365 in the *Oeuvres complètes*.

5. Du Châtelet had gone to court at Fontainebleau, the château favored by Louis XV in the fall because of the hunting. There, she met with ministers and courtiers to plead on Voltaire's behalf. She had come from Brussels where she had been dealing with her husband's contested inheritance.

leaving for Berlin! He notifies me of the news curtly, knowing full well that it will cut me to the heart, and he abandons me to unparalleled pain, of which other men have no idea, and that your heart alone can understand.

My blood is overheated from lying awake, I had chest pain; I am in the grip of fever, and I look forward to ending as the unfortunate Madame de Richelieu.[6] Except that I will end more quickly, and that I will have nothing to regret since your affection was an asset I could never enjoy. I return to end in Brussels a life where I have had more happiness than unhappiness, and which ends of its own accord, at a time when I could no longer bear it.

Can you believe that the idea that concerns me most in these dreadful moments is the frightful pain M. de Voltaire will be in when the elation of being at the court of Prussia diminishes. I cannot bear the idea that memory of me will one day be his torment. All those who have loved me must never reproach him for it.

In the name of pity and affection, write to me simply addressed "in Brussels." I will still receive your letter and if there is still life left in me, I will answer it and will tell you of the state of my soul, in these moments that appear so terrible to the unhappy and which I await with joy, as the end of an unhappiness I neither deserved nor foresaw.

Adieu, always remember me and be sure that you will never have a better friend.

To Johann Bernoulli II

In Brussels, this 28th April 1741

I do not know, monsieur, to what to attribute your silence ever since the letter I wrote to you when leaving Paris last winter. I fear that you may have suffered poor health, and you must not doubt the interest I take in it, as well as in the new victory that you just won at the Academy.

My absence from Paris has caused my orders about the *Foundations of Physics* to be rather badly executed. You were assuredly the first of those for whom I intended the book, and yet a copy has just been sent to you. I took advantage of this delay to add to it the letter that the last chapter of my book prompted from M. de Mairan, my response to this letter and the work of a M. the Abbé Deidier, a lost son of M. de Mairan, who put his name to a

---

6. Madame de Richelieu died of tuberculosis in August 1740.

7. No. 268 in Besterman collection, vol. 2, 48–50; D2468 in *Oeuvres complètes*.

work against M. your father and against me that M. de Mairan and he composed together. These are all the pieces of the proceedings up to now, and I believe that M. de Mairan has no ground for being so pleased about the success of his letter and that now he may to want to reply to mine.[8] Be that as it may, I feel how unworthy I am to defend the truth, but I also feel how strong one is with it. Undoubtedly, it is glorious to battle with the secretary of the Academy, but it is above all glorious for me to defend a truth that M. your father seemed to have put out of harm's reach. His memoir is like an impenetrable shield under the protection of which I do not fear any attack. That is the aegis of Minerva.[9]

The heart of the matter does not appear to be what interests M. de Mairan most in his letter, and I was obliged to follow him step by step in my response. However, you will see by the work by M. Deidier that it was not pointless to prove again the falsity of M. de Mairan's reasoning in his 1728 memoir and to demonstrate how odd a notion it is to attempt to assess the force of bodies by what they do not do.

You must remember, monsieur, that at Cirey, M. de Voltaire showed you a letter from M. de Mairan in which he said that, as no opponent had answered his memoir, he expected that he had ended the quarrel. It was a very different motive that had prevented M. your father answering it, but as for me, however easy it is to demonstrate the falsity of such pitiful reasoning, I believed that it was still glorious enough for me to destroy it and I see, by the effect that my letter produced, how necessary that was. I flatter myself, monsieur, that this little literary quarrel will restore your correspondence to me; you know how agreeable it is to me and how much I deserve your friendship by the sentiments with which I am, your very humble, and very obedient, servant.

Breteuil Du Châtelet

I have ordered that a copy of my book should also be sent to M. Daniel Bernoulli, and I beg you to give him a thousand compliments for me. M. de

8. Dortous de Mairan had sent Deidier's pamphlet with his own. It repeats many of the same arguments on the nature of collision. Mairan did not answer her reply to his criticism. She and her friends assumed that other members of the Academy insisted on this. Both her and Mairan's letters were reviewed in the *Mémoires de Trévoux* in August 1741. Although the reviewer did not agree with her position, he complimented her rhetoric and style, including her ironic tone and mocking rejoinders.

9. Du Châtelet is referring to Johann Bernoulli's memoirs on the laws of motion submitted for the Royal Academy of Sciences prize competitions of 1724 and 1726.

Maupertuis got himself lost at the battle of Neuss; one still has no news, and I am very much troubled by it.[10]

To F. M. Zanotti[11]

In Paris, this 1st June 1746

That the Academy of Bologna, monsieur, was pleased to honor me with admission as one of its academicians, is a prize that I would not have dared to hope to win, and this honor becomes yet more precious, if possible, by learning of it from you monsieur, for whom my esteem has long preceded the occasion that I have today to express it to you. I consider the choice that the Academy was pleased to make of me as an encouragement that it wanted to give to persons of my sex. For what more flattering motive could one propose to them, in order to engage them to cultivate the sciences from which prejudice had so far appeared to exclude them, than the hope of one day seeing themselves admitted to an illustrious body that is the glory of Italy and of the learned world?

For me, if something could add to my taste for the sciences it would be the desire to justify the choice of the Academy. The desire will stand in place of the genius I lack; and what must I not await from exchanges with you, which monsieur, if you permit me, I hope, to cultivate? I would see all my literary wishes fulfilled if I could one day thank the Academy and see myself able to attend its meetings and benefit from the enlightenment of its learned members. For me, it will be the holy city toward which my eyes will constantly be turned.

If something can diminish my regrets on this point, monsieur, it is the kindness that you have shown in undertaking to express my gratitude to them: it is equal to the gift, as my esteem for you equals the merit that an illustrious body so fairly made by choosing you to be its voice. It is with such sentiments that I profess to be, monsieur, your very humble and very obedient servant.

Breteuil Du Châtelet

⌘

---

10. Daniel Bernoulli was Johann's brother and also a mathematician and physicist. Maupertuis was then director of the Berlin Academy and had gone to the battle of Mollwitz at Frederick of Prussia's request. He was captured by the Austrians and taken to Vienna, and then released.

11. This letter is from Mauro De Zan, "Voltaire e Mme Du Châtelet," appendix, 156–57.

EXAMINATION OF THE BOOKS OF THE OLD TESTAMENT

Examination of Genesis

In Genesis Moses depicts God as a workman who does his work piece by piece, who spends six days creating the heavens and the earth, and who rests on the seventh, as if he were weary from doing much good work during his week. He represents him to us as capable of jealousy, anger, revenge, repentance, in other words, with all the faults of men. If God wanted to depict himself in an appealing way, he should at least have depicted himself with qualities that bring respect for men and not those that make them hated or viewed with contempt; let it not be said that it is in order to conform to the primitiveness of the Jews that God chose to act thus.

For first, God did not intend this book only for the Jews; it is still today the foundation of the religion that his son came to bring to men,[12] and still he has not corrected our [Christian] ideas on this story of the creation. And second, the more primitive the Jews were, the less judicious it was to present the deity to them in demeaning images they were unable to rise above. This answer that God conformed to the primitiveness of the Jews is also absurd for another reason. Most of the books of the Old Testament are impenetrably obscure on things that were essential for the Jews to know if redemption was announced to them. If those evils are predicted that in the end will punish their alleged obduracy, these books are little help, for they offer only allegories and symbols incomprehensible to the subtlest minds, let alone to a primitive people. Now, on the contrary, it was much more important to speak to them intelligibly when the purpose was to save them from total ruin and to make the most of the redeemer whom God was to send them, than to make them learn about how God created the heavens and the earth. So, this alleged relationship, which God is supposed to have made between the primitiveness of the Jews and the primitive ideas we are given about him, cannot excuse the way in which the supreme being is talked of here. . . .

He, who made the heavens and the earth, the creator of all things, is willing to give men a story of the way in which he created this great work, and he tells it to them as if he himself did not know what he had done. Each discovery men made in physics and astronomy revealed to us a new absurdity in this history of the creation. God, for example, did not make any other difference between the sun and the moon except that he calls one *luminare*

---

12. Although elsewhere in the *Examinations* Du Châtelet mentions Mohammed, she seems unaware of the importance of the Old Testament to Islam.

majus [major light] and the other *luminare minus* [minor light]. Yet, children know now, that the moon is an opaque body that often does not preside over the night and that never gives light other than that of the sun that it reflects to us. In Genesis 1:24 God says, *Producat terra animam viventem in genere suo. Let the earth produce all living souls, each according to his kind.* Yet, it is certain that the earth does not produce any animal or the water which, however, in verse 20 had produced the fish.[13]

In Genesis 1:4, God divides the light from the shadows, as if the shadows were something and could be separated from the light, with which they can never be mixed in the first place, since they are only the deprivation of it. It is very amusing to see 3 days and 3 nights marked by the morning and the evening before the sun was created. For it was only created on the 4th day, and the author takes care to say that God created the sun and the moon in order to divide the day from the night: this means that they were not so divided before their creation. . . .

The first commandment that God gave to man was not to eat of the fruit of a tree that was named the tree of the science of good and of evil;[14] it is difficult to ascertain whether the tree of life mentioned in Genesis 2:9 is the same as the tree of the science of good and of evil that God forbids man to eat from in Genesis 11:[17].

[Flavius] Josephus makes them two different trees.[15] It is also very difficult to ascertain what the virtue of the fruit of these two trees was. It seems, judging by the name of the one God forbade Adam to eat from, that the serpent was right when he said to Eve in order to incite her to eat from it that *then they would know good and evil;* but what the interpreters find even more puzzling is why God made this strange interdiction to man, and this appears

---

13. Du Châtelet uses the word *certain* in contrast to *probable.* Both words for her have a specific, "scientific" meaning; she sought *certain* as opposed to *probable* knowledge. In the *Examinations* she is applying these scientific criteria to the biblical account of creation. She continues to do this in other books of the Old Testament and in the Gospels of the New Testament. Note also that she is making her own translation of the Latin Vulgate throughout the *Examinations.*

14. Note that it is "science" in French, as in certain knowledge, versus simple "connaissance [knowledge]," which could be reasoned and only probable. See the entry "L'ame" in Voltaire's *Philosophical Letters.* Schwarzbach, ed., *Examens,* I, "Genèse," 157, n. 8. (Pagination is from an earlier draft, footnote number is correct.)

15. Flavius Josephus (37–ca. 101), of a priestly Jewish family, allied himself with Roman Emperors, but still was a military leader of the Jewish revolt of 66. Spared death, and released from imprisonment by the emperor Vespasian, he enjoyed numerous favors that allowed him the means and leisure to write his histories of the Jews. His *Ioudaike Archaiologia* [Jewish Antiquities] includes his account of the life and times of Jesus. Du Châtelet's knowledge of Josephus may have come from her reading of Dom Calmet, though Voltaire had a French translation of his history of the Jews from 1735–36 in his library at the time of his death, which she also might have consulted.

pure caprice on the part of God. But there is more; God, in creating man, had taken care to make an animal in order to tempt him, and this animal is the serpent. There are great disputes between the interpreters about the language in which the serpent spoke to Eve. Some believe that all this is only an allegory, but why not say all at once that it is a dream. Indeed, it is in this story of the creation that one must have recourse to allegory, and to supposing fallen angels, who take the form of the serpent in order to come to tempt the woman, although surely this was of all forms the least tempting. One is obliged to make this supposition of the fallen angels, because if it were only the animal that we know under the name of serpent that tempted Eve, it would be necessary for God to have given this animal reason and speech expressly in order to come and tempt man.[16]

But I leave it to every sensible mind to judge what one should think of this gratuitous supposition of the bad angels, necessary, however, to explain a book given by God to men to instruct them, a supposition of which not the least trace is to be found in this book. This fall of the bad angels, their hatred of God, their envy of man, are one of the fundamentals of the christian religion, and not a word about it is found in the book on which this religion is based: the word *Satan* even is not Hebrew. This fall of the angels is a tale that occurred to people very late, and to which the Epistle of St. Jude, verse 6, which seems to speak of something similar, gave birth. But this Epistle of St. Jude is, of all the books, the one whose authority is most disputed; for a very long time it was not accepted by the Church. Yet, I bet that of one hundred christians there are at least ninety-nine who believe that this story of the bad angels, on which the promise of a redeemer is founded and which is the reason for his mission, is to be found in the Old Testament. This makes one feel how necessary the examination of this book is.

So, the woman allows herself to be persuaded and eats this forbidden fruit; she makes her husband eat it, and there and then, they are fallen from their glory and their innocence. But the most absurd is that the punishment of the serpent is the first thing God thinks of. Although surely he was very innocent in all ways, since he was only the instrument of God or that of the devil, and that as he was a brute one cannot see the basis of the anger of an intelligent being nor of the punishment this being imposed on it. If it is said that the punishments imposed on the serpent must be suffered by the bad angels which it represents, one falls into another absurdity, for the bad an-

16. Du Châtelet considered this suggestion of Locke's that animals might have the qualities of men, in particular reason, which indicated a soul, one of the most dangerous that Voltaire played with in his writings. She found occasions to deny it herself in her writings, if obliquely, and consistently told him to take it out of his, for example, the proposed last chapter of his *Treatise on Metaphysics*.

gels, being damned, had nothing more to lose. This punishing of the serpent by forcing it to walk on its belly—which I believe it was only moderately troubled by—a punishment that unfortunately cannot be applied to the bad angels, no more than that of taking dirt as its nourishment, which is added here to the first, and which puzzles all the interpreters very much, 3:14. . . . [Latin omitted][17] *You will walk on your stomach and you will eat of the earth all the days of your life*—but of these punishments inflicted on the serpent, it only suffers half; for it is well known that the serpent eats no dirt, but fruits and plants like other animals. . . .

When God had seen to the punishment of the serpent, he thought of that of the man and the woman. They had already begun to feel his anger in perceiving themselves naked (Genesis 3:7), which seems to me a small misfortune that God remedied on the spot by making them garments of animal skins (Genesis 3:7).

In Genesis 3:8, Adam and Eve are seen hiding from God behind some trees, and hearing his voice walking in the garden; God asks Adam where he is, as if he did not know (verse 9) and Adam answers him (Genesis 3:10) that, seeing himself naked, he went to hide himself, when he heard God's voice walking in the garden. Finally, Adam justifies himself and tries to excuse his crime, as if he were answering to a man he could deceive. But what is most remarkable is the way God speaks to Adam and Eve after their disobedience; he demeans himself to mock Adam and to insult him in his misfortune. After having imposed his punishment on him, and having cursed him, he says to him (Genesis 3:22): *Ecce Adam quasi unus ex nobis factus, sciens bonum et malum. There is Adam become as one of us knowing the good and the bad;* it is to be noted here that God does not say, *There is Adam become similar to me,* but *as one of us.* And that when the serpent tempts the woman he does not say, *God knows that if you eat of this fruit you will be similar to him;* but *you will be as gods,* which proves what I said before, that whoever wrote this book did not recognize the unity of God; the christians find the Trinity in the *nobis* that God used in mocking Adam, and it is on this that it is founded.[18]

The place where the earthly paradise was situated is an inexhaustible subject of disputes in which I do not want to enter; on that, one can read Dom Calmet, in whose writings will also be found a part of what the Fathers

17. Customarily, Du Châtelet gives the Latin version of the passage from the Vulgate and her own translation. For purposes of continuity and brevity, the Latin is sometimes omitted with the omission indicated in this way in the text.

18. Du Châtelet means the presumption that God is manifested as god, man, and holy spirit. Note that she does not capitalize "christian" anywhere in her critique. This is yet another subtle act of disrespect.

said on the cherub armed with a flaming sword that God put at the gate of the earthly paradise to prevent Adam and Eve from entering there; for they were expelled from this place of delights and condemned to earn their bread by the sweat of their brow and etc.

So far God had behaved like a strange king who wanted to test the obedience of his subjects on ridiculous things, and who punished them afterward well beyond their crimes. If only Adam and Eve had been punished, one could strive to save the conduct of God from the reproach of malignity and injustice, and this story would only be the tale of a ridiculous old woman; but it is claimed that the entire race of men was cursed for this mistake, and that we all carry the burden of Adam's sin, a burden that is no less than our eternal damnation; . . .[19] so, this disobedience of Adam is what is called the *original sin*, by which we are all born in the power of the demon. From that moment God resolved to make a glorious deed and to send his only son in order to free us from this power, as we saw that he promised to the woman. But, having formed this design he keeps it *in petto* [to himself] for four thousand years, and instead of sending this redeemer who was so necessary—since 1,700 years after the creation, all men had corrupted his rule (Genesis 6:12)—he repents having created man (Genesis 6:6) and, to console himself, he resolved to drown him. We will soon see how he executed this resolution. . . .

But now we have reached the flood . . .

God saved Noah and all his family and with them all the animals of the earth and the sky. . . .

But where did Noah take these animals from in order to confine them in the ark? Those of America, for example, and of other countries separated by the sea from the place where Noah was? Some say that they swam across the sea in order to come and find him; others, that God brought them to him himself: one has a choice.

The same difficulty occurs in explaining how these countries were repopulated with animals after the flood. But the same answer is given, and it seems that God carried them back to the same place he had taken them from when he brought them to Noah.

It has been demonstrated that the ark, the dimensions of which are provided in scripture, could not contain a quarter of all that God ordered Noah to put there; moreover, why drown all the animals of the sky and the earth

19. Du Châtelet notes that this puts Pascal into ecstasies, an observation that probably comes from Du Châtelet's reading of Voltaire's *Philosophical Letters*, from the section he added on Pascal.

because men were bad? This seems neither consistent nor just, and God did not lack the power to save them; since surely there can be no greater miracle than a universal flood. . . .

The flood is demonstrably physically impossible. It has been calculated that it would take at least eight times as much water as the ocean contains to carry it into effect, and it cannot be said that the waters of the sea and the atmosphere were sufficient for the flood, because they rarefied. . . .[20]

Finally God, after having assuaged his vengeance in so cruel a way, orders the waters to recede; as it is not known where they came from, it is not known where they went either. . . .

The dove that Noah released to know if the waters had receded brought him back a twig of an olive tree with its leaves on it, which is impossible. The long time the trees were under water must have made most of them die and delayed by several months the rising of the sap of others.

It is surprising that Noah, his children, and most of the beasts did not die of the infection the silt must have caused in the air.[21] A land that had suffered a flood could not be habitable for animals of our kind for a year at least after its end. . . .

So, God chooses Abraham to be the father of his cherished people. The conversations that he has with this father of the believers are one of the most interesting things in the Old Testament.

God begins by making an alliance with Abraham. God and Abraham reciprocally pledge themselves and make mutual promises, Abraham to be circumcised and God to multiply his race as the sand of the sea. It is to be noted that God, in ordering Abraham to be circumcised does not explain to him what it is, which indicates that Abraham knew about it, and that the Hebrews did not invent it and were not the only people who used it. . . .

[Du Châtelet describes the destruction of Sodom.]

The rest of the story of Lot is curious. His wife who had been saved by the angels was changed into a statue of salt for turning her head toward Sodom; the angels allow Lot to retire to a small town and promise him that they will not destroy it. However, he next has the whim to leave and to retire to a cavern, and then his two daughters imagine that they are alone in the world with their father, although they had just left the small town of

20. Du Châtelet was familiar with the process of rarefaction, the becoming less dense, more refined, scarcer, from her consideration of the properties of fire for her submission to the 1738 prize competition of the Royal Academy of Sciences on the nature and propagation of fire.

21. In the eighteenth century many believed that infections rose from the mists of silted areas carrying diseases such as malaria and yellow fever.

Zo'ar, that had been saved for them. But finally under this strange miscon-
ception they made their father drunk and abused him.[22] It must be confessed
that this makes a pretty family to save over all others.

The care that God takes with Abraham's domestic arrangements is
laughable, he mixes in their squabbles, and sometimes settles them; thus,
if the mistress of the family beats the servant and dismisses her, God goes
and finds her, brings her back, and reconciles them. But next, on God's order, Abraham and Sarah use this same servant very harshly; Sarah being
sterile had given Hagar (this is the name of the servant) to Abraham so that
he could make her pregnant, and he sired a son named Ishmael; on a whim
Sarah decides to expel Hagar and her son. . . .

God, after using Hagar and Ishmael so harshly, scarcely treats his servant Abraham better; apparently fearing that Abraham might impose on his
kindness, he sets out to test him, and suggests the most revolting thing in
the world, to immolate his son Isaac himself, the son who had been given
to him miraculously in his old age. But happily God contented himself with
Abraham's obedience and did not want the sacrifice. . . .[23]

Examination of Exodus

. . . Moses had begun his mission to lead the Israelites from Egypt, but Pharaoh did not want to dismiss the people and increased their labors. Moses
complained to God about this, and God encouraged him and said to him,
Sed indurabo cor Pharaonis et non audiet vos et mittam manum meam super Aegyptum, "I
will harden the heart of Pharaoh, he will not release you and I will spread
my hand over Egypt."

Now, certainly, God's conduct is the most revolting. What would one
say of a man who ordered one of his servants to do something he had made
impossible to carry out. for that is what *Indurabo cor ejus. I will harden his heart*
means, and of a man who made his servant suffer a thousand evils and killed
his firstborn son to punish him for not carrying out his commands. Such is
the conduct of God toward Pharaoh. . . .

Moses and Aaron, seeing that Pharaoh would not allow the Israelites to
go and sacrifice in the desert,[24] changed all the waters of Egypt into blood;
the rivers, fountains, lakes, in a word, all the waters of Egypt. . . .

22. Lot's daughters have sexual intercourse with him.
23. This paragraph is a clear example of Du Châtelet's use of sarcasm.
24. Du Châtelet previously noted this request was a lie that God instructed Moses to tell Pharaoh when actually he meant the Israelites to flee.

[After questioning why the Egyptians did not die by this scourge, Du Châtelet continues puzzling over the events described.]

God, seeing that Pharaoh, despite this terrible scourge, would not give in, ordered Moses to make all the frogs come out of the river. The animals scattered into the houses and the countryside. Apparently they were not killed by the waters turned into blood like the other fish, and God saved them expressly for this second scourge. Some say they were dead, but he resuscitated them. . . . Pharaoh, more affected by this second scourge than by the first, although it was much less terrible, promised to send away the Israelites if Moses would send away the frogs; Moses drove the frogs away by killing them. These frogs that Moses had die in order to free Egypt must have ruined them with their infection. It is not explained why Moses only delivered Egypt from this scourge by delivering them from the stench. But these poor frogs, why make them die? This is what puzzles me.

The king did not keep his word and did not release the Hebrews *because God*, scripture repeats again (Exodus 8:15), *had hardened his heart;* which brought on himself the third plague, a kind of fly. . . . next a prodigious quantity of locusts. The poor Pharaoh whose heart God had hardened (Exodus 8:12) (scripture repeats this diligently at each scourge) was careful not to surrender. Thus, Moses struck all the animals with a plague.

This fifth scourge of the plague of the animals made them all die, except those of the Hebrews. . . . However, at the sixth scourge, which is that of the ulcers, it is said that the men and the animals were struck (Exodus 9:9). This is difficult, if they had all died from the plague, unless God had resuscitated them in order to strike them with ulcers and in order to kill them with the hail that Moses next unleashed (Exodus 9:25). It is to be noted that nowhere is it said that the plague of the animals and the ulcers had ceased, and because of this neither men nor beasts must have remained in Egypt, as the waters were blood and the earth infected by the frogs, the dead fish, the infection of ulcers, and that of the cadavers of the beasts killed by the plague.

Thus, it so happens that God miraculously preserved the Egyptians in the midst of all these plagues, to have the pleasure of sending them new ones. . . .

I confess that I find Pharaoh very patient; during all the time that these scourges lasted in Egypt he did not harm Moses or Aaron, and granted them an audience as soon as they asked for one, although it was certain that they were going to strike his country with some new plague. . . .

But the reason that God gives Moses and Aaron (Exodus 10:1–2) for continuing this comedy or rather this tragedy for so long, is remarkable. Here it is as given in the scriptures . . . [Latin omitted]: "And the Lord says to Moses, go and see Pharaoh, for I have hardened his heart and that of his servants, so that I might be able to make all marvels and that you could tell your sons and your grandsons how many times I struck Egypt and you would know that I am the Lord."

Is that not a beautiful reason for killing a whole people, and did Nero, the cruelest of the emperors, ever show a more revolting barbarity? Yet, this is given as the work of God himself. In truth, if men could offend God, would it not be the worst offense to charge him with such conduct?

Finally, the tenth and last scourge with which God struck Egypt, after which he allowed Pharaoh to let his people go, was the death of all the firstborn. In all the other scourges, God had given the choice to Pharaoh either to release the people or to suffer the punishment; but he sent this tenth scourge without giving him any warning of it, foreseeing apparently that Pharaoh would prefer to release the Hebrews and save his own son.[25]

Before sending this last scourge, God takes a funny precaution. As he did not want to kill the firstborn of the Hebrews and apparently feared to make mistakes, he orders them to mark their door with lamb's blood; . . .

. . . so that if an Israelite had unfortunately forgotten to mark his door with blood his son was doomed, not as a punishment for this forgetfulness, which would only have been unjust, but because God would have made a mistake and would not have been able to tell the house of a Hebrew from that of an Egyptian.

Further, it must be noted that the Hebrews lived apart in a separate district, which makes God's precaution even more ludicrous. . . .

The animals also experienced this tenth plague and their firstborn died; *Omne primogenitum jumentorum* (Exodus 12:29), and yet, all the beasts, *Omnia animantia* (Exodus 9:6) were dead of the plague of the fifth scourge, the ulcers of the sixth, and the hail of the seventh. Apparently children had not been included. The scourges of Egypt must have been only a conjurer's sleight of hand. For if they had been real, there should have been nobody left in Egypt.

The tenth scourge was the most serious and struck Pharaoh's mind most forcefully; for he allowed the Israelites to leave and announced it to Aaron and to Moses (Exodus 12:31), although Moses had said to him in Exodus 10:29 that he would not see his face again. . . .

Let us stop here for a little to see, in relation to the scourges of Egypt, the nature of God's miracles in the Old Testament so far. These miracles are

25. Du Châtelet clearly means to suggest that God intended the Egyptians to suffer this final scourge and so prevented Pharaoh from complying.

of two kinds: pointless or cruel. What is more pointless and more puerile, for example, than to send an angel to console Hagar (the servant of Sarah that she had dismissed), to send one to Pharaoh to warn him that he had made Abraham a cuckold, to send one to Jacob to order him to build an altar, and another to order him to repeat to him that instead of being called Jacob, he will be called Israel etc.?

But the miracles that bear the mark of cruelty are many more! Waters that changed into blood to poison an entire people, frogs, locusts, flies sent to lay them waste, contagious illnesses that attack some animals, ulcers that eat into men, storms, hail, darkness over an entire kingdom, children in their cradles killed by an angel, and all this in favor of the miserable little people of Israel, in order to make them wander forty years in a desert. For God did not lead them to Palestine by a direct path, fearing (Exodus 13:17) that they might regret having left Egypt if they at first encountered enemies they would have to fight, and fearing that they might return to Egypt . . . [Latin omitted].

But was it more agreeable for them to die of hunger, of thirst, and of all the miseries they experienced in the desert for forty years than to fight the enemies whose defeat at least would have gained a country for them whose land they could have cultivated and towns they would have inhabited? Or, was it more difficult for God to make them triumph over the Philistines than to make all the miracles that one is going to see that accompanied them constantly in the desert.

In the end, God gained nothing by this detour; for one will see that the Israelites never stopped complaining in the desert and regretting Egypt. . . .

God next enters into the greatest detail about the household of the Israelites.[26] For example, he orders the slave to leave with the clothes he brought (Exodus 21:3). He describes Aaron's breeches (Exodus 23:42). In Leviticus he tells the Israelites (Leviticus 19:27) how they must cut their hair and shave their beard. . . . In Numbers 15:38 God orders the Israelites to put fringes and hyacinth-colored ribbons on their cloaks. Finally, he prescribes to them the manner in which they must go to the toilet (Deuteronomy 23:13), *Gerens baccillum in balteo [cumque sederis] fodies per circuitum et ejecta humo opertes. You will make a hole with your staff and when you have relieved yourself, you will fill it with earth.* Only cats observe this ordinance. All of this is not worthy of God nor very useful to the republic.

Some of these laws given by God to the Israelites are very unjust. For

---

26. In this section Du Châtelet takes examples from different books of the Old Testament.

example, he punishes with the same torment the man who stole and the man who murdered his father and his mother, and the one who cursed them. The one who beat his slave to death was punished with death if the slave died on the spot. But he is not punished at all if the slave survives a day or two. . . . In Numbers 15:30 a sin of pride is punished by death, and it is not explained what the sins of pride that merited death were. It is ludicrous to impose the penalty of death and not to say precisely why. A poor man who was gathering wood on the day of the Sabbath was stoned, as is seen in Numbers 15:36, and he who used an oil similar to that which God indicated to Moses for anointing the tabernacle was punished with death. . . .

Examination of Leviticus

Chapter 12 is devoted to prescribing what women must observe to purify themselves after giving birth. God insists that the discharges after the birth of a girl last much longer than those after the birth of a son (this is certainly false) and yet for this reason the woman who gave birth to a girl was considered unclean for seven more days.

Examination of Numbers

. . . But here is perhaps the most ridiculous place in the laws of Moses. When a wife was suspected of adultery, her husband led her before the priest, chapter 5 from verse 19 to the end of the chapter, who took water from a bronze basin with a cup and threw in it dust from the temple square, which made the water bitter, and the priest wrote curses in a book, and then erased this writing in the water, which he next made this woman drink. If she was guilty, her thigh rotted, her womb swelled, and she died; if she was innocent, she had children. I find her very kind to sleep again with her husband after that; this is not as ludicrous as the priest's enchanted cup, but it is just as probable.

Moreover, this precaution seemed hardly necessary among the Jews where divorce was permitted. It is to be noted that not only was there no punishment for a husband who unjustly accused his wife, but that he could even bring this accusation more than one time. Besides, I must confess that this ceremony seemed so ridiculous even to the Jews that it is nowhere mentioned in their books that it was ever observed. Thus, God did not find any people stupid enough to carry out all that he had ordered. . . .

Examination of the Book of Joshua

[Du Châtelet discusses Joshua's war against the five Kings of the Amorites.]
   . . . The Lord frightens them at the sight of Israel, who pursue them and

make a great carnage, and while they fled before Israel, God caused stones to rain on them (Joshua 10:11) that killed still more than Joshua had.

Who would believe after that, that on this occasion God still had something more to do for the Israelites against the Amorites? However, Joshua was not yet satisfied, and while his enemies fled before him and God caused stones to rain on them, he set himself to command the sun to stop over Gibeon and the moon to rest on Ai'jalon, so that he could have time to take revenge on his enemies, and the sun and the moon obliged, according to what is written in the book of the Just.[27] Thus, the sun stopped for a whole day in the middle of the sky . . . [Latin omitted].

It was, according to the interpreters, about 10 o'clock in the morning, and at most twelve noon (according to the text: *in medio coeli*) when Joshua commanded the sun to stop in order to give him time to defeat enemies who had already been defeated. It must be admitted that he was very cautious.

I say nothing about the physical absurdity of this miracle, it is too obvious,[28] and this physical absurdity is moreover a true objection; since God who made the laws of nature cannot violate them without contradicting himself, especially when he violates them without necessity.[29] For nothing was so pointless as this miracle. Since, according to the text itself, the enemies were fleeing, and it was only 10 in the morning, and, besides, one does not see why these five kings and all their people had to be killed and the town of Makke'dah taken precisely on this day.

Besides, it is difficult to understand what the moon is doing there, and why Joshua needed it; or where Ai'jalon is, where he orders it to stay.

One is also a little puzzled to explain this order that Joshua gives to the sun, since it was proved that what turns is the earth, not the sun. Apparently the H. Spirit did not know the true system of the world.[30]

Again, it turns out that the sun, in stopping, must have done more harm than good to the Israelites. For it stopped *in medio coeli*. Now the heat must have enervated them, and it required a new miracle for the Israelites to

27. Calmet identifies this as a book of canticles. Schwarzbach, ed., I, *Examens*, I, "Joshué," 291 n. 42. (Pagination is from an earlier draft, footnote number is correct.)

28. Du Châtelet means that it went against the natural laws of physics.

29. Here Du Châtelet is drawing on her description of the Creator in chapter two of the *Foundations*, where she states that God cannot break the epistemological laws of reasoning established in her chapter one, such as the law of contradiction, and that his actions must be "necessary," rather than contingent, subject to changing circumstances.

30. Du Châtelet or her copyist abbreviated terms when referring to God and Jesus. "H. Spirit" indicates the Holy Spirit, the third manifestation of the Trinity.

have enough strength to chase their enemies and to follow them for 24 hours without drinking or eating; I say 24 hours: for the inspired author of Ecclesiastes says that this day lasted twice as long as the others . . . [Latin omitted].

The author, to give credence to so extraordinary a fact, says that it is written in the book of the Just: *The sun stopped in the middle of the sky.* This book of the *Just,* incidentally, is so lost that one knows neither who wrote it nor what was in it. It is the same with the book of the *Wars* of the Lord, which I discussed elsewhere. It is pleasing that the inspired books become lost; for as soon as they are cited to confirm what is written in a canonical book, they are considered inspired. The H. Spirit chooses not to strengthen his testimony with that of men. . . .

Examination of the Book of Job

Neither the person, nor the author of this book, nor the country, nor the time when it was written is known. Some say that it is only an allegory, others that what is told happens in a dream. Others assert that it is a piece of poetry, and a few have even wanted to find an exposition, dialogues, action, and a denouement, in other words a kind of dramatic poem. . . .

It is not known if Job was a Jew or a pagan. . . .

Be that as it may, here is how the book that bears the name of Job tells us his story.

(Chapter 1) Job was very rich, he lived as a gentleman; he was good to everybody; he had seven sons and three daughters. One day God on a spree saw Satan in the middle of the Children of God. The Children of God means the angels, and Satan is the Devil.

God amuses himself chatting with Satan and asking him where he comes from; apparently he had no idea. Satan answered him that he had just gone around the earth. Well, God said to him, have you not seen my servant Job, who has not his equal on the earth? Satan answered him, that this was a fine miracle, that he showered goods on Job and that he, Satan, was not surprised that Job was attached to God. But in place of doing good to him, do harm to him, added the Devil, and you will see that he will curse you to your face. God accepts the challenge and in order to see who will win the wager, says to Satan he can harm Job as much as he likes, but forbids him to touch Job's person. . . .

[Du Châtelet describes the afflictions, the visit of the comforters, and Job's turn away from God; she mocks the injustice of the punishments.]

And in chapter 42 Job asks God's pardon for all that he gave vent to against God during his misfortune. So, God did not triumph this time against the Devil as he had in 2:2, when the Devil had taken from Job only his goods and his children. . . .

Although Job caused God to lose his wager, it ended magnificently for him. God gave him back twice the amount of goods and the same quantity of children. He did not resuscitate the first, so, the poor children died in the bloom of youth, from a practical joke devised by God and Satan.

The friends of Job also had their turn. God scolds them (Job 42:7), for they did not speak the truth, as his servant Job did; yet, they had always taken the part of God against Job, who, in contrast, had always maintained that God punished him too severely for his sins. It is handsome of God thus to confess his error; but he should give a little thought to the fact that the torment of hell is one hundred times more unjust than all he had made Job suffer. In the end, God also pardoned the friends of Job in return for a sacrifice. Thus, in this story everyone ended by being satisfied.[31]

Examination of the Book of Judith

[Du Châtelet explains that this book has been canonical since the Council of Trent, but that it is not accepted by Protestants or Jews. She then describes the besieging of Bethulia by the Assyrian commander, Holofernes.]

. . . Judith, seeing the plight to which Bethulia was reduced took a strange resolution, which was to find a way into the camp of Holofernes like an honorable whore to try to insinuate herself into his good graces and to cut off his head during the night. She prepares herself for this beautiful action by fasting and prayer; next she takes off the garments of her widowhood, for she was one, arranges her hair as best she can, and adorns herself with her most beautiful finery in order to try to please Holofernes.

The prayer she addressed to God (Judith 9:2), before leaving for the camp, is remarkable . . . [Latin omitted].

Judith 9:2, *God of my father Simeon to whom you gave the sword in order to punish the strangers who had dishonored a virgin,* verse 3, *and to whom you gave their wives for prey, their daughters to be captives, and all their goods to recompense your servants who had been inspired by your zeal, aid me in the same way, oh my God, I am I who am only a woman widowed.* Verse 4, *For it is you who have made all that was made; you have foreseen all things one after another; and they have accomplished what you had resolved.*

It is difficult, after this, to justify God for having approved, recompensed, ordered the action of Simeon . . . which is surely the most abomi-

---

31. There is, in fact, no place where God, explicitly, admits a mistake. Du Châtelet must have come to this conclusion because of the way God rewards Job.

nable and the most detestable action of which one has ever heard. It may be seen that Judith did not reject such an origin; for what she did is scarcely inferior to the horror of the action of her ancestors that the H. S. celebrates here with her own words.

This cruel widow continues thus her prayer: . . . [Latin omitted].

Judith 9:12, *Make, Lord, his pride be punished by his own sword . . .* Verse 13, *let him be caught by his eyes in the snares of my charms, so that you might deceive him by the sweetness of my words.*

God made himself a party to this unworthy maneuver; for it is said in Judith 10:4 that after Judith adorned herself, *God added a new brilliance to her beauty because all her adornment had virtue as its object . . .* [Latin omitted].

Judith, thus prepared, sets off with her servant to go and find Holofernes; she did not tell anyone her plan, fearing apparently that people might be horrified by it and that she would be prevented from carrying it out. The soldiers of Holofernes who came across her saw immediately what this was all about and told her that she would certainly win the heart of Holofernes. They admired her beauty very much, and there is even some indication that they admired it at close range. For soldiers are not so respectful, especially of a person whose profession Judith appears to practice.[32]

Judith is introduced to Holofernes, who asks her what brings her there. In chapter 11 she tells him a hundred extremely clumsy lies. She tells him, for example, in 9:14–15, that her God will tell her when he is angry enough with the Jews, and that she will go and inform Holofernes. Surely, it was fortunate that she was pretty, for I do not believe she would have seduced Holofernes by her words. . . .

Judith spent four days in Holofernes's camp, God knows what she did there or rather what was done to her. On the fourth day Holofernes sends for her. The text seems to indicate that Holofernes had not yet had his pleasure with her; this scarcely seems probable.[33]

Judith goes to him with all her heart and when Holofernes falls asleep, she takes his sword and cuts off his head. . . .

There are some contradictions in this account . . . for example, Judith, in cutting off the head of Holofernes, had to strike him twice, verse 10; however, normally he would have woken up at the 1st. . . .

The Israelites must have been a very cruel people; for among all the

32. In her "Remarks" on Bernard Mandeville's *The Fable of the Bees*, Du Châtelet comments on this kind of rough behavior, but of sailors, not soldiers.

33. *Probable* had a specific scientific meaning for Du Châtelet, which she meant to indicate here, in the sense of a *fact*. She viewed Judith's chastity as not only uncertain but improbable.

people to whom Judith recounted her action, nobody expressed horror at it. Everyone praised it, everyone blessed her. Thus, it may be seen that she had been wrong to hide her plan and that they would have encouraged it. . . .

Examination of the Book of Isaiah

Isaiah is one of the prophets who have written the most, and he was of the race of David and prophesied in Judea. . . . His visions are no more reasonable than those of the others. He even saw the Lord sitting on his throne (Isaiah 6:1–2), with seven seraphim at his side. One of these seraphim took a piece of charcoal and touched it to his lips. The cherub had taken the precaution of handling the charcoal with fire tongs (Isaiah 6:6). Next, Isaiah 6:8, the Lord said, *Whom will I send and who is going to carry my words?* Isaiah took him from his misery and offered himself. The Lord sent him and here is the commission that he gave him (Isaiah 6:8–9). *You say to this people: Listen, you who listen and do not understand, and see this vision and do not recognize it* . . . [Latin omitted].

I am not surprised that God had trouble finding someone who wanted to carry out this commission. . . .

Isaiah 6:14 has the famous prophecy of a virgin who is to conceive and bear a son who will be called Imman'u-el. This passage being, with the seventy weeks of Daniel,[34] the clearest prophecy, of the messiah, it deserves a little examination.

Isaiah 7 should be read from the beginning to understand the literal meaning of this verse 14, which is one of the clearest of the prophets. Ahaz was besieged by the king of Israel and the king of Syria (7:1). The prophet Isaiah comes to comfort him on God's behalf and to give him hope (7:3–4). He says to him (7:11) to ask if he wishes for a sign from the Lord, be it in heaven, or on earth.

In Isaiah 7:12 Ahaz answers that *he will not ask for one, and that he does not want to test the Lord.* And then Isaiah says to him (7:14), *Oh well, the Lord will give you one himself; can you see this woman? She will conceive and bear a son who will be named Immanuel;* and verse 16, *and before this child is able to discern good from evil, your kingdom will be delivered from the two kings leagued against you.*

Isaiah 7:14: *Propter hoc Dominus dabit ipse vobis signum; ecce virgo concipiet et pâriet filium, et vocabitur nomen ejus Emanuel.* . . . In the Latin version there is *virgo,* which means *virgin;* but the Hebrew word *alma* that St. Jerome translated as

---

34. Du Châtelet also wrote at length on this passage in order to show the contradictions and scientific inaccuracies.

virgo (it is easy to understand why) means equally a young woman or a young girl. This can be proved by many texts of scripture that can be found in Dom Calmet. The woman of whom Isaiah speaks to Ahaz was his. This is proved in chapter 8 where the prophecy is fulfilled. For in 8:2 Isaiah says, *I took two faithful witnesses Uriah, high priest, and Zechari'ah, son of Jeberechi'ah.* And Isaiah 8:3, *I approached the prophetess and she conceived and bore a son: and the Lord said to me, Give him his name: make haste to take these spoils, for,* 8:4, *before this infant knows how to name his father and his mother, the force will have been taken from Damascus and the spoils from Sama'ria* . . . [Latin omitted].

And finally Isaiah adds, in speaking to the people, 8:18, *There are my children that the Lord gave me as a sign* . . . [Latin omitted].

How can such a clear meaning be twisted into a prophecy of Christ?

First, one must begin by changing the word *woman* into that of *virgin*, as St. Jerome did in the Vulgate: the name of *Imman'u-el* into that of *Jesus* for Jesus never bore the name of *Imman'u-el*, but how can one, even after doing this, find the fulfillment of this prophecy in *Jesus*? . . . Also, Father Calmet admits in his commentary that if one had only this single passage for convincing the Jews of the mission of JC it would be difficult to convert them. One has recourse to the mystical sense when one is forced to abandon the literal, and then anything one wishes may be seen in these verses. This is the same as with the clouds.

Examination of the Book of the Twelve Minor Prophets: Jonah

But surely here is the best of all.

[Du Châtelet tells the story of Jonah, his effort to flee the Lord, his three days and nights in the whale, and then his effort to fulfill God's order to prophesy in Nineveh.]

. . . It is difficult to understand what might have determined God to send Jonah to prophesy in a town that was plunged in idolatry and that lived in perfect ignorance of the true God and of his laws, which knew neither Jonah nor the one who sent him, and to answer to this that it was a whim of God's. But there is still another good answer, which is that as Jonah must have been a figure of JC, like him, he had to preach to the gentiles. In truth JC never left Judea. But that does not matter. . . .[35]

But where the interpreters shine is in the explanation of the fish that

---

35. Du Châtelet suggests this idea of a prefiguration of Jesus Christ. Calmet did not. Schwarzbach, ed., *Examens*, I, "Les Douzes Prophètes," 71 n. 34. (Pagination is from an earlier draft, footnote number is correct.)

swallowed Jonah. For assuredly there is nothing so difficult, as to save this adventure from absurdity. A man fully dressed thrown into the sea without being drowned, who finds in the nick of time a fish that had the courtesy to eat him, and this, so honorably as not to hurt him; who swallows him without suffocating him; who keeps him alive in his stomach all shod and clothed for three days and three nights; that this man breathes, eats, and lives at the bottom of the sea in this living prison; that he composes a canticle here, and that finally he escapes scot free, and why such a great, impossible miracle? . . . All these difficulties have not stopped the commentators who are, of all the species, the most intrepid. They say, first, that when one attacks a miracle from scripture, one should attack them all, and that it is no more unbelievable that Jonah was three days and three nights in the belly of a fish than it is that JC was resurrected. But some people would not be stopped by this response. They all agree that this fish is a whale, but unfortunately the biggest whales have extremely narrow gullets. And they also say that this miracle was necessary to give the world a telling prophecy of the resurrection of the savior. It is easy to feel that there is no response to that. Finally, it is said that what fables tell us is no longer believable, and that, it seems to me, is the best response.[36] See D. Calmet's dissertation on Jonah's fish.

EXAMINATION OF THE BOOKS OF THE NEW TESTAMENT

We have seen by the analysis that I just made of the books of the Old Testament what a sensible man must think of them. But the books that compose the New Testament, being the direct basis of the christian religion, surely it must be judged by what they contain. So I am going to examine them with exactness.

One of the things on which the defenders of the christian religion rely most to silence those that they call the incredulous is the authenticity of their books, and, yet, when one closely examines their origin, it is impossible to see the motives which caused some books to be chosen and declared to be *canonical*, as opposed to those they regarded as apocryphal.

It is above all with the gospels that it is difficult to clarify this point of criticism; for in the first centuries of the Church, there was an infinity of gospels composed under the names of all the apostles and all the disciples, and even Judas Iscariot had his own.

---

36. Du Châtelet takes all of these arguments from Dom Calmet. Schwarzbach, ed., *Examens*, I, "Les Douzes Prophètes," Jonah, 71 nn. 33–36. (Pagination is from an earlier draft, footnote numbers are correct.)

Father Calmet gives a list of thirty-nine of these apocryphal gospels, which are all cited as inspired in the writings of the Fathers, some of which are still extant. But one does not know what, among this number, determined the choice of the four that the Church accepts today.

F. Calmet gives some excerpts of the gospels that have been declared apocryphal. He reports, for example, that in the gospel of the Infancy, it is told that, when Jesus made little clay birds with other children, his drank, ate, and flew away; that when he worked with St. Joseph in his craft as a carpenter, he miraculously lengthened what was too short and shortened what was too long; that little Jesus having played with a child who died in a fall, and being accused of having made him fall, raised him from the dead out of fear of a whipping, and made the boy declare that he was not the cause of his fall. After which, the child died again right away.

In the gospel of the Birth, it is said that the midwife that Joseph went to fetch to attend Mary's lying-in examined her in order to know if she was a virgin, as [Mary] boasted, and found her to be a virgin, etc.

But one book that one must infinitely regret the loss of is the Acts of Pilate in which the motives for the condemnation of Jesus were exposed; but as for this one, it may be surmised that it was carefully suppressed.[37]

Having said a word about the apocryphal gospels, it is right to examine the canonical ones, not because they are more reasonable, but because they are more respected.

Examination of the Evangelist St. Matthew

CHAPTER 1.

. . . It is said in Matthew 1:18 that Mary found herself pregnant before Joseph and she had lived together; which with several other places in the gospel that tell of the brothers of JC[38] and with those where JC is called *the first born of Mary*, has led some to believe that Mary was not a virgin *parte post* [after] the incarnation, and that after having had Jesus by the Holy Spirit, she had other children by St. Joseph.

But a very strong objection that the Fathers themselves raise is why JC chose a married woman for his mother, since it is claimed that, according to the prophecies, the messiah must be born of a virgin, and that this virgin mother later became an article of faith, and one of the proofs of his divinity.

---

37. Du Châtelet uses sarcasm in the New Testament commentary even more aggressively than in that for the Old Testament.

38. As in the Old Testament, Du Châtelet or her copyist made a number of abbreviations in the text, all of which suggest disrespect, as here, JC for Jesus Christ.

Was this not, say the Fathers, to set a trap for us and to expose the mystery of the incarnation to calumny?

After raising this objection, the Fathers attempt to resolve it. Some say that God acted thus to save Mary's honor; others, that it was to give her and her son a guide for the flight into Egypt; others finally among whom the martyr St. Ignatius, St. Jerome, and St. Chrysostom[39] believe that it was in order to trap the Devil and to hide from him the incarnation of his son that God made him be born of a married woman. For the Devil, knowing that the messiah was to be born of a virgin, could not guess that the son of a married woman was the messiah. Thus, by means of this ruse Jesus was born, grew, and began his mission, without the Devil dreaming of crossing him. But I would ask St. Jerome and St. Chrysostom how the Jews could recognize JC for the messiah, since the Devil who is so sly could not divine it?

In Matthew 1:20, 23, Gabriel appears to Joseph to order him to keep Mary with him, whom he wanted to send away because of her pregnancy. Gabriel told him at the same time that he had the honor of being the cuckold of the Holy Spirit, and Gabriel announced to him that the child Mary would give birth to would be called Emman'u-el, a name that JC never used.

CHAPTER 2.

This chapter tells of the journey of the magi who came from the orient[40] to Jerusalem to look for the newly born king of the Jews, whom they came to adore because they had seen his star (Matthew 2:1–2).

It is a true pleasure to see the pains the interpreters take in order to know who these magi were, and where they were from, and how many they were, and what this star that appeared to them was.

It is certain that under the name of *magi* Orientals understood all sorts of diviners, priests, and philosophers. Now if they all came to Jerusalem, this

---

39. By "Fathers," Du Châtelet means the men designated by the Catholic Church as the "Church Fathers." St. Ignatius of Loyola (1491–1556) was the founder of the Jesuit order, a pillar of the Catholic Reformation. St. Jerome (340–420), an early Church Father, was the translator of the first Latin Vulgate. St. John Chrysostom (347–407) was an early Church Father, and, as Bishop of Constantinople, a prominent leader of the Greek Church. He is known for his commentaries on the gospels of Matthew and John, the Psalms, and Paul's Epistles. It is in this kind of discussion that Du Châtelet relies on Dom Calmet. She did not read these Catholic authors, but uses Calmet's descriptions and quotations, and then reacts to what they had written. Note that she does not usually accept the resolution that Calmet makes of contradictions and confusions.

40. Du Châtelet uses the contemporary terms *orient* and *occident* to denote *east* and *west* respectively.

would have made an army, and some must have been traveling a very long time. As scripture says only *the magi* in general, there is no more reason to name one than another: it has been said that there were three of them, apparently in honor of the Trinity, and henceforth it became customary to believe that there were three of them, doubtlessly in the same way they were made kings, in order to honor JC. . . .

But what was this star which they and they alone saw, since no historian mentions it? How did they divine that this star meant that the king of the Jews was born *and newly born?* Assuredly, only judiciary astrology can lead one to such a conclusion, and I dare anybody after this to mock this science. . . .[41]

The magi, after speaking to Herod in Jerusalem, saw their star again, which conducted them to Bethlehem, and only they saw it.

Although the magi spoke to Herod in Jerusalem, it is proved by Flavius Josephus[42] and by St. Luke (*nota* that St. Luke says nothing of it) that Herod was then in Jericho; but it is proved by reason that this whole history is absurd. For, to say nothing of the star, how likely is it that the magi, if they were not madmen, would come to say publicly in Jerusalem, where Herod reigned, that they came to find and adore *the king of the Jews,* and that they said this to Herod himself? Did they not sense such a statement must at the least cause their being locked up in an asylum? Thus, Herod, says the evangelist (Matthew 2:3), was troubled; yet, he was trustful on a point that concerned his interest (for whether true or false, the discourse of the magi could make an impression on the minds of a people naturally prone to credulity and fanaticism). Herod, I say, trusted these magi and contented himself with inviting them, once they had seen what was going on, to come and give him the news.

But in Matthew 2:16 this same Herod, who was so careless, is seen to take suspicion and caution to an incredibly barbaric excess; for it is told that he had killed in the city of Bethlehem and its environs all the male children two years old and younger. And it is notable that St. Matthew is the only evangelist who tells of this event, and that no contemporary writer, not [Flavius] Josephus, this historian who was so exact and such a critic of Herod,

41. "Judiciary astrology" would mean astrology used in a court. Du Châtelet is mocking the idea of this as a valid authority, or a science.

42. This is the same historian cited as an authority in her critique of the Old Testament, see note 15 above.

says a word of this hideous carnage, which must, more than anything else, have made Herod's memory execrable.

But can one believe, unless one has lost one's mind, that a king could give such an order, and that he could find people to execute it, and that on the report of some madmen who say that they saw a star that told them the king of the Jews was newly born? This story has all the signs of a fable that, nonetheless, has been made a beautiful festival in the Church.

The magi received in a dream a warning from heaven and went away without speaking to Herod and without Herod dreaming to look for them or to follow them; but they did not go without having opened their treasures and without having offered Jesus gold, incense, and myrrh. These presents are ridiculous and full of mysteries, for they represent the divinity of JC, his kingship, and his humanity. But is it not ridiculous to give incense and myrrh to a poor infant who only needed swaddling clothes and pabulum?[43] It is true that they will give him a bit of gold; but they must have given him very little, for there is no indication that from that time his parents or he were comfortably off. One sees, on the contrary, that Mary at the time of her purification is seen to offer only two doves, like the poorest.

Besides, it seems that this journey of the magi only served to have the innocents killed. For it does not seem to have been very useful to Jesus, who was nonetheless ignored for thirty years.

Joseph too received a warning in a dream: to flee to Egypt with the divine child in order to save him from Herod's fury and from the massacre of the innocents. What was the point of being the son of God if one had to flee to Egypt for fear of being killed? And what also pleases me is the abbé d'Houteville's joke about the hejira, or flight, of Mohammed. *It was pleasing*, says he, *that the favorite of God was obliged to flee*, but if this is what becomes of his favorite, what can be said of what happened to his son?[44]

It may be seen by this story, supposing it is true, that this son of God, who had come into the world to save all men, begins by costing the lives of the14,000 small children Herod had murdered; it must be confessed that such a savior more resembles a curse. . . .

[Du Châtelet considers the prophesies from Micah 5:2, Hosea 11:1, Jer-

---

43. Although a version of this objection can be found in Thomas Woolston's writings, Du Châtelet added the "swaddling clothes" and "pabulum," from her own experiences with her children.

44. The joke is from Claude-François-Alexandre Houteville's (1686–1742) *La vérité de la religion chrétienne prouvée par les faits* [The Christian Religion Proved by Facts] (1722), which earned him election to the French Academy. Du Châtelet and Voltaire probably read it together as part of their biblical studies at Cirey.

emiah 31:15 mentioned in St. Matthew's gospel and declares them untrue or doubtful. She concludes with a statement on the fourth prophesy in the gospel (Matthew 2:23) that *"He will be called a Nazarene"*].

But what is surprising is that this prophesy, *He will be called a Nazarene,* which caused Joseph to travel expressly to Nazareth to fulfill it, is not found in any prophet, as all the commentators have to admit.

This last citation provides a reflection applicable to all the passages cited in the gospel; this is that it seems Jesus directs all his actions to have it believed that he was predicted; thus, he flees to Egypt, goes to Nazareth. When he does such and such a thing; it is always in order to fulfill what is written in such and such a prophet, and the evangelist never fails to remark on it. But it may be seen from this last citation that zeal sometimes carries them away to cite prophecies that are not to be found anywhere, in order to have it believed that all his actions have been predicted . . . .

CHAPTER 4.

. . . Matthew 4:18 tells of the vocation of Simon and Andrew, who were fishermen and to whom Jesus promised he would make them *fishers of men.* Apparently they believed that this fish was worth more than the other, for they left their nets without hesitation. Jesus took two more fishermen named James and John into his service; but besides their nets, he made them abandon their father (Matthew 4:21–22), who was in their care, and it was not a very good action to have them abandon him . . .

CHAPTER 5.

The sermon on the mount spans three chapters (Matthew 5–7). This sermon contains a fairly pure morality, but very trite, and mixed, from time to time, with fairly ridiculous things.

It is in this sermon that Jesus sets forth the eight beatitudes; the first is not very tempting, for it is to be *poor in spirit;*[45] it is true that Jesus (Matthew 5:3) affirms that the Kingdom of Heaven is for them. This gives a poor idea of the Kingdom of Heaven; it is quite likely to make one not want to go there.

I place more value on his second beatitude, which is to be meek; yet, their reward is much less than that of the poor in spirit, for those who are meek only inherit the earth. Thus, according to the gospel, it is much better to be silly than to be meek.

---

45. Du Châtelet uses the phrase "pauvre d'esprit," meaning poor in intelligence. *Esprit* can have many meanings in French, of which *spirit* is only the most literal translation. For example, it is also synonymous with *mind* and *wit.*

Jesus (Matthew 5:13) says to his apostles that *they are the salt of the earth*, and he adds that *if the salt loses its force one would no longer know with what it will be salted* . . . [Latin omitted].[46] But it seems that the earth is never salted; however, a comparison must always have a true thing as its basis . . . .

In Matthew 5:22 Jesus says that *whoever becomes angry with his brother shall be in danger of judgment*, which means, *death*. For he pronounced the same punishment in the preceding verse against those who will kill. Now, this verse, 5:22, is very unjust. For often your brother is very wrong and deserves your anger; and if mere anger is punished with death, with what will one punish the effects of anger? . . .

In Matthew 5:39 Jesus orders his disciples to offer the other cheek if somebody smacks them on their face. However, D. Calmet very judiciously notes that when the soldiers of Pilate smacked Jesus during his passion, he did not turn the other cheek. Thus, christians are excused by his example from following this precept.

Again Jesus orders his disciples that if someone institutes a suit in order to obtain their robe, they should give him their robe and their cloak. It is said that all this should not be executed to the letter, and I believe that one need not trouble oneself saying it, but why then did Jesus order it? . . .

CHAPTER 7.

. . . In Matthew 7:29 the evangelist says, speaking of the sermon on the mount, which ends with this verse, that Jesus instructed the peoples on his own authority, and not as the scribes and the Pharisees did; the interpreters are very perplexed over finding the sense of this verse, for one could scarcely have less authority than the son of a carpenter followed by four fishermen.

CHAPTER 8.

In Matthew 8:3–4 Jesus cures a leper and forbids him to speak of it. It is pleasing to come to convert the world and to hide one's miracles.

Jesus touches this leper in order to cure him, which seems unnecessary for the miracle, and was against the law, which, as is clear from his whole life, he prided himself on observing. . . .

The evangelist says, Matthew 8:17, that Jesus cured the sick so that this word of Isaiah was fulfilled, *He took on himself our infirmities*. But it is not the case that JC was sick in the place of those he cured, as the capuchin who took on himself the pains of women in childbirth. This is what this prophecy of the passion of Jesus is commonly understood to mean. But St. Matthew, who ap-

---

46. Du Châtelet then gives the Latin. As in the Old Testament, deletions have been made for purposes of brevity and continuity.

plies it to the curing of the sick, did not understand it in that way, and I do not believe that the interpreters flatter themselves to understand scripture better than the Holy Spirit who dictated it. . . .

CHAPTER 9.

Matthew 9:9 tells of the conversion of St. Matthew, who very recklessly leaves his counting house[47] in order to follow a man who must have looked like a madman. One can understand that poor fishermen who had nothing to lose should do that; but for Matthew this seems to me completely reckless. . . .

CHAPTER 10.

. . . In Matthew 10:10 Jesus forbids his disciples when they travel to take with them a bag, tunics, shoes, or a walking stick. It must be admitted that he did not like taking precautions and it was a little harsh to have to go barefoot; so, St. Mark states that he allows them sandals, which would at least be some kind of comfort. But unfortunately, the words that St. Jerome translated in St. Matthew as *shoes* and St. Mark as *sandals*, in order to save it from contradiction, can be said to signify the same in Hebrew. But St. Luke places all in doubt on this; for, in Luke 22:25 [35] he says that Jesus forbade his disciples to wear *shoes*. This leaves no room for ambiguity; thus, according to St. Matthew and St. Luke, Jesus forbade his disciples to wear shoes when they travel, and he allowed them to do so according to St. Mark (Mark 6:9).[48]

In Matthew 10:34–35 Jesus gives a very revolting idea of his mission. *Do not believe*, he says, *that I am come to bring peace on earth; I am not come to bring peace but the sword: for I am come to separate the sons from the father, the mother from the daughter, and the daughter-in-law from her mother-in-law* . . . [Latin omitted]. And unfortunately this is all too true. God might just as well not have been born if he was going to bring so much harm to men. Yet, Jesus is presented to us as a prince of peace who reconciles man to God, and only inspires his disciples to peace, meekness, and harmony; who wants them to turn the other cheek to him who gave them a smack on the face and to give their tunics to those who only ask for their cloaks. All this is scarcely consistent, but that is not my fault. . . .

47. Matthew was a tax collector.

48. This is a perfect example of the way in which Du Châtelet mocks the text by showing contradictions among the gospels even on such a minor point as the prohibitions about shoes and sandals. Note how she uses the term *ambiguity*. Schwarzbach, in his edition of the *Examinations of the Bible*, quotes Calmet's discussion of this, which Du Châtelet follows here, but with a different conclusion. See Schwarzbach, ed., *Examens*, II, "Matthieu," 11, n. 110. (Pagination is from an early draft, footnote number is correct.)

CHAPTER 13.

In this chapter the evangelist tells how Jesus sat on the seashore and that a great crowd gathered around him to listen to him, and that he told some parables that no one understood, which led his disciples to represent to him (Matthew 13:10) that it was a mockery to say such things to the people, *Et accedentes discipuli dixerunt ei, quare in parabolam loqueris eis?* But Jesus gives them a very good reason. *I act thus*, he says to them (Matthew 13:11–13), *because it has been given to you to know the mysteries of the Kingdom of Heaven, but, this has been refused to them; for*, he adds, *more will be given to those who have; but as for those who have nothing, even what they have will be taken from them.* Surely, this is not difficult. *This is why*, continues Jesus, *I speak to them in parables, because, by this means, in seeing they do not see, and in listening they do not understand* . . . [Latin omitted].

It is difficult to understand after this why Jesus was given to wasting his breath speaking so long to people who could not understand if he did not give the reason in Matthew 13:14, where he says that *it is in order to fulfill a prophecy of Isaiah* . . . [Latin omitted]. But he gives still another reason in verse 15, *I speak to them obscurely*, says he, *for fear that if they understood me, they might convert and that I would be obliged to heal them: Incrassatum est enim cor populi hujus, et auribus graviter audierunt et oculos suos clauserunt, ne quando videant oculis et auribus audiant, et corde intelligant et convertantur, et sanem eos.*

It must be admitted that those poor Jews were much to be pitied. Jesus, in these parables that he told expressly in order that he would not be understood, compares the Kingdom of Heaven as such: to the seed thrown in different soils, a field full of good grain and tares;[49] a grain of mustard; and the yeast that a woman puts in the dough. Is all this not very much like the Kingdom of Heaven, and had not Jesus a fine imagination?

When this poor multitude to whom Jesus was saying all this had withdrawn, all edified and all silly, Jesus explains to his disciples all those parables. He says to them (Matthew 13:23) that the seed that is thrown in good earth represents those who listen to his words and that yields one hundred or sixty or thirty for one.

St. Augustine, in explaining this parable, says that the martyrs produce one hundred for one, virgins sixty, and married persons thirty.[50]

The comparison of the grain of mustard that Jesus also made in this chapter to the Kingdom of Heaven, rests on something false, for it is not true that a grain of mustard is the smallest of all the grains. The grain of the

---

49. "Ivraie" can be translated as *tares*, as in the St. James Bible, or by the more modern term, *weeds*.

50. St. Augustine, the Church Father (354–430), formulated the doctrine of Original Sin, including the relative merits of one group over another for redemption.

poppy and that of the fir tree, for example, are much smaller than that of mustard; and spruces are bigger than the plant that produces a grain of mustard. This parable is not one of those that Jesus explains to his apostles, and assuredly, this is a pity. . . .

CHAPTER 14.

Herod, King of the country where Jesus raised the dead and healed all the sick, had not yet heard of him according to the first verse of the gospel, and when he heard of him, he took him for John Bap whose head he had had cut off.[51]

The manner in which the death of this precursor of Jesus was reported in this chapter surely is scarcely probable.

Matthew 14:6–8, 11, claims that Herod had somebody bring a platter to Salome, his daughter, with the head of J Bap that she asked of him as a reward for her having danced well. But is it very likely that a little girl would be given the head of a man on a platter as a reward for dancing well? . . .

It is noted in the same verse, Matthew 14:13, that five thousand men followed Jesus, not counting the women and children; this was not the best way of staying *incognito*. . . .

Matthew 14:15: When it was late, his disciples, who were much more prudent than he, warned him that it was time to send away this multitude who listened, so that they could go and eat, and Jesus answers them (Matthew 14:16–17) that they should feed the people. To which his disciples replied that this was difficult, since they only had with them five loaves and two fish; but Jesus (Matthew 14:19–20) takes these five loaves and these two fish and shares them among this multitude—they, their wives, and their children—whose hunger was satisfied with it, and twelve baskets full of bread remained. The apostles had apparently brought the baskets expressly for the miracle.

One may be surprised that no fish remained; some interpreters say that is because there was less of it than of bread, before the multiplication. As for me, I believe that they had no more baskets to collect it. . . .

When the disciples were at sea, a storm arose and the boat was about to be lost. Then Jesus came to their aid by walking on the water (Matthew 14:25). But unfortunately this was at night, thus, no one noticed anything except for the apostles. Jesus would have done better to make this miracle in broad daylight before the Pharisees, in place of speaking to them of Jonah when they asked a marvel of him. But I always forget, when I reason thus, that the intention of God was not that they might convert. . . .

51. "John Bap," "Jean Bap" in the French, is Du Châtelet's abbreviation for John the Baptist.

CHAPTER 16.

. . . In Matthew 16:13 Jesus asks his disciples what was being said of him in the world. Apparently he did not know. They answered him (Matthew 16:14) that some said he was John Bap, others Elijah, others Jeremiah; but I believe they did not tell him all. Jesus asked them next (Matthew 16:15) what they themselves thought of him, and Simon who later was called Peter, answered in the name of all, that *they believed that he was the christ, son of God.* And Jesus, to reward Simon for having answered so well, says to him (Matthew 16:18), *You are Peter and on this rock I will build my Church* . . . [Latin omitted].[52]

Now up to then Peter was called Simon; but Jesus changed his name, expressly, I believe, in order to make this neat point (John 1:42). Some even say that in the Hebraic language that Jesus spoke, this play on words is even better: the word *peter* and the name Peter are the same gender. . . .[53]

CHAPTER 19.

. . . In Matthew 19:16–22 a young man comes to ask Jesus what he must do to enter the Kingdom of Heaven; and Jesus tells him that he must keep the commandments of God, which he enumerates for him, without speaking of the love of God. This young man having answered him that he had kept all these commandments from his youth, Jesus told him that he must sell all that he had and give it to the poor, and this young man left him very dissatisfied, in the firm resolve to do nothing. When Jesus saw that this young man did not want to do what he had ordered, he began to inveigh against the rich and say (Matthew 19:24) that it is much easier for a camel to pass through the eye of a needle than for a rich man to enter the Kingdom of Heaven. The successors of the apostles do not appear very concerned about entering the Kingdom of Heaven. But Jesus, in passing this sentence, sentenced half the world. For there he gave a precept on poverty that all the world cannot follow; for it is very necessary that riches be in the hands of someone, and it is equally impossible for all the world to be poor as for all the world to be rich. . . .[54]

---

52. In Greek, *petros* means *stone* or *rock*. Du Châtelet has noted an additional play on words.

53. According to Schwarzbach, Du Châtelet makes the decision to have Jesus speak "hebraic"; Calmet calls it Syriac. Schwarzbach, ed., *Examens*, II, "Matthieu," 16, n. 171. (Pagination is from an early draft, footnote number is correct.)

54. This passage reflects Du Châtelet's views of society, and echoes her comments on Bernard Mandeville's *The Fable of the Bees*, the text she and Voltaire were also reading when they began their biblical critique.

CHAPTER 21.

In this chapter Jesus sends his disciples to find a she-ass and its colt, he tells them (Matthew 21:2–3) that if someone asks them why they are taking away this colt and its mother, they only need answer: *The Lord needs it*. But I very much doubt, despite all that Jesus can say, that in his lifetime he was called *the Lord*.

In Matthew 21:4 Jesus explains the reason he sent them to seek this she-ass *and* its colt; as he intended to ride the ass, the colt seems rather useless. But one thinks very differently when one knows this colt was necessary to fulfill a prophecy of Zechariah (Zechariah 9:9) where this prophet says to a daughter of Sion *to rejoice because her king comes to her mounted on an ass and on her colt* . . . [Latin omitted].

In this fine equipage followed by a dozen ragamuffins who threw tree branches in his path, Jesus enters Jerusalem, everybody pointing at him (Matthew 21:10) and saying to one another, *Quis est hic? Who is that?* And they were told (Matthew 21:11), *It is Jesus of Nazareth*, which apparently, was to say it all.

When Jesus had entered Jerusalem he went to the temple and began to chase away the merchants (Matthew 21:12) who were in the forecourt; he then began to knock over their goods and to cite prophets to them. But those prophets he cites so opportunely, he cites wrong; for the passage he cites (Matthew 21:13), *My house is called a house of prayer and you make it a cave of thieves* . . . [Latin omitted]; this passage, I say, is not to be found in any prophet.[55]

Surely there is nothing so surprising as this action of Jesus, to see a lone man without authority, a man scorned, chasing on his own this crowd of merchants who were in the exterior galleries of the temple. So, St. Jerome says that of all the miracles of Jesus this one appears to him to be the greatest. I believe that this means the most absurd. But there is no sign that this practice that Jesus wanted to reform was contrary to the law of Moses; for if it had been, the Pharisees and the scribes were too rigid observers of the law to allow it.

But it can be asked why Jesus was so zealous about a house he had come to destroy?[56] One cannot answer this except by saying that this was not the

---

55. According to Schwarzbach, Calmet explains it by merging Isaiah 56:1 and Jeremiah 7:11. See Schwarzbach, ed., *Examens*, II, "Matthieu," 21 n. 219. (Pagination is from an early draft, footnote number is correct.)

56. Much of what follows was taken from Du Châtelet's summaries of Woolston.

intention of Jesus and that he was a very good Jew all his life; but what are we then, we, who claimed to be his disciples and who burn the people who follow this law he observed all his life. . . .

Jesus left Jerusalem in the evening and went to sleep in Bethany (Matthew 21:17), but on the next morning while returning from Bethany to Jerusalem, he was hungry. (Apparently, he had forgotten to have breakfast before leaving.) *He saw a fig tree on the way and he approached it to take some figs. But, not having found any, he cursed the fig tree which dried up in an instant* (Matthew 21:19) . . . [Latin omitted]. St. Mark, in telling this same miracle, adds a remarkable circumstance, that it was not the time for figs (Mark 11:13), *Non enim erat tempus ficorum.*

This miracle is so ridiculous and so revolting that several Fathers believed that this story is only an allegory and that it did not happen. Origen says it clearly,[57] and St. Augustine confesses that if Jesus did this action, he did a crazy thing; but if this miracle is taken as real, which one can hardly help doing if one follows the text, there are three terribly puzzling things.

First, the absurdity of this miracle, second, its uselessness, and third, its injustice.

There is surely nothing more absurd than to go looking for figs on a fig tree when it is not the season for figs. And there is nothing more unjust than to curse it for this reason.

But this curse of the fig tree includes yet another absurdity; for a tree being an insensible thing is susceptible neither to cursing nor blessing. . . .

As regards the uselessness of the miracle, it is obvious: for Jesus did not for all that have anything to eat. Miracle for miracle, it would have been better to make this fig tree bear figs that very instant. For, at the least, Jesus would have had his wish. . . . [58]

CHAPTER 27.

. . . From Matthew 27:27–31 exclusively, we can read a repetition of the insults Pilate's soldiers hurled at JC, which are a clear indication that his miracles had not made much noise and that he was not much feared.

Jesus was so harassed by them that he had not the strength to carry his cross, and a Cyrene who happened to be there by accident (Matthew 27:32)

---

57. Origen of Alexandria (185–254) was an early Christian theologian and philosopher, known for his refutations of sects considered heretical, such as Gnosticism.

58. Du Châtelet, in her summaries of Woolston, wrote pages on this miracle. She goes on at length here also, drawing on Woolston and on Calmet, citing the versions in the different gospels, and condemning Jesus's action.

was forced to carry it with him. It would have been an ideal task for a legion of angels. When Jesus arrived at the mountain of the calvary and was crucified, he was thirsty; and they gave him a sponge at the end of a stick, wine mixed with gall, a drink usually given to criminals on their crosses in order to benumb them, and so that they might suffer less. Jesus (Matthew 27:34) tasted it; for he did not know what this was; but he did not want to drink it. . . .

It is said in Matthew 27:45 that Jesus was crucified at the sixth hour; and St. Mark says (Mark 15:25) that it was the third hour when he was put on the cross.

St. Matthew adds in this (Matthew 27:45) that from the sixth to the ninth hour all the earth was in darkness; yet no historian, not even [Flavius] Josephus, speaks of this darkness that cannot have been caused by an eclipse because there can never be one when the moon is full; and it is known by the time of Passover that the moon was full. D. Calmet believes this darkness was caused by a big spot on the sun, and this is as good a thing to believe as any other.

In Matthew 27:46, at the ninth hour Jesus began to cry out; and it must be admitted that it would take rather less for one to cry out; and after this cry he expired. . . .

The centurion who guarded the crucified and who saw all that happened when Jesus died, concluded from it that he really was the son of God. This again proves that the Jews did not understand anything by this name other than the *favorite of God;* for, assuredly, it is not natural to imagine that the son of God might die.

CHAPTER 28.

This chapter tells the story of the resurrection of Jesus. It is certain, according to the four evangelists, that Jesus was only one day and two nights in the tomb, and yet he had said very precisely that *he would be there three days and three nights, as Jonah had been three days and three nights in the belly of the whale.* He had said, *Destroy this temple and I will rebuild it in three days;* and the evangelist adds that *he intended this to mean the temple of his body.* And one sees in Matthew 27:63 that the princes of the priests say to Pilate that *Jesus had said during his life that he would be resurrected three days after his death; Ille dixit ad huc vivens: post tres dies resurgam.* Thus, following his own prediction, he should have to be three days and three nights in the tomb; and the interpreters have made the greatest efforts in order to find the three days and three nights in the one day and two nights the evangelists say that he was in the sepulcher. . . . It seems that the disciples of Jesus had been up to a trick and wanted to win the game point when it came, and that they preferred him to be thought to be resurrected earlier than he had said, rather than not to be resurrected at all.[59]

The holy women, for Jesus always had some in his retinue, were the first informed that he was resurrected; they came very early on the morning of the first day of the week to visit the sepulcher.

The evangelist tells that when Jesus rose, there was a great earthquake, and that an angel descended from heaven and removed the stone that closed the sepulcher; apparently without this aid, Jesus would not have been able to go out of it.

This angel then sat on the stone. Its face was resplendent and its clothes white like the snow; and it said to the women who came to visit the sepulcher that they should not be afraid and that Jesus, whom they sought, was resurrected.

These women did not witness the earthquake mentioned in Matthew 28:2, and they only arrived after the angel had removed the stone from the sepulcher and Jesus was resurrected, and according to the evangelists no witness to the moment of resurrection is known. How and from whom did they learn then how this resurrection had happened? That there had been an earthquake and that it was the angel who overturned the stone that closed the entrance to the sepulcher? For, nowhere is it said where Jesus or the angel told this to anyone; and this alone must make the rest of the account very suspect.

If the evangelists had not agreed to compose a story, and had simply told what they had seen and heard, they would not have spoken of this earthquake, or of this stone removed by the angel. They would have been content to say that he rose without saying how, since they could not have known it.

This stone moved from the entrance of the sepulcher is another great mark of duplicity. For Jesus, who passed through walls after his resurrection—at least this is what the Fathers claim about what St. John tells of this (John 21:[20]26), that he entered into the room where the apostles were, *When the doors were closed. Clausis januis*—would have been able to leave his

59. This idea of the "trick" comes from a seventeenth-century clandestine text called *Traité des trois imposteurs* [Treatise on the Three Impostors] that Schwarzbach indicates was published circa 1700. It charged Moses, Jesus, and Mohammed as impostors only out for political power. Schwarzbach notes that Calmet mentions the text in his notes, Schwarzbach, ed., *Examens*, II, "Matthieu," 26 n. 290. (Pagination is from an early draft, footnote number is correct.) An abridged version of a 1719 edition is included in Margaret C. Jacob, ed., *The Enlightenment: A Brief History with Documents* (New York: Bedford/St. Martin's, 2001), 94–114.

tomb without this stone being removed, as it had been sealed by the princes of the priests (Matthew 27:66); if he had left the sepulcher without breaking the seals, this would have made his resurrection undeniable. . . .[60]

The princes of the priests were to have returned on the third day to see him resurrected, that is to say, to finish proving to the people that he was an impostor, and it was on the third day before all the princes of the priests, and before the people, on which Jesus should have risen. This would have been glorious for him and useful for the others. But judging by the way the evangelists tell that he rose, it seems that he wanted his resurrection to be taken as a skillful trick by his disciples. Indeed, they may well have won over the guards or made them drunk. It may even be the case that Pilate, who gained political advantage from the divisions among the Jews, favored this fraud. What is sure is that none of his guards saw him rise from the dead; no one bore witness to it. They were probably sleeping when his disciples stole his body, and this is all the more probable as the time of Passover was a time Roman soldiers who were in Judea took advantage of to become drunk. Besides, these guards, who were only there against fraud and not against violence, were probably few in number and consequently easy to win over or to make drunk.

Finally, an argument against this resurrection, to which there is no answer, is that if Jesus was resurrected, he should have shown himself to Pilate, to the princes of the priests, to all the people. For without this he lost the benefit of the resurrection, and put himself in the same position as so many impostors. If he only had as a reason for doing this to furnish a ground for doubt to all those who would read his story, it is certain that he should have given his resurrection as much authenticity as had been given to his passion. Now, since he did not do that, one has just cause for concluding that he was not resurrected.

The witness of the apostles was not admissible; for the example of Lazarus showed what he was able to do. . . .[61]

The only apparition of Jesus that Matthew tells of happened on a mountain of Galilee where JC had sent his apostles (Matthew 28:16), . . . and he adds, that when they saw him, they worshiped him, although some of them doubted that it was he (Matthew 28:17). Now, if those who saw him doubted, one may speculate about what those who did not see him thought.

60. Du Châtelet then describes the stone and the seals as precautions taken by the princes of the priests.

61. In her critique of chapter 11 of the Gospel of St. John, Du Châtelet describes this raising from the dead as a hoax arranged by Jesus with his friend, Lazarus.

It is to be noticed about this resurrection that Joseph of Aramathea, who is called a disciple of Jesus in Matthew 27:57 and who was probably in the conspiracy of the resurrection, had plenty of time to bury Jesus as he wished, and to use any trickery he wanted. For Matthew 27:62 states that it was only on the day after the death of Jesus that the princes of the priests decided to place some guards at his sepulcher. And it must be admitted that it is pleasing that the son of God should come into this world to be crucified publicly, and that he should rise incognito in order to save all men by the faith that they would have in this resurrection. According to the evangelists, nothing was more public and certain than his martyrdom; all Jerusalem witnessed it, but nothing is more doubtful than his resurrection; for no one saw it other than those who had an interest in seeing it, and who do not even agree in their account of the way or on the place where they saw it.

Resurrections present a very simple dilemma. Either they have happened in public and it is impossible that all the world does not believe in them, or they have happened in secret, and then no one is obliged to believe in them, which all at once determines the faith one must add to give credence to that of JC.

Examination of the Gospel of St. Luke

The same uncertainty prevails about the identity of St. Luke as about St. Mark; Luke like Mark is only a kind of abridger of Matthew.

CHAPTER 1.

His first chapter opens in a curious way; the Holy Spirit, for it is he who is the real author of St. Luke's gospel, dedicates his book to somebody named *Theophilus,* and he says to him that *several having undertaken to write what had happened, he took a fancy to write too.* For that matter, nobody knows who this Theophilus is whom the Holy Spirit condescended to honor in this way.[62]

In verse 32 St. Luke says that the angel Gabriel said to Mary that the son she would conceive *would be called son of the Most High: Et filius altissimi vocabitur;* this means that that is not who he really was, unlike what the christians believe, but that he was so only in name. Father Calmet, in order to save this,[63] has recourse to Hebrew; but he forgets then that it is uncertain

---

62. Neither Du Châtelet nor her contemporaries knew of this individual. Current biblical scholarship identifies him merely as "an unknown Christian." See *The New Oxford Annotated Bible,* ed. Bruce M. Metzger and Roland E. Murphy (New York: Oxford University Press, 1977), 1240 n. 1.3.

63. To make the passage conform to Christian belief, Calmet translates the Latin to mean, "he will be the Son of the Most High." See Schwarzbach, ed., *Examens,* "Luc," 30 n. 4. (Pagination is from an early draft, the footnote number is correct.)

whether St. Luke wrote in Latin or Greek, but it is very certain he did not write in Hebrew. . . .

There are great disputes among the Fathers about whether the incarnation happened little by little, or if the divinity of the word united hypostatically all at once, to the soul of Jesus and to his body in the womb of Mary.[64]

St. Luke is the only one who has given us the story of the incarnation. The three other evangelists say nothing of it. It is claimed that Luke was a relative of Mary, and that with this story he wanted to try to save her honor. . . .

CHAPTER 2.

Luke 2:1 speaks of a census, . . . and because of this census Joseph goes to Bethlehem from Judea with his very pregnant wife because he was of the tribe of Judea (Luke 2:4–5). But, surely, it would have been a pretty commotion if it had been necessary for everybody to leave his home to come to be registered in a certain place. So all the commentators say this was a whim of Joseph's that was suggested to him by the HS.[65] in order to fulfill the prophecy of Micah, which I discussed in St. Matthew about Bethlehem in Judea. . . .

One thing is sure: if it was the HS who made Joseph undertake this trip, this was very bad advice, for Mary very nearly gave birth in the middle of the street and was only too happy to find a manger for her lying-in.

It must be confessed that it is very unfortunate to come to one's tribe and to find no one, among a very hospitable people, who is prepared to give you shelter.

Further, no evangelist mentions the ox and the ass that are usually supposed to have been company for Jesus and his mother in the manger. That is an idea of some Fathers to show the fulfillment of a prophesy of Habakkuk that says (Habakkuk 3:2), *The Lord appeared in the midst of two animals*, and of the place in Isaiah where this prophet says (Isaiah 1:3), *The ox recognized his master, and the ass the stable of him who nourished him*; but, unfortunately, the evangelists did not notice this. For it would not have cost them very much to put an ox and an ass in a manger.

St. Luke says (Luke 2:7) that Mary was delivered in this manger *of her*

---

64. *Hypostatically* refers to the Trinity of three essences in one Godhead; here it is applied to the question of how Jesus's divine and human natures were created: all at once, or separately and then joined.

65. This is Du Châtelet's or her copyist's abbreviation for the Holy Spirit.

firstborn son; Et peperit filium suum progenitum; which supports the opinion of those who believe that Mary had other children than Jesus. . . .

All the accounts of St. Luke report equal tenderness and attentions from Joseph and from Mary for Jesus, which demonstrates that this Joseph was the best husband in the world. Mary says to Jesus in speaking of Joseph (Luke 2:48), *Your father and I have been looking for you anxiously;* the Holy Virgin lied then; for she very well knew that Joseph was not the father of Jesus. . . .

CHAPTER 15.

In Luke 15:7 Jesus says there is more joy in heaven for a sinner who is penitent than for the righteous who have no need of penitence. The pleasure of triumph over the devil must be very lively for God, and assuredly there is pleasure in having been a sinner when one enters paradise.

The rest of this chapter is taken up by the parable of the prodigal son that I find charming as a comedy.[66]

CHAPTER 24.

In Luke 24:15 Jesus appears after his resurrection to two of his disciples who do not know him, neither when seeing him nor when speaking to him. In Luke 24:28 Jesus, talking to his disciples, pretends that he wants to leave: . . . [Latin omitted] all in order that they might ask him to have supper with them, which turned out well for him.

It was very essential to him that they should ask him to have supper; for they were to recognize him only by his manner of breaking bread (Luke 24:30–31), which, however, is not difficult to imitate. As soon as his disciples had recognized him by the breaking of bread, he went nobody knew where, nor how he left, nor why his disciples let him go so easily; but since they had known it was he, they must have been curious to talk to him.

In Luke 24:36 Jesus appears to the apostles and to the two disciples who had already seen him, and, as his apostles did not want to believe that it was he (Luke 24:37), and as they believed that he was a spirit, he suggested that they touch him and said to them that spirits have neither flesh nor bone, which is not true. For the angels who appeared to the patriarchs had real bodies (Genesis 18:4–5), and Abraham washed the feet of three angels who came to announce to him that Sarah would have a son. Moreover, if the senses of sight and hearing deceived the apostles, why could the sense

---

66. Du Châtelet is referring to Voltaire's comedy, "L'enfant prodigue [The Prodigal Son]," and by implication means that the parable is comic as well. Voltaire wrote the play while they were at Cirey and reading critiques of the Bible.

of touch not deceive them as well? Finally, Jesus, in order to convince them more and more, eats in front of them (Luke 24:41–43), which was still not convincing. For all the angels of the Old Testament ate with those whom they came to visit. But it is amusing that the apostles should find it so simple to see a spirit when seeing Jesus; and I believe that it is very easy to make people who see spirits believe that a man rose from the dead.

Examination of the Gospel of St. John

The style in which this gospel is written is very different from that of the other gospels; and the statements that Jesus makes here are in an altogether different tone. Here these statements are full of repetitions and full of reasonings whose connections are often elusive, whereas in the other gospels his manner of speaking is sententious, and quite clear. Also, the authenticity of this gospel has been much challenged, as well as the authorship of St. John. But supposing that this apostle had written the gospel that bears his name, it is certain and not disputed that he wrote it when he was over ninety-five and consequently when most of the witnesses of things that he recounts were no longer alive. . . .

CHAPTER 1.

This chapter begins with Platonic gibberish on the *word*, which I strongly doubt anyone understands, and which is, however, one of the strongest proofs of the divinity of JC, for JC is that word.

CHAPTER 2.

In this chapter is told the story of the wedding at Cana where Jesus changed water into wine. He began by speaking very harshly to his mother, who had merely told him that the wine had run out (John 2:3), and Jesus answers, *Woman, what is there in common between you and me?* Father Calmet confesses that *Whatever turn one gives to these words, their harshness is undeniable.*

All that is told in the gospels proves that Jesus was a very bad son; for it may be seen that he again denied his mother when she came with his brothers to take him away. Finally, he never called Mary his *mother*, but *woman*. To excuse the harsh manner in which Jesus speaks to his mother, here it is said that he was a little drunk; this is not a good excuse.

But several people have wanted to infer from this name of *woman* that Jesus always gives to his mother, that she was not a virgin; for Jesus must have known that she was one. Thus, he should not have misled the others by calling her *woman*.

But one has trouble excusing the miracle that Jesus then made, for the evangelist suggests (John 2:10) that the guests were half drunk. At the least it is certain that it was at the end of the meal. Now not only is there no necessity for changing water into wine for people who had already drunk too much, but it is even indecent.

Moreover, why have water put into these jugs in order to turn it into wine? Jesus, since he wanted to do it, should have done it without water; this would have removed all pretext of doubt from this miracle, and he would only be reproached for its inutility and his indecency. But the water spoils all; could he not make wine without water? And if he could, why did he not do it? For it may be said that on this occasion, Jesus made nothing other than some mixture that he knew, like punch, for example, which the half drunk guests took for excellent wine. . . .[67] I am surprised that the day of this story was not a very well known festival among drunks, and Jesus ought to have transmitted to his disciples and to their successors the gift of repeating this miracle. This would be worth more than that of transporting mountains, which he promised them. . . .

CHAPTER 8.

Chapter 8 tells the story of the adulterous wife. This story is one of the most sensible in the gospel. Jesus told those who accused this woman, *Let he among you who is without sin* (John 8:7), this means I believe, exempt of this very sin, *throw the first stone at her.* And there was no one, Jesus not excepted, in a position to throw it; thus, Jesus dismissed her. However, this jurisprudence should not be followed for all crimes; for it seems that Jesus believed this woman guilty since in dismissing her he tells her (John 8:11) *to sin no more in future.* But it must be admitted that on this occasion the jurisprudence was aptly applied. . . .

CHAPTER 9.

The disciples of Jesus ask him (John 9:2) if a man who was born blind was born such because of his sins, or because of those of his parents? An insane question, but which shows that the Jews believed God still visited the sins of the fathers on their children, despite the promise he made to them, through the words of Jeremiah, no longer to act in that way. But supposing that the apostles had this in mind, the question they asked (John 9:2) Jesus was not

---

67. Du Châtelet came across the English word, *punch,* when she translated Bernard Mandeville's *The Fable of the Bees* and found the concept mystifying. Her decision to use it here is part of her mockery of this miracle. See Du Châtelet's translation in Ira O. Wade, "Remarks," in *Studies on Voltaire,* ed. Wade (Princeton NJ: Princeton University Press, 1947), 185.

any more reasonable; for since this man was blind from birth, it could not be because of his sins that he was born blind . . . [Latin omitted]. But Jesus manages to give them an even more ridiculous answer, if that is possible, than their question. *Neither this man*, says he, *nor his parents sinned; but it is so in order that the works of God might be made manifest in him* (John 9:3) . . . [Latin omitted]. Thus, the poor man was blind from birth solely so that Jesus might have the glory of curing him; and it is to be noted that he asked for alms (John 9:8). So Jesus cured him, but to achieve that he used an amusing recipe; he put diluted clay under his eyes (John 9:6), mixed with his spittle. *A thing*, says St. Chrysostom, *which ought to have made him blind if he had not been so already*. Thus Jesus probably took this precaution better to establish the miracle. But it seems to me that if nothing had been put on the eyes of this blind man, the miracle would have been even better established, and Jesus was wrong to make this miracle ridiculous by this ceremony. For it could be believed that the man had not been born blind and that Jesus had a drug in his mouth that was good for giving clear vision. This blind man, moreover, was a beggar with whom Jesus would have found it easy to come to an understanding. . . .

CHAPTER 10.

. . . In John 10:33 the Jews want to stone Jesus to death *because*, they say to him, *being only a man you make yourself God;* but Jesus, to calm them down explains in John 10:34 that they did not understand the words they reproached him for; *Is it not written in your law*, he told them, *I said that you are gods?* Thus, all those to whom the word of God is addressed have the right to say they are gods and sons of God (John 10:34) [Latin omitted]. . . .

I do not know how one dares maintain after this that Jesus believed himself to be *son of God*, and God *like his father;* surely he will not be accused of concealing what he thought about it, because he feared the Jews, since this fear would prove better than all his speeches that he was not *God.* It must be admitted, moreover, that it would be pleasing to be a shamefaced god.[68]

[CHAPTER 11].

It is in this chapter that the resurrection of Lazarus is reported. There are many points to be made about this resurrection. First, St. John is the only evangelist who writes about it, and only after the dissolution of the State of the Jews, when the public registers were lost and burned and all witnesses, including Lazarus himself, were dead. For, I noted that, supposing the gos-

---

68. Du Châtelet uses the word *honteux*, which carries all the meanings of *shamefaced*: bashful, modest, ashamed.

pel which bears the name of St. John is by the apostle of this name, he only
wrote when he was over ninety-five years of age. Now, this circumstance,
being reported only by St. John, is enough to make this resurrection very
suspect. For how likely is it that the other evangelists would have neglected
to speak of such a magnificent miracle, whose object, being still alive, was
an extant miracle that attested to what they claimed, and ought to suffice to
make all the rest believable?

In John 11:4 Jesus says to his apostles that *the illness of Lazarus was not mor-
tal; but it was only for the glory of God and his son* . . . [Latin omitted]. So, either
Jesus lacked foreknowledge if he did not know that Lazarus was dead; or he
said this intentionally to his apostles, and in order to blur the time he took
to go and find Lazarus. For, it may be seen from John 11:6 that it was two
days before he set out . . . [Latin omitted]: apparently in order to give time
to those with whom he had agreed to prepare everything, and in order to
have longer since Lazarus was dead, so that the miracle of the resurrection
would be more magnificent.

In John 11:11 he next says to his apostles, once he had set out: *Our friend
Lazarus sleeps, but I am going to awaken him* . . . [Latin omitted]. But having apparently received news later on that everything had been well prepared, he
tells them clearly in John 11:14 that *Lazarus was dead; Lazarus mortuus est;* and
he adds next in John 11:15, *But for your sake I rejoice that I was not there so that you
may believe; but let us go to him* . . . [Latin omitted]; whence it follows that Jesus
was indeed able to prevent Lazarus from dying, but that he could not exercise this power from afar.

But Lazarus's death, of which we just saw that he rejoiced with his disciples, we are going to see him weep over in John 11:35, *Et lacrimatus est;* although, surely, it is not very natural to mourn for someone who one is going to resurrect. But it was apparently a prelude that he believed necessary
to this comedy. . . .

The sepulcher where Lazarus had been laid was a cave . . . ; thus it was
possible to live there and it may well be that his sisters brought him food to
eat at night, by some entrance that they alone knew; and it was easy for Lazarus, who was a friend of Jesus, as were his sisters, to have agreed with him
to put on this performance. . . .

Next Jesus, as a matter of form, makes a small prayer to his father (John
11:42), and next he cries out very loudly to Lazarus to come out (John
11:43) . . . [Latin omitted]; a very ridiculous thing if Lazarus was really dead;
since it is extremely pointless to speak to a dead man, but which was very
useful to Jesus in order to warn Lazarus to begin playing his part. The dead
man came out of the cave as soon as he had heard Jesus's voice. But what is

quite puzzling is what is noted in John 11:44, *He had his feet and hands bound when he came out.*[69]

. . . It is said in John 11:44 that Lazarus had his face covered, *Et facies illius sudario ligata erat*—a remarkable and very suspect circumstance. For if Lazarus had had his face uncovered before being resurrected, it would have been possible to see if he had the look of a dead man and of a four-day-old dead man as well. Finally, Lazarus was not in his burial cave long enough to remove all suspicion, and it is certain that the cloth that covered his face must create some. He, who was all powerful, had so many means to dissipate all suspicions; why did he not do it? For, it is certain that if this resurrection had been established, and that the witnesses had not perceived any fraud, it would not have been possible for a single incredulous one to remain among them. Yet St. John notes in 11:45–46 that *some of those who had been witnesses to this resurrection went to find the Pharisees and the princes of the priests to give them warning and to conspire against Jesus.*

. . . Now I ask whether there is the least likelihood that people who have been the witnesses of a nonsuspect resurrection should seek to ruin and bring about the death of the man who effected this resurrection? Can a man who resurrects the dead not bring about the death of the living? And if JC had done such a miracle before all the Jews who were then at Martha and Mary's house (it is noted in John 11:19 that *there were many of them* . . .) [Latin omitted] would all those who had before been his enemies not have been at his feet?

However, one sees that to the contrary (John 11:53), it was from that day that the princes of the priests and the Pharisees resolved to have Jesus killed, and that from this time that he had to flee to a desert with his disciples, and that he dared no longer show himself in public. . . .[70]

CHAPTER 13.

It is said in John 13:2 that the Devil took possession of the heart of Judas. It is pleasing that Jesus chased away so many demons and that he lived so peacefully with this one. . . .

It is noted in John 13:23 that John, whom Jesus loved, and who is, it is claimed, the author of this gospel, was lying in Jesus's arms during the last

---

69. Du Châtelet comments also on how puzzling it is that he could walk at all.

70. From this story of Lazarus Du Châtelet hypothesizes that the apostles learned to stage Jesus's resurrection. She uses much of her summary of Woolston. See Thomas Woolston, "Miracle 14e: Jésus ressuscite le Lazare," in his *Six Discours sur les miracles de notre sauveur*, trans. Mme Du Châtelet, ed. William Trapnell (Paris: Honoré Champion, 2004), 358–61.

supper, and there are those cunning ones who insist that this predilection of Jesus for John was not very honorable. Thus, without the women who were always following Jesus, his affection for John would have ruined his reputation. . . .

CHAPTER 14.

And in John 14:4 Jesus says to them: *You know where I am going, and you know the way. Et quo ego vado scitis, et viam scitis.* And when Thomas says to him: *Lord, we know nothing of it,* . . . [Latin omitted] Jesus answers him, *I am the way, and the truth and the life* (John 14:6), *Dicit ei Jesus: Ego sum via, veritas et vita: If you had known me, you would have known my father, henceforth you will know him, and you have already seen him* (John 14:7): *Si cognovissetis me, et patrem meum utique cognovissetis; et amodò cognocestis eum, et vidistis eum.* What gibberish! The whole chapter continues in this tone, which, however, appears difficult to sustain. . . .

CHAPTER 20.

After his resurrection Jesus appears to Mary Magdalen dressed as a gardener (John 20:14). She begins by seeing him and speaking to him without knowing him. It must be that in rising from the dead Jesus had changed his physical appearance and his face. However, this was not the way to be recognized; so, all those who see him do not recognize him at first, neither by his appearance nor his speech. Indeed, I believe that nothing changes one more than to be crucified. But if Jesus wanted his resurrection to be believed in, he should have found a remedy for that and presented the same appearance. For it can be said that there was someone who found it amusing to pass himself off as Jesus risen from the dead, but he only dared to show himself to people who had been forewarned, like his apostles and the holy women. So, initially Mary Magdalen (John 20:14) does not recognize Jesus when he asks her whom she is seeking . . . [Latin omitted]; and she answers him (John 20:15), *If you have taken him away, tell me where you have laid him and I will take him away* . . . [Latin omitted]; finally she recognized him (John 20:16) because he calls her Mary; *Dicit ei Jesus; Maria. Conversa illa dicit ei: Rabbi &.* And it is difficult to know why she recognized him rather sooner at this word *Mary* than at all he had said to her previously. Jesus then forbids her to touch him, *Because,* says he (John 20:17), *I am not yet ascended to the father* . . . [Latin omitted]. Is not that a good reason not to be touchable? But it was apparently only women he could not be touched by before ascending to his father. For we saw in Luke 24:39 that he had his apostles touch him, *Palpate et vidite;* and one will soon see (Luke 24:27) Thomas put his hand into his side. Besides, this verse proves that Mary Magdalen regularly touched him; for he forbids her to do that without her having said anything to him, which could indicate that she felt like doing it. It must be that being crucified changes other things than the appearance. . . .

In John 24:19 Jesus arrives in the midst of his apostles when the doors were shut. Some say he passed through the keyhole; others that he penetrated the walls, and that this is a proof of the Eucharist, in which bodies interpenetrate. Jesus had a real body after his resurrection, since we saw in Luke 24:39 that he had his apostles touch him and told them that he had flesh and bones and that spirits have none of that. . . .

CHAPTER 21.

In this chapter the apostles (21:1) see Jesus again on the shore of the sea of Tiberias (John 21:4), but again they did not recognize him (John 21:4–5), neither in seeing him nor in speaking to him. This is very extraordinary, since they had already seen him since his resurrection. But maybe these were not the same apostles.

The apostles had come to the coast to work at their first occupation of fishers of fish; apparently not having found their reward in being fishers of men.

Jesus (John 21:4) appears before them on the beach and in this apparition he makes three beautiful miracles: (1) to make them catch many fish (John 21:6); (2) to make them find lighted charcoal all lit on the shore to cook the fish for them (John 21:9); and (3) to have bread ready for them to eat (John 21:13). Is this not similar to the story of the three *aulnes* of blood sausage?[71]

For me, I believe that the evangelist's story was inaccurate, and that Jesus also made them find salt on the beach; for fish without salt is bad food.

Besides, the Fathers are very puzzled to know if the angels brought this lighted charcoal and this bread, or if Jesus created both. . . .

Examination of the Epistles of St. Paul: I Corinthians

. . . In I Corinthians 7:1 he says that *it is well for man not to touch any woman. Bonum est homini mulierem non tangere.* And what becomes of the first commandment of God to man after his creation, *Crescite et multiplicamini; Increase and multiply?* It is surely not the intention of the creator that women and men live

---

71. The *aulne* or *ell* was a unit of measurement, approximately 120 cm, used primarily for cloth. Du Châtelet is referring to the Charles Perrault fable of "Les souhaits ridicules [The Ridiculous Wishes]," in which a poor woodcutter, granted three wishes by Jupiter, wastes them wishing first for sausages, then that they be on his wife's nose, and finally that she be as she was. For the tale, see http://clpav.fr/lecture-souhaits.htm, accessed 9 December 2007.

apart, and it cannot be that of the redeemer. Since the redeemer is none other than God, he cannot have a will other than his own. For, one cannot say without absurdity that God willed two contradictory things;[72] yet, it is what follows from this verse of St. Paul . . . [the Latin passage from above is repeated] and from Genesis 2:18, *Non est bonum hominem esse solum.*[73] It is difficult to understand why St. Paul gives this precept to the Corinthians. For it is obvious from all of Jesus's life that he liked women very much, and we will see that St. Paul himself did not hate them. But he was completely mad, and that is a fact. In truth he says that people who are married must stay married, but he adds (I Corinthians 7:8), *As for people who are not married, or are widows, it is well for them to live in this state, as I live here* . . . [Latin omitted]. For he always gives himself as an example. But if since Adam his advice had been followed, JC would not have found anyone to die for and that would have been a shame. . . .

In I Corinthians 11:3, *Jesus,* he says, *is the head of man; man is the head of the woman and God the head of Jesus.* Jesus is not then more *one* with God than the man with the woman?[74]

In I Corinthians 11:6 he says that *it is shameful for a woman to have cut hair;* and moreover, he says that if the head is not covered, she must cut her hair. What is certain is that according to this chapter of St. Paul, nuns commit a great sin and a great indecency when they cut their hair as they do; for he says in I Corinthians 11:15 that *it is glorious in a woman to have long hair; because nature gave her hair as a covering for herself.* . . .

72. This is a reference to Du Châtelet's ideas about God to be found in chapter two of her *Foundations of Physics.* God, like all in the universe is subject to the law of contradiction.

73. It is not good that man should be alone.

74. Here Du Châtelet again questions the divinity of Jesus.

V

COMMENTARY ON NEWTON'S
PRINCIPIA

VOLUME EDITOR'S INTRODUCTION

This set of chapters also comes from Du Châtelet's scientific writings. They constitute the first section of her *Commentary* on Newton's *Principia*. Some have speculated that this "Brief Exposition" was intended originally as volume two of her *Institutions*. This may be so; in any case, when she returned to her studies in mathematics and physics in the 1740s, she had decided to undertake a much more ambitious task, translation of the *Principia* itself. Her translation of Sir Isaac Newton's *Philosophiae Naturalis Principia Mathematica* [Mathematical Principles of Natural Philosophy] remains the only complete rendition into French, and the only one in any language that the Newtonian expert, I. Bernard Cohen, acknowledged as a guide in his own English version of this influential work.[1] To translate the *Principia*, Du Châtelet read the important contemporary books and treatises in experimental physics, continued her study of analytic geometry, and mastered calculus. As such, she numbered among the twenty or so eighteenth-century men and women who understood this advanced mathematics, and who could manipulate its propositions to apply to other cases.

The decision to include a commentary meant that Du Châtelet gave her readers four different ways to understand Newton's ideas. There was the translation itself, and the abridged and simplified description of Newton's "system of the world," which is presented here. For those trained in the new mathematics, she went on to offer a section of analytical solutions, or algebraic equivalents, for the most controversial idea of the *Principia*, the

---

1. Du Châtelet used Newton's 1726 third edition, but also consulted the 1713 second edition, and the altered Latin version of Book III of the *Principia* published as *De Systemate mundi* [System of the World], the 1731 edition. The Jacquier and Le Seur edition of the *Principia* with its continuous commentary in the annotation was a valuable resource as well.

so-called problem of three-bodies (the interaction of *forces* exerted by three bodies moving through space; for example, a comet passing between two planets). In the fourth and last section she summarized two works considered to give key proofs of Newton's universal theory of the workings of the universe: Daniel Bernoulli's treatise on the Moon and the Earth's tides, and Alexis Clairaut's on the effects of attraction on the shape of the Earth.[2]

Du Châtelet completed the translation in the hectic year of 1745 when she and Voltaire were active courtiers at Versailles. In contrast, the work of the *Commentary* dragged on year after year. She kept expanding its scope as new scientific memoirs appeared that corrected one or another aspect of Newton's suppositions. Du Châtelet did not finish correcting the proofs until just days before the birth of her fourth child in September 1749. She never saw publication of the two volumes because of her death a week later of a complication occasioned by the birth. No one knows what happened to the manuscript after that, but fortunately, her text was not lost altogether. The publisher must have kept the proofs. As the end of the 1750s approached, there was much excitement in the scientific community about Newton's theories because he had used predictions of the return of comets as proof of his theory of universal attraction. Clairaut, Du Châtelet's mathematics mentor and her advisor in the final stages of this, her last project, though he refused to act in any editorial capacity, encouraged the publishers to bring out her work. It would be useful publicity for Newton's theories and for his, Clairaut's, own role in establishing them.[3] The official and complete version of Du Châtelet's translation and commentary appeared in 1759, ten years after her death. With its publication her learned reputation

2. For a more extensive discussion of these choices, see Antoinette Emch-Dériaz and Gérard Emch, "On Newton's French Translator," in *Emilie Du Châtelet: Rewriting Enlightenment Philosophy and Science*, ed. Judith P. Zinsser and Julie Candler Hayes, SVEC 1 (2006); Judith P. Zinsser, "Translating Newton's *Principia*: The Marquise Du Châtelet's Revisions and Additions for a French Audience," *Notes and Records of the Royal Society of London* 55, no. 2 (2001): 227–45. The four proofs hypothesized by Newton and subsequently proved by the end of the eighteenth century were the shape of the Earth, the precession of the equinoxes, the tides, and the orbit of comets and predictions of their visible return to the Sun's system.

3. This excitement may explain why the 1756 version was incomplete, and the reason for the odd title page of the 1759 edition with her name in capital letters and no mention of Newton. Her learned reputation and the notoriety occasioned by her pregnancy and death may have been viewed as more likely to encourage sales than the Englishman's name. On the events surrounding the return of the comet in 1758, see Simon Schaffer, "Halley, Delisle, and the Making of the Comet," in *Standing on the Shoulders of Giants: A Longer View of Newton and Halley*, ed. Norman J. W. Thrower (Berkeley: University of California Press, 1990), 254–92; Craig B. Waff, "Comet Halley's First Expected Return: English Public Apprehensions, 1755–58," *Journal for the History of Astronomy* 17, no. 1 (1986): 1–37.

was assured. The *Foundations* and now this massive project proved her abilities as a mathematician and her understanding of some of the most complex controversies in eighteenth-century physics. Thus, she played a role in the acceptance of Newton's explanations of the workings of the universe, and to the virtual canonization of the man and his method of reasoning by experiment and mathematical analogy.

The *Commentary* and translation have been reprinted in facsimile editions. In 1966, Blanchard brought out the 1756 incomplete version; in 1990, Jacques Gabay produced the complete 1759 edition.

The letters that precede her *Commentary* illustrate the intertwining nature of her major preoccupations from 1745 to 1749. There was her work on "my Newton" as she referred to it and her duties as a courtier both at Versailles and after 1748 in Lunéville, at the court of King Stanislas. Then, in the spring of 1748 she began her affair with Jean-François de Saint-Lambert. When she became pregnant, the complications multiplied. As she explained in her letters to her lover, she needed to be in Paris to have Clairaut's assistance with the complicated mathematical section of the *Commentary* and her description of his memoir on the shape of the Earth. She would have preferred to be in Lorraine and she hoped to have the birth there, far from the ridicule of Paris and Versailles. There, she also would have the comfort of her friends such as the King's official mistress, Mme de Boufflers; of Saint-Lambert, her lover; and her husband—whose welcome response to her pregnancy had been to hope that it would be another son.

The letters to Jacquier and Boufflers come from the Besterman collection. All of those to Saint-Lambert come from the edition of her love letters by Anne Soprani.

ৎৎ

RELATED LETTERS

To Father Jacquier[4]

In Paris, the 13th April 1747

I assure you that the irregularity of the post is a misfortune, monsieur, for M. Clairaut informs me that you complain about not receiving my news, and I was sorry not to receive a response to the last letter I wrote to you. I am in the

---

4. No. 360 in the Besterman collection, vol. 2, 155–56; D3522 in the *Oeuvres complètes*.

same situation with my daughter.[5] This is one of the inconveniences of the war, and although this is not the greatest, I endure it with much impatience when it deprives me of your letters and makes you doubt my affection.

I am always very busy with my Newton. It is in press. I am going over the proofs, which is very boring, and I work at the commentary, which is very difficult. Your excellent work is a great help to me and if I had had the courage to undertake a perpetual commentary, I would not have hesitated to translate yours. I am very sorry that we are deprived for so long of your work on integral calculus. I would very much like to know what delays it. I never received this Italian journal where you were so kind as to have my answer to Jurin printed. M. Cramer who is here and whom it gives me great pleasure to see, because of all you said to me of him, makes me think of it.[6]

It is with great pleasure that I congratulate you on the favor to your cousin, Mme du Hausay, about whom M. Clairaut told me. If you write to her, let her know, I pray you, that I would be very glad for her to come to see me when I am at Versailles, where she still is, because I would have the pleasure of speaking of you with her and would speak to her of an essential service she can render to M. Clairaut that I believe he finds it difficult to speak to her about himself.[7] I even beg you, when you write to M. Clairaut, not to speak to him of what I have told you on this subject.

Yesterday I attended the public session of the Academy where M. de Buffon read to us a memoir on the manner of burning by reflection at very great distances, by means of several mirrors on moving planes which brought the images of the Sun to the same focus.[8] He burned at one hundred fifty *pieds* and his reasoning tends to prove that with a great enough number of mirrors one could burn from six or seven hundred feet away, which vindicates Archimedes against Descartes; M. de Buffon's memoir is well written and very instructive.

5. Gabrielle Pauline had married the Duke de Montenaro-Carafa in 1743 and lived in Naples at the royal court. In the War of the Austrian Succession (1740–48), although the Kingdom of Naples and France were allies in 1747, much of the fighting was in the Italian peninsula.

6. In the course of doing her translation and commentary, Du Châtelet used Jacquier's and Le Seur's edition of the *Principia* with its "perpetual commentary," a commentary done through the footnotes. Their edition also included the essays by Daniel Bernoulli and Clairaut that Du Châtelet chose to summarize in her *Commentary*. Jacquier had arranged for publication in Italy of her answer on *forces vives* to the English mathematician and physicist, James Jurin. Gabriel Cramer was another Swiss mathematician with whom she corresponded.

7. Mme de Hausset was a relative of Jacquier's who held a court appointment in service to Mme de Pompadour, Louis XV's official mistress.

8. George Louis Leclerc, Count de Buffon (1707–1788), was a naturalist. The Royal Academy of Sciences held two public sessions a year which women were allowed to attend. Typically, as Du Châtelet describes here, a member gave a paper and did scientific demonstrations.

You saw my daughter in Naples. So you take an interest in her, and you will be very glad to learn that she has been named principal lady-in-waiting [*dame du palais*] to the queen of Naples, which she so strongly desired.

M. de Voltaire sends a thousand fond compliments to you, and I reiterate to you monsieur, assurances of an affection that will last as long as the life of your very humble and very obedient servant.

Breteuil Du Châtelet

To Saint-Lambert in Nancy[9]

[Cirey] This Saturday [11 May 1748]

Ah well, here I am then alone in this sad château, which was this morning for me the most delicious abode. But am I not alone in the world when I no longer see you? I intended to write to you all that I have thought since I said good-bye. You will see that this joy of the armistice, the pain of being far from you, fills my heart with so much pain and so much joy that it succumbs from being torn apart. But I am dying of lassitude, I do not have the strength to write, I have only that to love. Let me know on arriving if your trip is known. I believe that I will not be talked about on this side.[10] Let me know above all how you are feeling, for that is a concern that silences all other feeling.[11]

I am overwhelmed with business affairs. I want to leave on Tuesday, and I no longer want it when I think that this will increase the distance between us. But I will only think of the means to bring us together, be sure. I hope that you will not go to the regiment, that you will take advantage of your leave and my absence to cultivate Mme your mother, and convince her to do what you wish for, that you will have care for your health, that you will work, that above all you will write to me, that you will not leave me ignorant of what is going on in this heart that I adore, and without which I can no longer live. I find my letter very cold when I compare it with all that I feel for you. But I am dead, I exist only by my love. You will see on Mon-

9. No. 16 in the Soprani edition of *Lettres d'Amour*, 54–55; no. 373 in the Besterman collection, vol. 2, 169–70. This letter probably signified the beginning of Du Châtelet and Saint-Lambert's sexual liaison.

10. Du Châtelet and Saint-Lambert endeavored to keep the new level of their relationship secret from the court at Lunéville and from Voltaire and Du Châtelet's husband. An armistice in the War of Austrian Succession meant that her husband might be returning from active service in the German states.

11. Saint-Lambert had been ill and feared the journey from Lunéville to Cirey would be too much for him.

day if I love you and if I know how to say it to you. Write to me in Paris, rue Traversière, next to the Richelieu fountain.[12] But write to me as you look at me. Console a soul that is wholly yours; fill with joy a heart that worships you. If you love me, I can endure everything, even your absence, which is the greatest of evils.

Adieu, my heart goes in search of yours.

Think that I will only write to Mme de Boufflers when I have had your news.[13]

I reopened my letter to entreat you not to speak of *Zadig*. She would believe that it was I who read it to you in an amorous encounter. It would be better to say nothing of it, for this would perhaps bring about an explanation between them.[14]

Here is "Bernard" and four verses to Mme Pompadour, bad enough.[15] It is an act of infidelity that I commit for you: I require that you will not show them to anyone at all. You will receive the letter that is fit to be shown on Monday.

To Saint-Lambert[16]

[Commercy] [July 1748]

I am a lazy woman. I am getting out of bed. I have only a moment and I use it to say to you that I adore you, miss you and desire you. Come then as early as you can. You will drink and dine here. I hope that you will also love here.

12. Du Châtelet shared a *hôtel* [townhouse] in Paris with Voltaire; she had the first floor, he the second (the kitchen was on the ground floor).

13. Marie Françoise Catherine de Beauvau-Craon, Marquise de Boufflers-Remiencourt, was the sister of Saint-Lambert's patron, the Prince de Beauvau-Craon, and the official mistress of the Duke de Lorraine, King Stanislas of Poland. As such, she was the reigning influence at the court. She expected young men like Saint-Lambert to pay court to her, even to exchanging sexual favors. It is assumed that Saint-Lambert obliged. He certainly wrote poetry for her (*Chloe*). Du Châtelet flattered and attended her as well as part of her role as a courtier in Lorraine.

14. *Zadig*, a tale, or *conte*, written by Voltaire that mocked court life, was to be published in Nancy and already was rumored to lack royal approval. Thus, it would be a possible source of difficulty for the author. Du Châtelet is assuming that mention of it would prompt conversation between Voltaire and Boufflers and the revelation of this assignation at Cirey with Saint-Lambert.

15. Du Châtelet here has sent Saint-Lambert some of Voltaire's light verse: one to their Paris friend, the poet Pierre Joseph Bernard, known as Gentil Bernard, the other, one of many quatrains that he wrote for Mme de Pompadour.

16. No. 33 in the Soprani edition, 98; no. 397 in the Besterman collection, vol. 2, 203. Du Châtelet and Saint-Lambert were able to be in Commercy together, and on such occasions exchanged short notes like this, often two or three times a day.

To Saint-Lambert[17]

[Plombières] Saturday morning [24 August 1748]

It is unquestionably the most unhappy person in the world who writes to you. No one has spoken to me yet, but I see very clearly that we will stay here, and that the menstrual period of Mme de B and her supposed bleeding will be used as a pretext. The viscount no longer wishes her to go to Saverne; he prefers that she stay here. She is like a dog, like a pauper in hospital, she sleeps not a minute, and there is no doubt that if she is in poor health, she will not recover here. But she is hard on herself, weak and complacent, and she willingly stays where she is, however ill she is here. Moreover, she loves better being here with her bleeding than at Lunéville, where she would have neither the viscount nor the *comète*.[18] Lastly, one must know how to bear what one cannot avoid. Let us imagine that the ten or eleven days we must be separated will be all at once and not spread over different times.

You would pity me if you saw the extreme malaise and boredom of this place. If something had been able to soften the sadness of your absence, it would have been to go to Saverne and Strasbourg, which I very much want to see. I would have preferred to be in Lunéville. At least I would not have had all these misfortunes together. But imagine what it is to be in a stable all alone, all day long, to go out only to kill time with a damnable game of *comète*, which interests me not at all, and to think that I could spend delicious days with you at Cirey or at Lunéville. If the Chevalier de Listenay was not in Paris, and if it had been possible for him to come to Cirey, I believe that I would have killed myself for not being there.[19]

The viscount does not want to put on *Le Complaisant* [The Complaisant One].[20] So, do not hurry to learn your role. But do not spread this news. I fear that I may run out of work, for I work ten hours a day and I had not counted on being here for so long. God knows when this will end!

It would have been impracticable for you to come here. First, all here is terribly expensive, and that would have ruined you. Furthermore, we are lodged fifty to a house. I have a tax farmer who sleeps next to me. We are

17. No. 45 in the Soprani edition, 111–13; no. 430 in the Besterman collection, vol. 2, 220–21. Du Châtelet had been asked by King Stanislas to accompany Mme de Boufflers to Plombières, a popular spa in Lorraine. Boufflers had also brought her lover, Viscount Rohan-Chabot.

18. *Comète* was a very popular card game.

19. Du Châtelet is probably referring to Charles Roger de Bauffremont, Chevalier de Listenois; spellings often varied.

20. This was a play by Antoine Ferriol, Count de Pont-de-Veyle, the Count d'Argental's brother.

only separated by a hanging and, however softly one speaks, one hears all that is said; and when someone comes to see you, everyone knows it and sees you to the back wall of your room. We only eat at the princess's;[21] you are not yet intimate enough to spend your day there. Finally, this is impracticable in every way. We must consider this as a time of calamity and try to no longer endure a similar one. At the end of the day, if we do not go to Saverne we will not lose very much time.

If M. Du Châtelet keeps the command of Lorraine, I will have an establishment in Nancy, where I will stay in his absence, so I will no longer depend on others.[22] Consider then, that in order to spend my life with you, there is nothing that I would not do, and that there is no good fortune I would want if it separated me from you. I want to spend all the time of my life with you and I will only sacrifice this pleasure to decency; I sacrifice it now because I believe that there is more to gain than to lose, and that one of the things most necessary to avoid in life is remorse.[23] I do not know when you will receive this letter, but it is a great consolation for me to talk to you and to open my heart to you.

To Mme De Boufflers - Remiencourt[24]

[Thursday, 3 April (1749)]

Ah well, I must then tell you my unfortunate secret without waiting for your response to the other one I asked about. I feel that you will promise me it and that you will keep it, and you will see that it cannot be kept much longer.

I am pregnant, and you can imagine my affliction, how much I fear for my health, and even for my life, how ridiculous I find it to give birth at forty years of age after seventeen years without having given birth to a child, how

21. The Princess de La Roche-sur-Yon, a member of the court at Lunéville. She was present for the birth of Du Châtelet's child and stood as godmother to the infant at the baptism.

22. The marquis did not keep this post, but instead became Grand Marshall of the King's Lodgings, a post requiring constant attendance on his majesty, and thus Du Châtelet was assured a place at the Duke of Lorraine's court.

23. This admonition about "remorse" echoes the language of Du Châtelet's rules for happiness in her *Discourse on Happiness*, which she gave to Saint-Lambert, and which circulated after her death.

24. No. 454 in the Besterman collection, vol. 2, 247–48; D3901 in the *Oeuvres complètes*. With this letter Du Châtelet initiated the delicate negotiations surrounding her lying-in at Lunéville.

upset I am for my son.[25] I do not want to make the news public yet, fearing that this will prevent his establishment, supposing that some opportunity arose, which appears very unlikely to me. No one suspects anything, it shows very little. By my count, however, I am in the fourth month, but I have not yet felt it move. That will only happen at four months and a half. I am showing so little that if I did not feel some dizziness or some discomfort, and if my bosom was not very swollen, I would believe that it was some illness. But this is highly unlikely, in view of the good health that I have enjoyed up to the present.

You feel how much I count on your affection and how much I need it to console me and to help me to endure my state. It would be very hard to spend so much time without you, and to be deprived of you during my confinement. But how could I possibly be at Lunéville for my confinement and cause so much embarrassment? I do not know if I ought to rely on the kindness of the king enough to believe that he wishes it, and that he will allow me the small apartment of the queen that I occupied, for I could not give birth in the aisle wing, because of the smell of smoke, the noise, and the distance there would be between me, the king, and you.[26] I fear that the king will then be at Commercy and that he would not want to cut his journey short. I will probably be confined at the end of August or the beginning of September at the latest. I am ignorant of the king's traveling plans. It would be very hard to spend eight more months without you and perhaps more, for with the time of my lying-in it will be at least eight months and, if I am not fully recovered, I could not undertake such a long trip at the beginning of winter, just after my lying-in. This will be a time when your affection will be most agreeable and most necessary, and when the kindnesses of the king will be the greatest consolation. It would be very hard to be deprived of it; I hope that you will not permit it. You see, however, how many considerations must stop me! I do not want to take advantage of the king's kindnesses to me, nor of your affection.

M. Du Châtelet wants me to be in Lunéville for the birth, or at least

25. Aside from the inappropriateness of the pregnancy at her age, Du Châtelet is thinking of the possible consequences for her son and his inheritance. His sister had given up all claims to the Du Châtelet estates on her marriage, but a new brother or sister would have to be provided for. He was awaiting news of a regiment to command.

26. Visiting courtiers usually had rooms in the "aisle" portion of the palace. Du Châtelet was asking to have the suite of rooms of the former queen that was on the gardens, adjacent to the king's apartment and across a courtyard from Boufflers's. Stanislas was delighted with her pregnancy and granted all her requests, even to preparing one of the little summer houses in the gardens for her.

he wishes it strongly. I wish it more than he, but it is for you to see if this is possible and suitable. It is for you to tell me if you wish it, if the king wishes it, and what your counsel is. If I am to lie in at Lunéville, I will return there at the end of May, or at the beginning of June, because I will risk less then. I am not afraid of traveling, I will go slowly, I have never been hurt in the past and I am very strong.

Nothing would be more unhealthy than to do without you; decide then my fate and if you want it to be a happy one, see to it that I will be with you. I will await your response with impatience. You will tell the king as much as you like, I put my fate in your hands. I count on you to find a good *accoucheur* and a good attendant in Lorraine; it would be very expensive to lie in in Paris, and very sad to lie-in without you.[27]

To Saint-Lambert[28]

[Paris] 21 [May 1749]

No, I have no more suspicions, I only have love. It is so easy for you to remove these suspicions, that you are very guilty in leaving me with them. It is by writing me tender letters that you will destroy them, even more than by going to Nancy, where I will nonetheless be very glad to know you are going to be on the 1st.

You justify yourself very coldly about your letter to Chevalier de Listenay. You no longer tell me that it would be impossible to leave me, the idea has become familiar to you, it frightens you no longer, and I fear that this will become easier for you as it would become more difficult for me. You have not scolded me a single time about staying in Paris for the month of May, and you appear to fear equally that I may leave late and that you may spend a few days at Cirey with M. Du Châtelet.

My departure does not depend absolutely on me, but on Clairaut and the difficulty of what I do.[29] I sacrifice everything to that, even my looks. I beg you to remember that if you find me changed. Have you any idea of the life I have led since the departure of the king? I rise at 9 o'clock, sometimes at 8, I work until 3, I have my coffee at 3 o'clock; I take up work again at 4, I leave it at 10 in order to have a little to eat alone, I chat until midnight with

27. An *accoucheur* was a male midwife and a luxury in Du Châtelet's time. Most women relied on a *sage-femme*, or traditional midwife, when giving birth and immediately afterward.

28. No. 89 in the Soprani edition, 220–24; no. 476 in the Besterman collection, vol. 2, 293–96.

29. Clairaut was assisting her in the completion of the complex mathematical sections of her *Commentary*, including her summary of an essay of his.

M. de V., who attends my supper, and I take up work again from midnight to 5 o'clock. Sometimes I wait for M. Clairaut, and I attend to my affairs and read through my proofs. Mme Du Deffand, Mme de B, everybody without exception is denied for supper and I have made a rule for myself not to go out to supper in order to be able to finish my work. I admit that, had I led this life since I came to Paris, I would have finished by now. But I began by having many engagements; I gave myself up to society in the evenings. I believed that the day would suffice. Then, I saw that it would be necessary to give up lying in in Lunéville, or to lose all the fruit of my work in case I die in childbirth, or to have fetters here that would force me to return immediately after my lying in in case I did not die. And I felt that the only way to avoid all these intersecting inconveniences and to make the most of my trip to Paris, which cost me enough chagrin—for me to gain at least some advantage out of it—was to sequester myself absolutely, to stake my all, and to devote all my time to my book.

My health is keeping up marvelously. I am abstemious and I drink gallons of orgeat, which sustains me.[30] My baby moves very much and feels, I hope, as well as I. If I were able to enjoy M. Clairaut's company I would leave at the end of the month. I will still have many things to do there and I will only stay here for the time necessary to finish those for which M. Clairaut's counsel is indispensably necessary to me. If I could take him with me, I would have left a long time ago, but that is an impossible thing.

Mme B tells me about the journey of Mlle Dandreselle and the interest that the chevalier takes.[31] So, your plans are more distant than ever. I confess that if they could be carried out they would bring delight and tranquility to my life. I always send letters from V; but if I were very sure of you, I would oppose her taking him back. He is silly and querulous about details; it is not worth the trouble.

I must deal with the fear you have of being alone at Cirey with M. Du Châtelet. It does not depend absolutely on me to spare you that, and if you prefer to see me ten or twelve days later rather than risk that contingency, I have nothing to say to you. It seems to me that you are putting this in the balance scales, and you must feel the effect that this has on me. The fear of exposing you to this contingency prevents me from giving a fixed date to M. Du Châtelet for fear of missing it. My traveling depends on the good will of M. Clairaut and on the time he gives me. I am still not able to set a

30. Orgeat was barley water.

31. Adélaïde-Louis-Jeanne-Philippine Picon d'Andrezel was another member of the Lunéville court.

definite day; it would be necessary to set one too far away. The king's journey to Commercy distresses me. If I have not left when he leaves, I believe it will be necessary to urge M. Du Châtelet to take you to Commercy to take possession of the lodgings of the gracious curé, and it will not be long before I cause you to leave them.[32] I write by this post to the chevalier in order to press him to make an appointment with Mlle Dandreselle in Cirey, whom I will go and beg to do it if it is convenient for him, and whom I will even bring, if he wishes it. The king told me last year that it was not necessary to have permission to go to Commercy. So I see nothing that could oppose your journey. You will have spent about fifteen days in Nancy, which will be enough for you and for me. Ask M. Du Châtelet to impart to you the details that I give him on this; this will be easy for you, if only you speak of it to him.

I do not see any indication of Mme Boufflers' traveling; she treats me charmingly and I love her as much as I fear her; this is very rare. I promise you, whether she comes or not, not to wait for her. One returns the 12th to Marly up to [the] 20th, one goes on 1st July to Compiègne, I cannot see any time for a journey. I have been asked to act in *La Mère coquette* [The Coquettish Mother]:[33] do you agree? Will you play in it? I can love only what I share with you. For I do not love Newton. At the least I finish it out of a sense of duty and of honor, but I only love you and what relates to you.

Press the chevalier to give this appointment to Mlle Dandreselle at Cirey. In this case, he would come there with you, and you would no longer be afraid, I hope, to stay at Cirey for 24 hours. I would take her to Commercy; that would be very pleasant. I ask Mme B to keep my son under pretext of the plays; he would only be an encumbrance at Cirey. I hope that after having had so much leave for Lunéville you are taking measures to have one for Cirey. I will only be there as long as you want, and I promise myself to spend my life doing everything you want, for I would like to believe that you will want us never to leave each other.

The prince speaks of a journey to England next year, but I do not fear that you will sacrifice me to any prince.[34] He spoke to me of your affairs first;

32. In the fall of 1748 when Du Châtelet first spoke to Stanislas and asked him to allow Saint-Lambert to accompany the court to Commercy, the curé's lodging adjacent to the royal château was understood to be where he would stay.

33. This was a comedy by Charles Collé, who coined some of the most insulting jokes about Du Châtelet's pregnancy then circulating in Paris and at the courts of Versailles and Lunéville.

34. The Prince de Beauvau-Craon was Saint-Lambert's patron and had arranged for his post in the Lorraine Guards.

I had stopped discussing them with him, because I know the uselessness of it; it is always necessary to be on good terms with him, to enjoy the pleasure and the informality of his company and to expect nothing from him. I always have the affair of the regiment so vividly in my mind, and I believe it is the only feasible one. I am not satisfied with the answers I received from Vienna. It is not the last word, and I still have some hope.

Adieu, this is how one writes when one loves as much as I do. Adieu, I adore you, my soul leaves me to go and find you. I believe that I will die for joy when I see you again, if I find you as I left you. Your health makes me anxious. I hope that you do not deceive me about your regimen. I hope that the chevalier has not taken up again . . . [words missing in the original text] I do not like him to set you a bad example. Think of Mlle Dandreselle as in an invincible ignorance and I am only too much on the right track. You no longer speak to me either of *"Matin* [Morning]," of *"Saisons* [Seasons]," or of these verses on solitude with the one one loves. What are you doing then? I have not been able to see anything of the *"Saisons* [Seasons]" of the Abbé de Bernis; he is almost as lazy as you are.[35]

Answer all in this letter, your answer will be long, and the short letters drive me to despair; they are a sure proof that one has little to say and that consequently one is scarcely in love.

꒰꒱

FROM THE COMMENTARY ON NEWTON'S *PRINCIPIA*

INTRODUCTION

I. First ideas of philosophy on astronomy.

Philosophers began by having in astronomy, as in other matters, the same ideas as the common people, but they rectified them. Thus, it was initially believed that the Earth was flat, and that it was the center around which all celestial bodies turned.

---

35. Du Châtelet here is referring to poems that Saint-Lambert was working on. *Les Saisons* [The Seasons], a very long poem glorifying an economically productive countryside was finally finished long after her death and published in 1769. The Abbé François Joachim de Pierre de Bernis, an advisor to Mme de Pompadour, also composed verses on the seasons, which he read at the salons in Paris.

II

DISCOVERIES OF THE BABYLONIANS AND OF PYTHAGORAS.

The *Babylonians*, and then *Pythagoras* and his disciples, having examined these ideas derived from the senses, recognized that the Earth is round, and regarded the Sun as the center of the universe.(*a*)[36]

III

EFFORTS MADE TO MAINTAIN THAT THE EARTH IS AT REST.
PTOLEMY'S SYSTEM[37.]

One can only be surprised that the true system of the world having been discovered, the hypothesis that the Earth is the center of celestial movements still prevailed. Although this hypothesis agrees with appearances, and appears at first sight to be extremely simple, it does not easily account for celestial movements. Thus, *Ptolemy*, and those who have since his time wanted to sustain this opinion that the Earth is at rest, have had to clutter up the skies with different epicycles, and with countless circles that were very difficult to conceive of and to use, for there is nothing so difficult as to put error in the place of truth.

It is very likely that the authority of *Aristotle*, which was almost the only measure of truth in *Ptolemy*'s time, is that which led this great astronomer [Ptolemy] into error; but why did *Aristotle* himself not endorse the true system, which he must have known since he attacked it? This does not redound to the credit of the human mind. Be that as it may, up to *Copernicus* it was believed that the Earth was at rest and the center of celestial movements.

---

36. Du Châtelet offers this note: "(*a*) M. *Newton* in his Book *De Systemate mundi* [The System of the World], also attributed this opinion to *Numa Pompilius*, and he says (pag. 1.) that it was in order to represent the Sun in the center of the celestial spheres that *Numa* had built a round temple dedicated to Vesta, the Goddess of Fire, in the middle of which a perpetual fire was maintained." Du Châtelet is referring to Newton's book version of Book III of the *Philosophiae naturalis principia mathematica* [Mathematical Principles of Natural Philosophy]. Du Châtelet studied two of the three editions, 1713 and 1726, as well as the *System of the World*, published in both English and Latin (1728, 1731).

37. Du Châtelet proceeds to give a brief history of astronomy: from Ptolemy (after 83–161) and his system of concentric circular orbits of planets and the Sun around the Earth; to Nicolaus Copernicus (1473–1543) and his discovery of the Earth as the center of the solar system; to Tycho Brahe (1546–1601) and his extensive observations, which Johannes Kepler (1571–1630) then used to formulate the elliptical orbits of the planets around the Sun.

IV

COPERNICUS REVIVED THE ANCIENT SYSTEM OF PYTHAGORAS ON THE MOVEMENT OF THE EARTH.

This great man revived the ancient system of the *Babylonians* and of *Pythagoras*, and supported it with so many reasons and discoveries that the old fallacy could no longer prevail. Thus, Copernicus again placed the Sun at the center of the world, or, to put it more exactly, at the center of our planetary system.

V

TYCHO BRAHE'S SYSTEM.

Although celestial phenomena are very easily explained in *Copernicus*'s system, although observations and reasoning equally favor it, there was in his time a very able astronomer who wanted to reject the evidence of his discoveries. Tycho, deceived by a badly done experiment, (b)[38] and perhaps even more by the desire to create a system in between that of *Ptolemy* and that of Copernicus, supposed that the Earth was at rest, and that the other planets that turn around the Sun turn with it around the Earth in twenty-four hours. This leaves in place one of the great difficulties of the Ptolemaic system, the one occasioned by the excessive rapidity of movement of the first moving body. This proves how dangerous it is to misuse one's insights.

SERVICES THAT TYCHE RENDERED TO ASTRONOMY.

If *Tycho* lost his way in his assumptions about the movements of celestial bodies, he rendered great services to astronomy with the exactitude and the long sequence of his observations. He determined the position of a great number of stars with an exactitude unknown before him; he discovered the refraction of light through the air, which is key to astronomical phenomena; he was the first to prove by the parallax of the comets that they rise signifi-

---

38. Du Châtelet offers this note: "(b) *Copernicus* had to answer the objection that the movement of the Earth would have to produce effects that were not taking place; that, for example, if the Earth moves, a stone thrown from the top of a tower must not fall at the foot of this tower, because the Earth moved while the stone was in free fall, and that nonetheless [the stone] falls to the foot of the tower. Copernicus responded that the Earth stands in the same relationship to bodies that fall to its surface as a moving ship in relation to things dropped on its deck, and he affirmed that a stone thrown from the top of the mast of a ship in motion would fall down at the foot of its mast. This experiment, whose result is not contested at present, was badly done at the time, and was the cause or the pretext that prevented *Tycho* from acknowledging the discoveries of *Copernicus*."

cantly above the Moon; it is he who discovered what is known as *the variation of the Moon*. Finally, it is from his observations on the course of the planets that *Kepler*, with whom he spent the last years of his life near *Prague*, drew his admirable theory of the movement of the celestial bodies.

VI

AFTER COPERNICUS MUCH REMAINED TO BE DISCOVERED.
Copernicus had rendered a great service to astronomy and to reason, in reestablishing the true system of the world, and it was quite an achievement that human vanity had agreed to consider the Earth as just one of the planets, among many, but much remained to be discovered. As yet, nothing was known about either the curve that the planets describe in turning in their orbit, or about the laws that direct their courses, and it is to *Kepler* that we owe these important discoveries.[39]

DISCOVERIES OF *KEPLER*.
 THE ELLIPTICITY OF THE ORBITS. THE PROPORTIONALITY OF THE AREAS AND THE TIMES. This great astronomer found that the astronomers who had preceded him were wrong in supposing that the orbits of the planets were circular, and he discovered, in making use of *Tycho's* observations, that the planets move in ellipses of which the Sun occupies one of the foci, and that they travel through the different parts of their orbits at different speeds. Thus, the area described by a planet, that is to say, the space contained between the lines drawn from the Sun to any two places occupied by the planet, is always proportional to the time.
 THE RELATION BETWEEN THE PERIODIC TIMES AND THE DISTANCES. Some years later, in the course of comparing the time of the revolutions of several planets around the Sun with their different distances from this star, he found that the planets farther from the Sun move more slowly in their orbit; and by seeking to know if this proportion varies with distance, he finally found in 1618, after several attempts, that the duration of their revolutions is as the square root of the cube of their mean distances to the Sun.

VII

Not only did *Kepler* find these two laws that bear his name and that rule the courses of the planets and the curve they describe, but also he foresaw the force that made them describe it. The seeds of the idea of the power of at-

---

39. Kepler is credited with three laws. Du Châtelet takes the first for granted and describes only the second and third laws. I am grateful to Adam Apt for this clarification.

traction are to be found in the preface of his commentary on the planet of Mars. And he goes so far as to say that the tides are the result of the pull of the Moon's gravity on the water toward the Moon; but he did not derive from this principle what one would have thought such a great man as he could have derived. For, he next gives in his *Epitome d'astronomie* [Epitome of Astronomy] (c)[40] a physical reason for the movement of the planets derived from very different principles. In this same book on the planet Mars, he supposes a friendly side to the planets and an enemy side; and on the occasion of their aphelion and of their perihelion, he says, that the Sun attracts one of these sides and repulses the other.[41]

VIII

SINGULAR ANECDOTE ON ATTRACTION.

The concept of the attraction of celestial bodies is to be found much more clearly stated in a book on the movement of the Earth by *Hooke*, printed in 1674, that is to say, twelve years before the principles [*Principia*]. *Here is the translation of his words*, p. 27. "Then I will explain a system of the world that differs in many regards from all the others, and which complies in every way with all the ordinary rules of mechanics. It is founded on these three suppositions.[42]

"1. That all celestial bodies, without any exception, have an attraction or gravitation toward their own center, by which they attract their own parts and prevent them from moving away, as we can see from the Earth. But they also attract all the celestial bodies that are in the sphere of their activity. Consequently, not only do the Sun and the Moon have an influence on the body and the movement of the Earth, and the Earth an influence on the Sun and the Moon, but also Mercury, Venus, Mars, Jupiter, and Saturn have, by their attractive force, a considerable influence on the movement of the Earth; as also the reciprocal attraction of the Earth has a considerable influence on the movement of these planets.

---

40. Du Châtelet offers this footnote: "(c) See, Greg. Bk. I. Prop. 69." She is probably referring to David Gregory's (1666–1708) *Astronomiae physicae et geometricae elementa* [The Elements of Physical and Geometrical Astronomy] published in 1702, translated into English in a revised edition in 1726. It was here that she must have read of Kepler's *Epitome astronomiae copernicanae* [Epitome of Copernican Astronomy], published in 1621.

41. Du Châtelet defines *aphelion* and *perihelion* in chapter one, section VII of this Commentary.

42. Robert Hooke (1635–1703) was a chemist and mathematician. With this quotation, Du Châtelet is giving Hooke credit for the concept of universal attraction twelve years before the publication of Newton's *Principia*. She may be referring to his "An Attempt to Prove the Motion of the Earth by Observations," published in 1674. Hooke's argument with Newton over attribution for the inverse-square law may also have been reported in another work that she studied.

"2. That all the bodies that have received a simple and direct motion continue to move in a straight line, up to that point whereby some other effective force they might be diverted and forced to describe a circle, an ellipse, or some other more complex curve.

"3. That the attractive forces are the more powerful in their operations, as the body on which they operate is closer to their center.

"With regard to the proportion in which these forces diminish as the distance increases, I admit that I have not yet verified this by experiments, but it is an idea which, having been followed as it deserves to be, will be very useful to astronomers to reduce all the celestial movements to one certain rule, and I doubt that one might ever be able to find it without this. He who understands the nature of the circular pendulum and circular motion will easily comprehend the basis of this principle, and will know how to find the directions in nature for establishing it exactly. I give this beginning here for those who have the opportunity and the capacity for this research, etc."

IX

It should not be thought that this idea just thrown in the air in *Hooke's* book detracts from the greatness of Mr. Newton, who was so considerate as to mention it in his book *De Systemate mundi. (d)*[43] The examples of *Hooke* and *Kepler* serve to show the gap between a truth foreseen and a truth demonstrated, and of what little use the illuminations of genius are in the sciences, when they cease to be guided by geometry.

X

KEPLER'S STRANGE IDEAS.

Kepler, who made such beautiful and important discoveries as long as he followed this guide of geometry, provides one of the most striking proofs of the errors into which the best minds can fall when they abandon it to indulge in the pleasure of inventing systems.[44] For instance, who would have thought that this great man would indulge in the reveries of the Pythagoreans on numbers; that he could believe that the distances of the principal planets and their number were relative to the five regular solid bodies of geometry, *(e)*[45]

43. Du Châtelet offers this footnote: "*(d)* Page. 3. Edition of 1731."

44. "Systems," such as Descartes', or that of that of the Scholastics, apparently based solely on conjecture without verification by observation and experiment, were anathema to Du Châtelet and other followers of Newton.

45. Du Châtelet offers this footnote: "*(e) Mysterium cosmographicum.*" Kepler's work [The Cosmographic Mystery] was published in 1597.

which he believed were inscribed in them. Then, his observations having shown that the distances of the planets did not agree with this supposition, he imagined that celestial movements corresponded to the proportions by which one divides the string of a musical instrument, in order for it to emit the tones making up an octave. (*f*)[46]

VERY WISE COUNSEL GIVEN BY *TYCHO* TO *KEPLER*.

Kepler, having sent to *Tycho* a copy of the work in which he tried to establish these chimeras, *Tycho* answered him, that he (*g*)[47] advised him "to give up those speculations drawn from first principles and to apply himself rather to establishing his arguments on the solid basis of observations."

HUYGENS'S BIZARRE IDEA.

The great *Huygens* himself (*h*)[48] believed that the fourth satellite of Saturn, which bears his name, with our Moon and the four of Jupiter, made six, and thus the number of secondary planets was complete, and that it would be useless to seek to discover new ones, because there are also six principal planets, and the number six is termed *perfect*, because it is equal to the sum of its divisors, 1, 2, and 3.

XI

ADVANTAGE OF *NEWTON* OVER *KEPLER*; IN HIS TIME THE TRUE LAWS OF MOTION WERE BETTER KNOWN.

It is by never diverging from the most profound geometry that M. Newton found the proportion in which gravity acts and that the principle, suspected by *Kepler* and by *Hooke*, became in his hands such a fecund source of admirable and unexpected truths.

One of the things which had prevented *Kepler* from drawing from the principle of attraction all the truths that are a result of it is the ignorance in his day of the true laws of motion. M. *Newton* had the advantage over *Kepler* of benefiting from the laws of motion established by *Huygens*, and which he, in turn, pushed much further.

---

46. Du Châtelet offers this footnote: "(*f*) *Mysterium cosmographicum*."

47. Du Châtelet offers this footnote: "(*g*) 'Uti suspensis speculationibus a priori descendentibus animam potius ad observationes quas simul afferebat consierandas adjicerem.' (it is *Kepler* who speaks) 'Notae in secundam editionem mysterii Cosmographici.'"

48. Du Châtelet offers this footnote: "(*h*) Dedication in his system of Saturn." Christiaan Huygens (1629–1695), the Dutch astronomer who rejected Newton's theory of attraction. She is referring to his observations about Saturn's rings, in *Systema Saturnium* (1659).

XII

ANALYSIS OF THE *BOOK OF THE PRINCIPLES* [PRINCIPIA].

The Book of the Mathematical Principles of Natural Philosophy, the translation of which precedes this commentary, comprises three books, in addition to the definitions, the laws of motion, and their corollaries. The first book is composed of fourteen sections, the second has nine, and the third contains the application of the propositions of the first two to the system of the world.

XIII

DEFINITIONS.

The Book of the Principles begins with eight definitions. In the first two M. *Newton* explains how one ought to *measure the quantity of matter and the quantity of motion*. In the third he defines *the force of inertia*, or resistant force, with which all matter is endowed. In the fourth he explains what is understood by the term *active force*. In the fifth he defines *centripetal force;* and in the sixth, seventh, and eighth he explains how to measure *its [centripetal force] absolute quantity, its motive quantity*, and *its accelerative quantity*. Then he establishes the three following Laws of motion.

XIV

LAWS OF MOTION.

1. That all bodies persevere in the same state of rest or of uniform motion in a straight line.

2. That the change that occurs in motion is always proportional to the motor force, and happens in the direction of this force.

3. That the action and reaction are always equal and opposite.

XV

FIRST BOOK. THE FIRST SECTION CONTAINS THE PRINCIPLES OF THE GEOMETRY OF INFINITY.

After having explained these laws and having drawn from them several corollaries, M. *Newton* begins his first book with eleven lemmas that make up the first section; he sets forth in these eleven lemmas his method *of first and last ratios*. This method is the foundation of the geometry of the infinite, and with its help, this geometry has all the certainty of the old one.[49]

49. Du Châtelet is making the distinction between Euclidian geometry and the geometrical ratios and proportions Newton used for the *Principia*. See, for the English translation of the sections she refers to, Isaac Newton, *The Principia: Mathematical Principles of Natural Philosophy*, trans. I. Bernard Cohen and Anne Whitman, assisted by Julia Budenz (Berkeley: University of Cali-

AND THE OTHER THIRTEEN GENERAL PROPOSITIONS ON THE MOTION
OF BODIES.

The thirteen other sections of the first book of the Principles [*Principia*] are
devoted to demonstrating some general propositions on the motion of bodies, without having regard, either to the nature of these bodies, or the medium in which they move.

It is in this first Book that M. *Newton* gives all of his theory of the gravitation of the celestial bodies. But he has not limited himself to examining
the questions which are relevant to it, he has rendered his solutions general,
and has given a great number of applications of his solutions.

XVI

SECOND BOOK. HE DEALS WITH THE MOTION OF BODIES IN
RESISTANT MEDIA.

In the second Book M. *Newton* considers the motion of different kinds of
bodies in resistant media.

M. *NEWTON* COMPOSED THIS BOOK TO DESTROY THE VORTICES
OF *DESCARTES*.

This second Book, which contains a very profound theory of fluids and the
motion of bodies that are plunged therein, appears to have been intended
to destroy the system of vortices, although it is only in the scholium of the
last proposition that M. *Newton* openly challenges *Descartes*, and that he demonstrates that celestial movements are not compatible with his [system of]
vortices.[50]

XVII

THIRD BOOK. HE DEALS WITH THE SYSTEM OF THE WORLD.

Finally, the third book of the Principles [*Principia*] deals with the system of
the world. In this Book M. *Newton* applies the propositions of the first to the

---

fornia Press, 1999), 433ff. All references to the *Principia* will be to this English translation. The
base text for Cohen and Whitman's translation was Newton's third edition, but references are
made to the first and second in the footnotes. Du Châtelet worked from the second (1713) and
third (1726) editions in doing her translation and commentary.

50. René Descartes (1596–1650) hypothesized a universe consisting of a system of vortices
made up of particles that carried the planets in their orbits through impulsion, the impact of the
particles on each other. Newton, through a unique style of hypothesis, mathematical model and
experiment, and analogy proved the theory impossible. For example, he undermined the theory
of vortices by studying the motion of bodies in liquids, the subject of Book II of the *Principia*.
Similarly, by analogy, by experiment, and by his mathematical model, he proved his alternative explanation of the solar system.

explanation of celestial phenomena. It is in this application that I will try to follow M. *Newton* and show the logical sequence of his Principles, and how easily they explain astronomical phenomena.

XVIII

WHAT IS MEANT IN THIS TREATISE BY THE WORD *ATTRACTION.*
Moreover, I here declare, as M. *Newton* himself did, that in using the word *attraction*, I only take it to mean the force that makes bodies tend toward a center, without claiming to assign the cause of this tendency.

CHAPTER ONE:
PRINCIPAL PHENOMENA OF THE SYSTEM OF THE WORLD

I

It will be useful before giving an account of the way in which the theory of M. *Newton* explains celestial phenomena, to give a brief idea of our planetary system.

The truths discovered by M. *Newton* will necessarily be part of this exposition, but explanations of how he arrived at them will be deferred to subsequent chapters. This [chapter] will contain only the exposition of the phenomena as such.

II

FIRST DIVISION OF THE CELESTIAL BODIES OF OUR PLANETARY SYSTEM INTO *PRINCIPAL PLANETS* AND *SECONDARY PLANETS.*
The celestial bodies that compose our planetary system are divided into *principal planets*, that is to say, those that have the Sun as the center of their movement, and *secondary planets*, which are called *satellites;* the latter turn around the principal planet, which is their center.

NAMES AND CHARACTERS OF THE PRINCIPAL PLANETS.
There are six principal planets, whose characters and names are *Mercury, Venus, Earth, Mars, Jupiter, Saturn.*

This enumeration of the principal planets follows the order of their distances to the Sun, beginning with those closest to it.

WHICH ARE THE PLANETS THAT HAVE SATELLITES.
GENERAL ENUMERATION OF THE CELESTIAL BODIES THAT COMPOSE OUR PLANETARY SYSTEM. The Earth, Jupiter, and Saturn are the only planets for

which satellites have been discovered: the Earth only has one, the Moon; Jupiter has four; and Saturn five besides its ring. These compose our planetary system of eighteen celestial bodies, including the Sun, and the ring of Saturn.

III

SECOND DIVISION OF PLANETS INTO *SUPERIOR PLANETS* AND *INFERIOR PLANETS*.

WHICH ARE THE INFERIOR PLANETS AND WHAT THEIR ARRANGEMENT IS. The principal planets can be divided into *superior planets* and *inferior planets*. The inferior planets are closer to the Sun than the Earth; these planets are Mercury and Venus. The orbit (*a*)[51] of Venus encompasses the orbit of Mercury around the Sun, and the orbit of Earth is exterior to those of Mercury and of Venus, and encompasses them as well as the Sun.

HOW THIS ARRANGEMENT WAS DISCOVERED. This arrangement is known because Venus and Mercury sometimes appear to us to be between the Sun and us. This could not happen if these two planets were not closer to the Sun than the Earth; and it is clearly perceived that Venus moves farther away from the Sun than Mercury, and that as a consequence its orbit encompasses that of Mercury.

WHICH ARE THE SUPERIOR PLANETS AND WHAT THEIR ARRANGEMENT IS. The superior planets are those that are farther away from the Sun than the Earth; they are three in number, *Mars, Jupiter,* and *Saturn.*

It is known that the orbits of these planets encompass that of the Earth, because the Earth is sometimes between the Sun and them.

The orbit of Mars encompasses that of the Earth, the orbit of Jupiter that of Mars, and the orbit of Saturn that of Jupiter. Thus, of the three superior planets, Saturn is farthest from the Earth, and Mars is the closest.

HOW THIS WAS DISCOVERED. This arrangement is known because the planets closest to the Earth sometimes conceal from us (*b*)[52] those that are the most distant.

---

51. Du Châtelet offers this footnote: "(*a*) One calls *orb*, or orbit, the curve that a planet describes in turning around the body, which acts as its center."

52. Du Châtelet offers this footnote: "(*b*) Wolff, *Eléments d'Astronomie* [Elements of Astronomy]." She is referring to Christian Wolff (1679–1754), the German philosopher whose ideas she described in her *Foundations of Physics.* French translations of Wolff's writings were sent to her and Voltaire by Frederick of Prussia at the end of the 1730s.

IV

THE PLANETS ARE OPAQUE BODIES.

All the planets are opaque bodies. The opacity of Venus and of Mercury is certain because when these planets pass between the Sun and us, they appear on this star as small black spots, and they have what are called *phases*. That is to say, that the quantity of their illumination depends on their position in relation to the Sun and to us.

HOW THIS WAS NOTICED.

For the same reason we infer the opacity of Mars, that it also has *phases*, and the opacity of Jupiter and of Saturn can be assumed, because their satellites do not appear to be illuminated by these planets when they are between the Sun and these satellites. This proves that the hemisphere of these planets that is not illuminated by the Sun is opaque.

V

THE PLANETS ARE SPHERICAL. HOW THIS WAS DISCOVERED.

Finally, it is known that the planets are spherical bodies, because, however they are placed in relation to us, their surface always appears to us bounded by a curve.

It is assumed that the Earth is spherical, because in eclipses its shadow always appears bounded by a curve; because on the sea a ship is seen disappearing little by little in the distance, so that one begins by losing sight of the body of the ship, next the sails, and finally the masts. Lastly, the Earth must be spherical because no edge to the Earth's surface has been found, although several explorers have circumnavigated it, yet that is what would have to be if the Earth were flat.

VI

ALL THE BODIES OF OUR PLANETARY SYSTEM APPEAR TO BE OF THE SAME TYPE, EXCEPT FOR THE SUN.

All that we know of the principal planets proves to us then that these are spherical, opaque, solid bodies.

IT IS LIKELY THAT THE SUBSTANCE OF THE SUN IS FIRE.[53]

The Sun appears to be of an entirely different nature from the planets; we do not know if it is composed of solid or fluid particles. We know only that

---

53. Du Châtelet wrote about the nature of the Sun at greater length in her *Dissertation on the Nature and Propagation of Fire*; see part 2, section XIV.

its particles shine; that they heat, and that they burn when they are gathered together in a sufficient quantity. Thus, in all likelihood, the Sun is a fiery body roughly similar to fire on Earth, since its rays produce the same effects.

VII

THE CURVE IN WHICH THE CELESTIAL BODIES TURN AROUND THE SUN. WHAT THE *LINE OF THE APSIDES*, THE *APHELION*, AND THE *PERIHELION* ARE.

All the celestial bodies make their revolutions around the Sun in more or less elongated ellipses (c)[54] in which the Sun occupies one of the foci; thus, the planets, in turning around the Sun, are sometimes nearer and sometimes farther from it. The line that passes through the Sun, and that ends at two points on the ellipse, the one nearest to the Sun and the other the greatest distance from the Sun, is called *the line of the apsides*, the point of the orbit the farthest from the Sun is called the *aphelion* of the planet, and the point closest to it is called its *perihelion*.

IN WHAT SENSE THE PLANETS TURN AROUND THE SUN.

The principal planets carry with them in their revolution around the Sun the satellites of which they are the center. The revolution of the planets around the Sun goes from occident [west] to orient [east]. (d)[55]

OF COMETS.

From time to time, celestial bodies appear that move in every direction, with extreme rapidity when they are close enough to be visible: these are the comets.

THE COMETS ARE PLANETS.

We do not yet have enough observations to know the number of comets. We only know—and we have only been certain about this for a short time— that these planets turn around the Sun like the other bodies of our planetary world, and that they describe ellipses so elongated that they are only visible to us for a very small part of their orbit.

54. Du Châtelet offers this footnote: "(c) Type of curve that in ordinary speech is called *oval*. The foci are the two points at which gardeners place their stakes to trace this type of figure, which they often use."
55. Du Châtelet offers this footnote: "(d) In all that is said here, one supposes the spectator placed on the Earth."

VIII

ALL THE PLANETS AND THE COMETS OBSERVE *KEPLER'S* LAWS.

In turning around the Sun, all the planets observe *Kepler's* two laws, which were discussed in the Introduction.

We know that the comets observe the first of these laws, I mean, the law according to which celestial bodies describe (e)[56] equal areas in equal times. And it will be seen later that it is probable, from the observations that one has been able to make up to the present, that the comets also observe the second law, namely, that of periodic time (f)[57] in sesquiplicate [one and a half times] ratio to the distances.

It is, I believe, more useful to give an example of the sesquiplicate ratio than a definition. Suppose then, that the mean distance of Mercury to the Sun is 4, that of Venus 9, that the periodic time of Mercury is 40 days, and that when the periodic time of Venus is sought, one cubes the first two numbers 4 and 9, and the result is 64 and 729. One next takes the square root of these 2 numbers, and this is 8 for that of the first, 27 for that of the second. One next makes this rule of three 8:27::40:135, that is to say that the square root of the cube of the mean distance of Mercury to the Sun is to the square root of the cube of the mean distance of Venus to the Sun, as the periodic time of Mercury around the Sun is to the periodic time sought for Venus around the Sun, which turns out to be 135 on the basis of suppositions that one made, and it is this that one calls the ratio sesquiplicate.[58]

IX

PROOFS OF THE MOVEMENT OF THE EARTH.

If these two laws of *Kepler*, that all observations have confirmed, are accepted, they furnish very strong arguments to prove the movement of the Earth that was so obstinately disputed for so long. For if one takes the Earth as the center of celestial movements, these two laws are not observed at all; the planets do not describe areas proportional to the time around the Earth, and the time of the revolutions of the Sun and the Moon, for example,

---

56. Du Châtelet offers this footnote: "(e) The word *area* in general means a surface, here it signifies *a space enclosed between two lines pulled from the center to two points where the planet is located;* these areas are proportional to time, that is to say, they are greater or smaller, as the time in which they are described is longer or shorter."

57. Du Châtelet offers this footnote: "(f) The periodic time is the time that a planet takes to make its revolution on its orbit."

58. The formula she has created: $8:27 = 40:x$, $8x = 1080$, $x = 135$.

around this planet, are not as the square root of the cube of their mean distance to the Earth. For the periodic time of the Sun around the Earth being about 13 times greater than that of the Moon, its distance to the Earth would have to be, following *Kepler*'s rule, between 5 and 6 times greater than that of the Moon. Now, one knows that this distance is about 400 times greater, thus, if one accepts *Kepler*'s laws, the Earth is not the center of the celestial revolutions.

Moreover, the centripetal force (*g*)[59] that M. Newton made evident as the cause of the revolution of the planets, makes the curve they describe around their center concave (*h*)[60] toward it, since its effect is to draw them away from the tangent (*i*).[61] Now, the orbits of Mercury and Venus are in some parts convex to the Earth, so the inferior planets do not turn around the Earth.

It is easy to prove the same thing about the superior planets, for these planets appear sometimes (*k*)[62] *direct*, sometimes stationary, and sometimes retrograde, all apparent inequalities that would not take place for us, if the Earth were the center of celestial revolutions.

None of this would appear to a spectator placed on the Sun, since they are only an effect of the movement of the Earth in its orbit, combined with that of the planets in theirs.

This is why the Sun and the Moon are the only celestial bodies that appear to us to be always direct. For the Sun does not travel through an orbit at all, its movement cannot combine with that of the Earth, and the Earth being the center of the movements of the Moon, the Moon must always appear to us direct as all the planets would appear to a spectator placed on the Sun.

59. Du Châtelet offers this footnote: "(*g*) The phrase *centripetal force* is self-explanatory, it could not mean anything other than the force that makes a body tend to a center."

60. Du Châtelet offers this footnote: "(*h*) The two sides of the glass of a watch can serve to make the words *concave* and *convex* understood; the exterior side of the glass is *convex*, and the side next to the face is *concave*."

61. Du Châtelet offers this footnote: "(*i*) The tangent is the line that touches a curve, but which can never cross it."

62. Du Châtelet offers this footnote: "(*k*) A planet is said to be *direct* when it appears to go according to the order of the signs, that is to say, from *Aries* to *Taurus*, from *Taurus* to *Gemini*, and etc., this is also called *going consequentially*; it is *stationary* when it appears to correspond for some time to the same points of the sky. And finally, it is *retrograde* when it appears to go against the order of the signs; this is also called *going in antecedence*, that is to say, from *Gemini* to *Taurus*, from *Taurus* to *Aries*, etc."

OBJECTION MADE TO *COPERNICUS*, DRAWN FROM THE PLANET VENUS.
HIS ANSWER TO THIS OBJECTION.

The planet Venus furnished one of the objections made to *Copernicus*'s system. If Venus, others told him, turned around the Sun, it would have phases like the Moon. Indeed, said *Copernicus*, if your eyes were good enough to distinguish these phases, you would see them; and perhaps one day astronomers will find the means to perceive them.

DISCOVERY THAT CONFIRMED THIS ANSWER.

Galileo was the first to have verified *Copernicus*'s prediction, and each discovery that has been made since on the course of the planets has confirmed it.

X

AT WHAT ANGLE THE PLANES OF THE PLANETS INTERSECT.

The planes (l)[63] of the orbits of all the planets cut each other along lines that pass through the center of the Sun, so that a spectator placed at the center of the Sun would find himself in the planes of all these orbits.

WHAT ARE CALLED THE NODES AND THE LINE OF THE NODES OF
AN ORBIT.

The line in which the plane of each orbit cuts the plane of the ecliptic, that is to say, the plane in which the Earth moves, is called the *line* of the nodes, and the points of this section are called the *nodes* of the orbit.

INCLINATION OF THESE PLANES TO THE ECLIPTIC.[64] THESE
PROPOSITIONS ARE TAKEN FROM *GREGORY*, BK. I., PROP. 3.

All these planes are inclined to the plane of the ecliptic, at the following angles.

The plane of the orbit of Saturn makes an angle of 2°½ with the plane of the ecliptic, that of Jupiter is 1°⅓, that of Mars is a little less than 2°, that of Venus a little over 3°⅓, finally, that of Mercury is about 7°.

XI

The orbits of the principal planets being ellipses with the Sun as a focus, all these orbits are eccentric, and, more or less so, according to the distance between their center and the point where the Sun is placed.

63. Du Châtelet offers this footnote: "(l) The plane of the orbit of a planet is the surface in which it is supposed to move."

64. The ecliptic is the apparent path of the Sun, indicated by the plane of the Earth's orbit extended as if to meet the Sun, inclined to the Earth's equator by an angle of 23°27".

ECCENTRICITIES OF THE PLANETS MEASURED IN HALF-DIAMETERS OF
THE EARTH.

The eccentricity of all of these orbits has been measured, and found to be as
follows, the eccentricity of

Saturn: 54,207 parts,
Jupiter: 25,058
Mars: 14,115
Earth: 4,692
Venus: 500
Mercury: 8,149 parts,

taking the demi-axis of the big orbit of the Earth as a common measure, and
supposing it to be 100,000 parts.

ECCENTRICITY OF THE PLANETS IN HALF-DIAMETERS OF THEIR
BIG ORBIT.

Reducing the eccentricity of the planets to the half-diameter of their big
orbit, and supposing this half-diameter to be 100,000 parts, the eccentricities are

Saturn: 5,683 parts,
Jupiter: 4,822
Mars: 9,263
Earth: 5,700
Venus: 694
Mercury: 21,000 parts.

Thus, the eccentricity of Venus is almost imperceptible.

XII

PROPORTIONS OF THE DIAMETERS OF THE DIFFERENT PLANETS.

The planets are different in size; only the absolute diameter of the Earth is
known, as this planet is the only one the circumference of which it has been
possible to measure; but we know the ratio between the diameters of the
other planets, and taking that of the Sun as common measure and supposing it to be 1,000 parts,

that of Saturn to the Sun: 137
that of Jupiter: 181
that of Mars: 6
that of the Earth: 7
that of Venus: 12
finally that of Mercury: 4

From which one sees that Mercury is the smallest of all the planets, for we
know that the volumes of spheres are as the cubes of their diameters.

XIII

DISTANCES OF THE PLANETS TO THE SUN.

The planets are at different distances from the Sun. Taking the distance of the Earth to the Sun as the common measure, and supposing it to be 100,000 parts, the six principal planets are drawn up around the Sun in the following order, when they are at their mean distance from it:

Mercury is at	38,710
Venus at	72,333
the Earth at	100,000
Mars at	152,369
Jupiter at	520,110
finally, Saturn at	953,800.

DISTANCES OF THE PLANETS TO THE EARTH.

The mean distances from the Sun and from the planets to the Earth have been calculated in half-diameters of the Earth. Here are those given by M. *Cassini:* the Sun, Mercury, and Venus are almost equally far away at their mean distance, which is 22,000 half-diameters of the Earth; Mars is at 33,500, Jupiter at 115,000, and Saturn at 210,000.[65]

XIV

PERIODIC TIMES OF THE PLANETS AROUND THE SUN.

The times of the revolutions of the planets around the Sun are shorter, the closer they are to it. Thus, Mercury, which is the closest to the Sun, makes its revolution in 87 days; Venus, which is next, makes its in 224; the Earth in 365; Mars in 686; Jupiter in 4,332; and finally, Saturn, which is the farthest from the Sun, needs 10,759 days to turn around it. All of this is in rounded numbers.

XV

ROTATION OF THE PLANETS.

In addition to their movement of translation around the Sun, the planets have another movement around their axis that is called their *diurnal revolution*.

---

65. Jacques Cassini (1667–1756), the French astronomer, was famous for his geodesic and astronomical observations. He contested Newton's view that the Earth was flattened at the Poles, and thus was in conflict with the views of Du Châtelet's mentor, Pierre-Louis Moreau de Maupertuis (1698–1759), and the results of his expedition to Lapland.

MEANS USED TO DISCOVER IT. WHICH ARE THE PLANETS THE ROTATION
OF WHICH IS KNOWN.

TIMES OF THE ROTATIONS OF THE PLANETS AROUND THEIR AXES. Only
the diurnal revolution of the Sun and of four planets is known: the Earth,
Mars, Jupiter, and Venus. The spots that were noticed on their faces, (m)[66]
and that have been seen to appear and disappear successively, made discovery of this revolution possible. Mars, Jupiter, and Venus having spots
on their surface, we learned from the return of these same spots, and by
their successive disappearance, that these planets turn on themselves, and
we learned the time needed for these revolutions; thus, it has been observed
that Mars turns in 23h20', and Jupiter in 9h56'.

UNCERTAINTIES ABOUT THE TIME OF THE ROTATION OF VENUS. Astronomers are not in agreement on the time of the revolution of Venus around its
own axis. Most believe that it turns in approximately 23 hours; but M. *Bianchini*, who has made a very particular study of the appearances of this planet,
believes its revolution on its axis to be 24 days.[67] As he was obliged to trans-
port the instrument with which he was observing during the observation itself, because of a house that hid Venus from him, and as this operation took
almost one hour, it may be that during this time the spot he was observing
changed. Be that as it may, his authority in this matter warrants a suspension
of judgment until we have more complete observations.

M. *Delahire* with a sixteen-foot-long telescope has observed some mountains on Venus higher than those on the Moon.[68]

OBSERVATION ALONE CANNOT PROVE THE ROTATION OF MERCURY, OR
THAT OF SATURN, AND WHY

Mercury is so flooded by the rays of the Sun that one cannot be assured, by
observation alone, if it turns on its axis; it is the same for Saturn because it
is so far away.

In 1715 M. *Cassini* observed with a telescope 118 *pieds* long, three bands
on Saturn similar to those one sees on Jupiter. But apparently no one has
been able to make this observation with enough exactitude to conclude from
it the rotation of Saturn around its axis.

66. Du Châtelet offers this footnote: "(m) One calls the *face* of the planet that part of its surface that is visible to us."
67. Francesco Bianchini (1662–1729) was an Italian astronomer whose measurements Du
Châtelet probably read of in another work of astronomy.
68. Philippe de la Hire (1640–1718), a French mathematician and astronomer, best known for
trying to replicate Newton's optical experiments.

BY ANALOGY: THESE PLANETS ALSO TURN ON THEIR AXES

Mercury and Saturn being subject to the same laws that direct the courses of the other celestial bodies, and these planets, by all that we can know of them, appearing to us as bodies of the same type, analogy suggests that these two planets turn on their centers as the others, and that one day perhaps someone will succeed in perceiving this revolution, and how much time it takes.

XVI

HOW THE REVOLUTION OF THE SUN ON ITS AXIS WAS DISCOVERED. THE SPOTS OF THE SUN.

From time to time spots appear on the Sun, which have taught us that this star also turns on its axis.

There had to be many observations after the discovery of these spots before some were observed that lasted long enough to enable us to determine the time of the Sun's revolution on its axis. In his fifth astronomy lesson, *Keill*[69] reports that some spots took 13½ days to go from the occidental rim of the Sun to its oriental rim, and that at the end of 13½ more days, they reappeared on its occidental side. From which he concludes that the Sun turns on its axis in about 27 days from occident to orient, that is to say, turning in the same direction as the planets.

Using the same spots, it was ascertained that the Sun's axis of rotation makes, with the plane of the ecliptic, an angle of about 7 degrees.

In his commentary Father *Jacquier* made a point about these spots that deserves reporting.[70] Seeing that no observation proves the equality of the time of occultation, and, that on the contrary, from all the observations he studied these times appeared to be unequal, and that the time of occultation during which they are hidden has always been longer than that during which they are visible, he concluded (as did M. *Wolff*, art. 413 of his *Astron.*) that these spots are not inherent in the Sun, but are at a distance above its surface.

---

69. John Keill (1671–1721), professor of astronomy at Oxford, was a defender of Newtonian ideas of the universe against those of Leibniz. His *Introduction to Natural Philosophy* appeared in Latin in 1701 and English in 1720.

70. Father [Père] François Jacquier (1711–1788) with Father Thomas Le Seur (1703–1770) did a profusely annotated edition of Newton's *Principia*. The annotation was so extensive that it constituted a commentary on the work. This was known as the Geneva edition of the *Principia* (1739–42).

Jean *Fabrice* (*n*)[71] was the first to discover these spots (in Germany, in 1611) and from this deduced the diurnal revolution of the Sun. Next, the Jesuit *Scheiner* (*o*)[72] observed them and also gave his observations. *Galileo*, about the same time, made the same discovery in Italy.

In *Scheiner's* time more than fifty spots were observed on the surface of the Sun, whence one can assign the cause of a phenomenon reported by some historians, that the Sun appeared very pale, sometimes for an entire year. For it is only necessary that some spots be big enough, and last long enough to cause this phenomenon.

Today we no longer doubt that the Earth turns on its axis in 23h56', which makes up our astronomic day and causes the alternation of days and nights, which all regions of the Earth enjoy.

XVII

THE EFFECT OF THE ROTATIONAL MOVEMENT OF THE PLANETS IS TO BROADEN THE PLANET AT ITS EQUATOR.

OF CENTRIFUGAL FORCE. This movement of celestial bodies around their center alters their form; for it is known that a circular motion makes bodies that turn acquire a force, which, with the time of their revolutions remaining the same, increases as the circle it describe increases, and one calls this force, *centrifugal force*, that is to say, a force *that moves away from the center*. Thus, by the rotation, the central parts of the planets acquire a centrifugal force all the greater, the closer they are to the equator of these planets—since the equator is the great circle of the sphere—and smaller, the closer they are to the poles of these planets (*p*).[73] Supposing then, the celestial bodies to have been spherical in the state of rest, their rotation around their axes must have raised the regions of the equator, and lowered those of the poles, and as a consequence changed the spherical form of the planet to a spheroid flattened at the poles.

71. Du Châtelet offers this footnote: "(*n*) *Wolff Elementa Astron.* Chap. I."

72. Du Châtelet offers this footnote: "(*o*) This Jesuit told his Superior that he had discovered the spots on the Sun, to which his Superior responded gravely, *That is impossible, I have read Aristotle two or three times, and I found nothing of this kind there.*" Christoph Scheiner (1573–1650), the Jesuit astronomer, published his observations anonymously because of Jesuit opposition to these discoveries, and their confirmation of similar observations made by Galileo Galilei (1564–1642).

73. Du Châtelet offers this footnote: "(*p*) The *poles* are the points around which the revolving body turns, and the *equator* the circle parallel to these points that divides the revolving sphere in two equal parts."

WHICH ARE THE PLANETS IN WHICH THE ELEVATION OF THE EQUATOR
CAN BE PERCEIVED.

Thus, the theory demonstrates that all the planets must be flattened toward their poles by their rotation, but this flattening is only perceptible in Jupiter and in our globe. It will be seen later on that the theory makes it possible to determine the quantity of this flattening in the Sun, but that it is too small to be perceivable by observation.

The measures taken at the polar circle, in France and at the equator, have given us the proportion of the axes (q)[74] of the Earth as being about 173 to 174.[75]

Telescopes show us the flattening of Jupiter, whose flattening is much greater than that of the Earth, because this planet is much bigger, and it revolves much more rapidly on its axis; the ratio of the axes of Jupiter is estimated as 13 to 14.

XVIII

THE OBSERVATIONS THAT DEMONSTRATE THAT THE EARTH, MARS,
JUPITER, VENUS, AND THE SUN HAVE ATMOSPHERES.

The spots of Venus, of Mars, and of Jupiter being variable and often changing form, make it very probable that these planets are, like ours, enclosed in an atmosphere, the changes in which produce these appearances.

As regards the Sun, as its spots are not inherent in its face, and as they appear and disappear very often, there is no doubt that it has an atmosphere immediately surrounding it in which these spots form and dissipate by turns.

XIX

MASSES OF THE SUN, OF JUPITER, OF SATURN, AND OF THE EARTH.
THEIR DENSITIES. WEIGHTS OF THE SAME BODIES ON THEIR DIFFERENT
SURFACES.

All that I have just set forth was known before M. *Newton*, but before him no one believed it was possible to know the mass of the planets, their density, and what the same body would weigh if it were transported succes-

---

74. Du Châtelet offers this footnote: "(q) In general, *axis* or *diameter* is every line that passes through the center of a body and ends at the circumference; in this case, the axes are two lines that pass through the center, one of which ends at the poles and the other at the equator."

75. Du Châtelet is referring to Maupertuis' expedition to Lapland in the Arctic Circle and a similar expedition to the equator in Peru and Brazil by Charles-Marie de La Condamine (1701–1774). Their measurements proved the flattening of the Earth at the poles and thus, one of the four hypotheses Newton presented to confirm his theory of attraction.

sively to the surface of different planets. In the following chapter, it will be shown how M. *Newton* arrived at these strange discoveries. Suffice it to say here that he found that the masses of the Sun, of Jupiter, of Saturn, and of the Earth, that is to say, the quantities of matter they contain, are 1, 1/1067, 1/3025, and 1/169282, supposing the parallax (r)[76] of the Sun to be 10'3". He found that their densities are respectively: 100, 94, 67, and 400; and that the weight of the same body transported successively to the surface of the Sun, of Jupiter, of Saturn, and of the Earth, would be 10,000, 943, 529, and 435, respectively.

To determine these proportions M. *Newton* supposed the half-diameters of the Sun, of Jupiter, of Saturn, and of the Earth, to be 10,000, 997, 791, and 109, respectively.

WHY THESE PROPORTIONS CANNOT BE KNOWN FOR THE
OTHER PLANETS.

It will be seen in the following chapter why one can know neither the density nor the quantity of matter of Mercury, Venus, and Mars, nor what bodies weigh on these three planets.

XX

PROPORTIONS OF THE SIZES AND THE MASSES OF THE THREE PLANETS
AND OF THE SUN.

It follows from all these proportions that Saturn is about 500 times smaller than the Sun, and that it contains 3,000 times less matter; that Jupiter is 1,000 times smaller than the Sun, and that it contains 1,033 times less matter; that the Earth is but a small dot in relation to the Sun, as it is 1,000,000 times smaller; and finally that the Sun is 116 times bigger than all the planets taken together.

XXI

PROPORTIONS OF THE SIZES AND THE MASSES OF THE PLANETS
AND OF THE EARTH, AND THE OTHER PLANETS IN RELATIONSHIP TO
ONE ANOTHER.

In comparing the planets with one another, one finds that only Mercury and Mars are smaller than the Earth; that Jupiter is not only the biggest of the planets, but that it is also bigger than all the other planets taken to-

76. Du Châtelet offers this footnote: "(r) *The parallax* of the Sun is the angle at which the radius of the Earth is seen from the Sun. Thus, the parallax of a star, whatever its relationship to the Earth, is the angle at which the radius of the Earth could be seen from this star."

gether, and that this planet is more than two thousand times bigger than the Earth.

XXII

OF THE PRECESSION OF THE EQUINOXES. IN WHAT DIRECTION IT
HAPPENS, AND HOW LONG IT TAKES TO ACCOMPLISH IT. ITS ANNUAL
DURATION.

The Earth, aside from its annual and its diurnal movement, has yet another movement by which its axis disturbs its (s)[77] parallelism and corresponds at the end of a certain time to different points in the sky. This movement causes what is called *the precession of the equinoxes*, that is to say, the retrogradation of equinoctial points, or of the points at which the equator of the Earth cuts the ecliptic. The movement of the equinoctial points happens against the order of the signs, and it is so slow that is takes 25,920 years to complete, that is one degree in 72 years, and 50" in about a year. As will be seen later on, M. *Newton* found the cause of this movement in the attraction of the Sun and the Moon on the protuberance of the Earth at its equator.

TROPICAL YEAR AND SIDEREAL YEAR.

Because of the precession of the equinoxes astronomers distinguish between the *tropical* year and the *sidereal* year. They call the tropical year the interval of time that elapses between the same two equinoxes in two annual revolutions of the Earth, and this year is a little shorter than the sidereal year, which is the time the Earth takes to return from any given point of its orbit to that same point.

XXIII

OF THE SECONDARY PLANETS.

We still need to speak of the secondary planets. They number 10, without counting the ring of Saturn; these 10 planets are the 5 moons of Saturn, the 4 of Jupiter, and the one that accompanies the Earth.[78]

THEY FOLLOW *KEPLER'S* LAWS.

Observations have shown us that the secondary planets observe *Kepler's* laws in turning around their principal planet.

---

77. Du Châtelet offers this footnote: "(s) One calls *parallel* a line that always keeps the same position in relation to some supposed fixed point."

78. Astronomers have now identified many more secondary planets, twenty-four for Jupiter alone.

DISCOVERY OF THE SATELLITES OF JUPITER.
Only recently were the satellites of Jupiter and of Saturn discovered, and this discovery was impossible before the telescope. $(t)^{79}$ *Galileo* discovered the 4 satellites of Jupiter, which he named the *stars of the Medicis;* and they have been of great use in geography and astronomy.

AND OF THOSE OF SATURN.
M. *Huygens* was the first to discover a satellite of Saturn, and the 4th is named for him. M. *Cassini*, the elder, discovered the other four.

XXIV

DISTANCES OF THE MOONS OF JUPITER TO THE PLANET, AND THEIR
PERIODIC TIMES AROUND JUPITER.
If we take the half-diameter of Jupiter as the common measure, its 4 satellites are placed at the following distances from it, beginning with the closest one.

The first is at 5, the second at 9, the third at 14, and finally the fourth at 25 in round numbers, according to the observations of M. *Cassini* on the eclipses of these satellites.

Their periodic times around Jupiter are longer, the farther they are from the planet. The first revolves in 42h, the second in 85, the third in 171, and the fourth in 400, discounting any minutes.

We know neither the diurnal revolution, nor the diameter, nor the size, mass, density, nor the quantity of the attractive force of these satellites, and up to the present, even with the best telescopes, they are so small that these discoveries can scarcely be hoped for. It is the same with the five moons that revolve around Saturn.

XXV

DISTANCES OF THE SATELLITES OF SATURN FROM SATURN AND THEIR
PERIODIC TIMES AROUND THIS PLANET.
Taking the half-diameter of the ring of Saturn as a common measure, the distances of the satellites of Saturn from this planet are in the following proportions, beginning with the one closest to the planet.

The first is at 1, the second at 2, the third at 3, the fourth at 8, and the fifth at 24 as a round number, and their periodic times are, according to

79. Du Châtelet offers this footnote: "(t) M. *Wolff* in his Astronomie, Chap. II. asserts that in Germany Simon *Marius*, a Brandenburg mathematician, discovered three satellites of Jupiter in the same year that *Galileo* discovered them in Italy."

M. *Cassini*, 45 hours, 65 hours, 109 hours, 382 hours, and 1,903 hours, respectively.

The satellites of Saturn all make their revolution in the plane of the equator of this planet, only the fifth is off from it by 15 or 16 degrees.

CONJECTURES OF M. *HUYGENS* ABOUT A SIXTH SATELLITE OF SATURN.

Several astronomers, among them M. *Huygens*, have surmised that a sixth satellite of Saturn between the fourth and the fifth, may perhaps be discovered one day, the distance between these satellites being proportionately too big in relation to the distances between the others; but then there would be another difficulty. This satellite, which would be the fifth, would be so very much smaller than the four between it and the planet that better telescopes would need to be perfected to perceive it.

The orbits of the satellites of Jupiter and Saturn are almost concentric with these planets.

OBSERVATIONS OF M. *MIRALDI* ABOUT THE SATELLITES OF JUPITER.[80]

M. *Miraldi* has observed spots on the satellites of Jupiter; but no one has yet been able to deduce any consequence from this observation, which could, if it were investigated further, teach us many things about the movements of the satellites.

XXVI

THE RING OF SATURN. IT DOES NOT TOUCH THE BODY OF THIS PLANET.
ITS DISTANCE TO THE BODY OF THE PLANET. ITS DIAMETER. ITS WIDTH.
ITS THICKNESS.

Saturn, in addition to its five moons, is also enclosed by a ring. This ring does not touch Saturn at any point; the fixed stars can be seen through the space that separates it from the body of this planet. According to M. *Huygens*, the diameter of this ring is to the diameter of Saturn as about 9 to 4, thus it is more than double the diameter of Saturn. The distance of the body of Saturn to its ring is about half of this diameter, so that the width of the ring is almost equal to the distance between its interior rim and the globe of Saturn.

THIS IS AN OPAQUE BODY, WHICH HAS PHASES.

Its thickness is very little, for when its edge is presented to us, it is not visible to us, and it only appears then as a black line that crosses the globe

---

80. Giancomo Miraldi, an Italian astronomer, noted for his observations of the planet Mars in the first decades of the eighteenth century.

of Saturn; further, this ring has phases according to Saturn's position in its orbit, which proves that it is an opaque body, and that just like the other bodies of our planetary system, it only shines by reflecting the light of the Sun to us.

One cannot tell if the ring of Saturn revolves, for there is no change in its appearance from which one might conclude this rotation.

The plane of this ring always makes an angle of 23°½ with the plane of the ecliptic, thus its axis always stays parallel to itself in its translation around the Sun.

OF THE DISCOVERY OF THIS RING. WHAT WAS BELIEVED BEFORE M. HUYGENS.

It is to M. *Huygens* that we owe the discovery of Saturn's ring, which is a unique phenomenon in the sky. Before him, astronomers had observed phases in Saturn, for they confused this planet with its ring; but those phases were so different from those of the other planets they could not be explained. One can see in *Hevelius* the names he gives to these appearances of Saturn, and how (u)[81] far he was from suspecting the truth.

M. *Huygens*, in comparing the different appearances of Saturn, found that they were caused by a ring that encircled the planet, and this supposition corresponded so well with all that the telescopes discover there that, at present, no astronomer doubts the existence of this ring.

GREGORY'S IDEA ABOUT THIS RING.

Gregory, in speaking of the idea of M. *Halley* that the terrestrial globe could well be only an assemblage of concentric crusts over an interior stone, has conjectured that the ring of Saturn was formed of many concentric crusts that have become separated from the body of the planet, whose earlier diameter was equal to the sum of its current diameter, and of the width of the ring.[82]

It is still being conjectured that the ring of Saturn may only be an assemblage of moons the great distance makes us see as contiguous; but all this is not founded on any observation.

---

81. Du Châtelet offers this footnote: "(u) *Hevelius* in *Opusculo de Saturni nativa facie* distinguishes between the different aspects of Saturn by name: *monaspericum, trispbericum, spherico-ansatum, ellipti-coansatum, sphaeri-cocuspitatum,* and he subdivided these into yet others." Johannes Hevelius (1611–1687), the Danzig astronomer, published a number of celestial atlases and catalogues.

82. Edmund Halley (1656–1742), the English astronomer and professor of geometry at Oxford, was known particularly for his observations of comets.

THE SATELLITES OF JUPITER AND SATURN ARE SPHERICAL BODIES.
HOW THIS WAS ASCERTAINED.

It is known by the shadows of the satellites of Jupiter and Saturn on their principal planets that these satellites are spherical bodies.

XXVII

OF THE MOON.

Our Earth has only one satellite, which is the Moon, but its proximity has meant that much more could be discovered about this satellite than about the others.

THE CURVE IT DESCRIBES AROUND THE EARTH.

The Moon makes its revolution around the Earth in an ellipse of which the Earth occupies one of the foci. This ellipse changes position and type continuously, and one will see in the following chapters that the Sun is the cause of these variations.

ITS PERIODIC MONTH.

The Moon follows *Kepler's* rules in turning around the Earth, and it is only disturbed by the action of the Sun on it. It makes its revolution around the Earth, from occident to orient, in 27 days 7h43', and this is called its *periodic month*.

ITS PHASES. ITS SYNODIC MONTH.

The face of the Moon that we see is sometimes entirely illuminated by the Sun and sometimes only in part; this illuminated part appears bigger or smaller according to its position in relation to the Sun and the Earth, this is what are called *its phases*. It goes through all its phases in the space of one revolution, termed *synodic*, and this is composed of the time that [the Moon] takes to go from its conjunction with the Sun to its next conjunction; this synodic month of the Moon is about 29½ days.

THE MOON IS AN OPAQUE AND SPHERICAL BODY.

The phases of the Moon prove that it is an opaque body, and that it only shines by reflecting to us the light of the Sun.

HOW THIS WAS DISCOVERED. THE EARTH LIGHTS THE MOON DURING ITS NIGHTS.

The Moon is known to be a spherical body, because it always appears to us as curved.

Our Earth lights the Moon during its nights just as the Moon lights us during ours, and it is by the light reflected from the Earth that one sees the Moon when it is not illuminated by the light of the Sun.

PROPORTION OF THIS ILLUMINATION.

As the surface of the Earth is about 14 times greater than that of the Moon, the Earth seen from the Moon must appear 14 times brighter, and send 14 times more rays to the Moon than the Moon sends to us, supposing merely that these two planets are equally able to reflect light.

INCLINATION OF THE PLANE OF THE ORBIT OF THE MOON.

The plane of the orbit of the Moon is inclined to the plane of the ecliptic at an angle of about 5°.

The big axis of the ellipse that the Moon describes in turning around the Earth is called the *line of the apsides* (x)[83] of the Moon.

The Moon accompanies the Earth in its annual revolution around the Sun.

WHAT THE *PERIGEE* AND *APOGEE* ARE. THE LINE OF THE *APSIDES* OF THE MOON MOVES. TIME OF THE REVOLUTION OF THIS LINE.

If the orbit of the Moon only consisted of its circuit around the Sun with the Earth, the axis of this orbit would always stay parallel, and the Moon, being in its *apogee* and in its *perigee*, would always be the same distance from the Earth, and would always correspond to the same points in the sky. But the line of the apsides of the Moon moves in an angular movement around the Earth according to the order of the signs, and the apogee and the perigee of the Moon only return to the same points at the end of about 9 years, which is the duration of the revolution of the line of the apsides of the Moon.

REVOLUTION OF THE NODES OF THE MOON. TIME OF THIS REVOLUTION.

The orbit of the Moon cuts the orbit of the Earth at two points, called *its nodes;* these points are not always the same. They perpetually change by a retrograde movement, that is to say, against the order of the signs, and this movement is such, that in the space of nineteen years the nodes have made

---

83. Du Châtelet offers this footnote: "(x) The *line of the apsides* of the Moon is the line that passes through the *apogee* and the *perigee*. The *apogee* is the point of the orbit farthest from the Earth and the *perigee* is the point of this orbit that is the least distant. As a rule, apsides are, for all orbits, the points the most distant and the closest to the central point."

one entire revolution, after which they return to intersect with the orbit of the Earth, or the ecliptic, at the same points.

ECCENTRICITY OF THE MOON.
The eccentricity of the orbit of the Moon also changes continually; this eccentricity is sometimes greater and sometimes smaller, so that the difference between the smallest and the greatest eccentricity exceeds half the amount of the smallest.

In the following chapters it will be seen how M. *Newton* found the cause of all of these eccentricities of the Moon.

ITS MOVEMENT AROUND ITS AXIS. THE DURATION OF THIS MOVEMENT.
The only movement of the Moon that is even is its rotation around its axis; this movement lasts precisely the same time as its revolution around the Earth, thus its day is 27 of our days, 7h43'.

This equality between the day and the periodic month of the Moon means that it always displays nearly the same face to us.

LIBRATION OF THE MOON.
The evenness of movement of the Moon around its axis, combined with the unevenness of its movement around the Earth, causes the Moon to seem to us to oscillate on its axis, sometimes towards the orient, and sometimes toward the occident; this is called its *libration*. By this movement it sometimes displays to us parts that were hidden, and hides from us parts that were visible.

ITS CAUSE.
This libration comes from the elliptic movement of the Moon, for if this planet moved in a circle, of which the Earth occupied the center, and turned on its axis in the time taken by its periodic movement around the Earth, it would always display the same face to the Earth, without any variation.

Nothing is known about the shape of the part of the Moon that is on its other side in relation to us; some astronomers even want to explain this libration by giving a conical form to the part of the Moon's surface we do not see, and they deny its rotation on its axis.

The surface of the Moon is full of heights and cavities, which causes it to reflect the light of the Sun from every direction. If it were smooth like one of our mirrors, it would only reflect the image of the Sun to us.

DISTANCE OF THE MOON TO THE EARTH.

The mean distance of the Moon from the Earth is about 60½ half-diameters of the Earth.

ITS DIAMETER. ITS MASS. ITS DENSITY.

The diameter of the Moon is to the diameter of the Earth as 100 to 365, its mass is to the mass of the Earth as 1 to 39,788, and its density is to the density of the Earth as 11 to 9.

WHAT BODIES WEIGH ON THE MOON.

Finally, the same body that weighs three pounds on the surface of the Earth would weigh about one pound on the surface of the Moon.

All these proportions are known for the Moon, and not for the other satellites, because this planet has a particular characteristic; it is its action on the waters of the sea that M. *Newton* was able to measure and to use in the determination of its mass. In one of the following chapters we shall give an account of the method he followed to arrive at this measurement.

CHAPTER TWO:
HOW M. NEWTON'S THEORY EXPLAINS
THE PHENOMENA OF THE PRINCIPAL PLANETS

I

The first phenomenon that must be explained when one wants to account for celestial movements is the perpetual circulation of the planets around the center of their revolution.

By the first law of motion, a body follows of its own accord the straight line in which it has begun to move; so, in order for a planet to be deflected from the small straight line it tends to describe at each instant, there must be a different force acting continuously on it in order to divert it. This is the same as the rope that, in the hand of one who makes a body revolve in a circle, at each moment prevents this body from escaping along the tangent of the circle one is making it describe.

HOW ANCIENT PHILOSOPHERS AND MOST RECENTLY *DESCARTES*
EXPLAINED THE CIRCULATION OF THE PLANETS IN THEIR ORBITS.

In order to explain this phenomenon, the Ancients had imagined solid heavens and *Descartes* vortices; but both of these explanations were pure hypotheses devoid of proofs, and if *Descartes'* was more philosophical, it was no more solidly established.

II

In the first proposition (a),[84] M. *Newton* begins by proving that the areas a body describes around an immobile center toward which it tends continuously are proportional to the time; and reciprocally in the second proposition, that if a body in turning around a center describes areas proportional to the time, this body is attracted by a force that carries it toward this center. Thus, since according to *Kepler's* discovery, the planets describe around the Sun areas proportional to the time, they have a centripetal force that makes them tend toward the Sun and keeps them in their orbit.

IT IS CENTRIPETAL FORCE THAT PREVENTS THE PLANETS FROM ESCAPING ALONG THE TANGENT.
Further, M. *Newton* has demonstrated (prop. 2, cor. 1) that if the force that acts on the body, making it tend toward various points, would accelerate it or slow it down, the sweep of the areas would then no longer be proportional to the time. So if the areas are proportional to the time, not only is the body driven by a centripetal force that carries it toward the central body, but this force also makes it tend to one and the same point.

III

AND PROJECTILE FORCE PREVENTS THEM FROM FALLING TOWARD THEIR CENTER. PROP. 36.[85]
Just as the revolution of the planets in their orbits proves a centripetal force that pulls them back from the tangent, so, from the fact that they do not fall in a straight line toward the center of their revolution, it can be concluded that a force other than centripetal force acts on them. M. *Newton* tried to determine (b)[86] the time each planet, placed where it is, would need to fall into the Sun if it only obeyed the action of the Sun on it. He found that the different planets, to fall into the Sun, would need half the periodic time a body took to make its revolution around the same center at a distance twice smaller than its own, and that consequently this time must be to their pe-

---

84. Du Châtelet offers this footnote: "(a) When propositions are cited, without citing the Book, the propositions are from Book One [of the *Principia*]."

85. See Cohen translation, *Principia*, Book I, proposition 36, 523.

86. Du Châtelet offers this footnote: "(b) *De Systemate mundi*, p. 31. edition of 1731."

riodic time as 1 to 4√2. Thus Venus, for example, would take around forty days to arrive at the Sun, for, roughly, 40:224::1:4√2; Jupiter would need two years and one month, and the Earth and the Moon sixty-six days and nineteen hours, etc. So, since the planets do not fall into the Sun, there must be some force that opposes the force that makes them tend toward their center, and this force is called *projectile force*.

IV

OF THE CENTRIFUGAL FORCE OF THE PLANETS.
The effort that causes the planets, by virtue of this force to move away from the center of their motion, is called their *centrifugal force*. Thus, in the planets, the centrifugal force is the part of their projectile force that moves them directly away from the center of their revolution.

V

The projectile force goes in the same direction in all the planets, for they all turn around the Sun from occident to orient.

Supposing the resistance of the medium in which the planets move to be nonexistent, the reason for the conservation of the projectile motion of the planets is found in the inertia of matter, and in the first law of motion, but its physical cause and the reason for its direction are still hidden from us.

VI

M. *NEWTON* MANAGED TO DISCOVER THAT THE FORCE CARRYING THE PLANETS TOWARD THE SUN FOLLOWS THE INVERSE PROPORTION OF THE SQUARE OF THE DISTANCES, THAT IS, TO THEIR DISTANCES TO THE SUN AND THEIR PERIODIC TIMES. PROP. 4. COR. 6.[87]AND FIRST SUPPOSING THEIR ORBITS TO BE CIRCULAR.
After proving that the planets are kept in their orbits by a force that tends toward the Sun, M. *Newton* demonstrates in prop. 4 that the centripetal forces of bodies that describe circles are to one another as the squares of the arcs of these circles described in equal times, divided by their radius; from which he deduces that, if the periodic times of bodies revolving in circles are in sesquiplicate [one and a half times] ratio to their radii, the centripetal force that carries them toward the center of these circles is in a reciprocal ratio to the squares of these same radii, that is to say, to the distances of

87. See Cohen translation, *Principia*, Book I, proposition 4, corollary 6, 450.

these bodies to the center. Now, by *Kepler's* second law [third law in modern terms] that all the planets observe, the times of their revolutions are in sesquiplicate proportion to their distances to their center, so the force that carries the planets toward the Sun decreases in inverse proportion to the square of their distances to this star, supposing that they turn in concentric circles around the Sun.

VII

BEFORE *KEPLER* IT WAS BELIEVED THAT THE PLANETS TURNED AROUND THE SUN IN ECCENTRIC CIRCLES.

The idea that most naturally springs to mind, as far as the orbits of the planets are concerned, is that they make their revolutions in concentric circles; but their different observed diameters and greater exactitude in observations made us acknowledge long ago that their orbits could not be concentric to the Sun. Before *Kepler* their courses were explained by eccentric circles that corresponded well enough with the observations for the Sun and the planets, except for Mercury and Mars.

BUT *KEPLER* SHOWED THAT THEY TURN IN ELLIPSES.

The path of this last planet made *Kepler* suspect that the orbits of the planets might be better described as an ellipse of which the Sun was one of the foci, and this curve agrees so perfectly with the phenomena that now all astronomers recognize that the planets turn around the Sun in ellipses, and that this star is one of the foci of these ellipses.

VIII

Starting from this discovery, M. *Newton* sought the law of centripetal force necessary to make the planets describe an ellipse, and he found in prop. 11 that this force must be inversely proportional to the square of the distances of the body to the focus of this ellipse.[88] But we have just seen that he had found in prop. 4, cor. 6 that in circles the periodic times of revolving bodies being in a ratio of one and a half times the distances, the force would be inversely proportional to the square of these same distances. It only remained—in order to be entirely sure that the centripetal force that directs celestial bodies in their paths follows the proportion of the inverse square of the distances—to examine whether or not the periodic times follow the same proportion in ellipses as in circles.

88. Ibid., proposition 11, 462–63.

M. *NEWTON* DEMONSTRATED THAT IN THE ELLIPSES THE PERIODIC TIMES
ARE IN THE SAME PROPORTION AS IN CIRCLES.

Now, M. *Newton* demonstrated in prop. 15 that the periodic times in ellipses
are in a ratio of one and a half times to their major axis, that is to say, that
these times are in the same ratio in ellipses and in circles, the diameters of
which were equal to the major axes of the ellipses.[89]

This curve that the planets describe in their revolution has this property, that if one takes small arcs described in equal time, the space between
the line drawn from one of the extremities of this arc and the tangent of the
other extremity increases in proportion as the square of the distance to the
focus diminishes, and this in the same proportion, from which it follows
that the attractive power proportional to this space also has the same proportion.

IX

AND THAT CONSEQUENTLY THE CENTRIPETAL FORCE THAT KEEPS THE
PLANETS IN THEIR ORBITS DECREASES AS THE SQUARE OF THE DISTANCE.
THE CENTRIPETAL FORCE BEING IN THIS PROPORTION, THE PLANETS CAN
ONLY DESCRIBE CONIC SECTIONS, OF WHICH THE SUN IS ONE OF THE
FOCI. PROP. 10.[90]

M. *Newton*, not content with just examining the law that causes the planets
to describe ellipses, considered whether this same law could not cause bodies to describe other curves, and he found (prop. 13, cor. 1) that whatever
the projectile force was, the law never made them describe anything but a
conic section, the focus of which was the center of force.[91]

The other laws that would cause bodies to describe conic sections
would cause these sections to be around other points than the focus. M.
Newton found, e.g., that if the force is as the distance to the center, it causes
the body to describe a conic section the center of which will be the center
of the forces. Thus, M. *Newton* not only found the law that centripetal force
follows in our planetary system, but he also demonstrated that another law
could not exist in our world as it is.

---

89. Ibid., proposition 15, 468.
90. Ibid., proposition 10, 459–60.
91. Ibid., proposition 13, corollary 1, 467.

X

MANNER OF DETERMINING THE ORBIT OF A PLANET, TAKING THE LAW OF
CENTRIPETAL FORCE AS A GIVEN.

M. *Newton* next sought in prop. 17 the curve that a body must describe, the
centripetal force of which decreases in inverse ratio to the square of the
distances, supposing that this body leaves a given point with a speed and a
direction chosen at will.

For the solution to this problem, he took from the remark he made in
prop. 16 that the speeds of bodies that describe conic sections are at each
point of these curves inversely proportional to the perpendiculars dropped
from the focus on the tangents, and directly proportional to the square root
of the parameters.[92]

This proposition not only creates an interesting problem for geometry,
it is also very useful in astronomy. For having discovered by some observations the speed and the direction of a planet in some part of its orbit, one
can, by means of this proportion, find the rest of the orbit, and the determination of the orbits of comets can, to a great extent, be founded on the
same proposition.

XI

WHAT CURVES OTHER LAWS OF CENTRIPETAL FORCE WOULD CAUSE
BODIES TO DESCRIBE. THE PROPORTION BETWEEN THE CENTRIPETAL
FORCE AND THE PROJECTILE FORCE CAUSES THE PERPETUAL
CIRCULATION OF THE PLANETS IN THEIR ORBIT.

It is easy to see that different laws of centripetal force than that of the square
of the distances would cause bodies to describe different curves, and there
could be such a law by which the planets, despite the projectile force, fell
toward the Sun, and another by which, despite their centripetal force, they
went off into the infinity of celestial space; yet another would make them describe spirals, etc. In prop. 42, M. *Newton* looks into all the curves that could
be described in all sorts of hypothetical centripetal force.[93]

XII

It is obvious from all that has just been said that the perpetual circulation
of the planets in their orbits depends on the proportion between the centripetal and projectile force, and that those who ask why, when the plan-

92. Ibid., proposition 17, 470–72; proposition 16, 468.
93. Ibid., proposition 42, 532–33.

ets have reached their perihelion, they rise again to their aphelion, do not know this proportion. For in the highest apsis, centripetal force surpasses centrifugal force, since then the body approaches its orbital center, and in the lowest apsis centrifugal force exceeds in turn the centripetal force, since in rising the body moves away from the center. So some combination of the centripetal and the centrifugal force was necessary, so that these forces alternatively might exceed each other, and that they might cause the body to go perpetually from the highest to the lowest apsis and from the lowest to the highest.

Another objection is made about the continuation of celestial movement, based on the resistance bodies must experience in the medium in which they move. M. *Newton* responded to this objection (Bk 3, prop. 10), where he shows that the resistance of mediums diminishes in relation to their weight and density;[94] now he had demonstrated in the Scholium of Bk 2, prop. 22 that at the height of 200 miles above the surface of the Earth, the air is thinner than that at the surface in the ratio of approximately 30 to 0.000,000,000,000,003,998 or 75,000,000,000,000 to 1.[95] From which he concludes (Bk 3, prop. 10) that, supposing this density for the medium in which Jupiter moves, this planet traveling five of its half-diameters in thirty days, would in 1,000,000 years, scarcely lose a 1,000,000th part of its motion to the resistance of such a medium. Thus, it is plain that the medium in which the planets move is perhaps so subtle that its resistance may be regarded as nonexistent and that the constant proportionality constantly observed between the areas and the times assures us that, indeed, its effect is imperceptible.[96]

XIII

HOW THE PLANETS CAN CONSERVE THEIR MOTION DESPITE THE RESISTANCE OF THE MEDIUM IN WHICH THEY MOVE.

Since one has seen above that the proportionality of the times and the areas that the planets describe around the Sun proves that they tend toward this star as toward their center, and that the ratio between their periodic times

94. Ibid., Book III, proposition 10, 815–16.

95. Ibid., proposition 22, Scholium, 696. This is described in Book III, proposition 10, theorem 10, but there appears to have been a printer's error. The text incorrectly reads "approximately 300,000000000000003998." I am grateful to Gérard G. Emch and Antoinette Dériaz Emch for bringing this to my attention and directing me to the correct reading.

96. See Cohen translation, *Principia*, Book III, proposition 10, 815–16. Significantly, Du Châtelet has accepted the idea of some medium in space, a point of much controversy between Newtonians and Cartesians in her day.

and their distances reveals that this force acts in a ratio twice the inverse of the distances; if the planets that make their revolution around the Sun find themselves surrounded by other bodies that turn around them and obey these same proportions in their revolutions, it will be proved that these revolving bodies experience a centripetal force that carries them toward these planets, and that this force decreases as that of the Sun in proportion to the square of the distance.

We only know three planets that have bodies revolving around them, Jupiter, the Earth, and Saturn. One knows that the satellites of these three planets describe areas around them proportional to the time, and consequently that they are driven by a force that tends toward these planets.

XIV

THE COMPARISON OF THE PERIODIC TIMES AND DISTANCES OF THE SATELLITES OF JUPITER AND SATURN DEMONSTRATES THAT THE FORCE THAT CARRIES THE SATELLITES OF THESE PLANETS TOWARD THEIR PRINCIPAL PLANET ALSO FOLLOWS THE INVERSE PROPORTION TO THE SQUARE OF THE DISTANCES.

Of Jupiter and Saturn, each having many satellites for which one knows the periodic times and the distances, it is easy to know if the times of their revolution around their planet are to their distance in the proportion discovered by *Kepler*. Observations demonstrate that Jupiter and Saturn's satellites also observe *Kepler's* second law in turning around their planet, and that, consequently, the centripetal force in Jupiter and Saturn decreases in inverse proportion to the square of the distance of the bodies to the center of these planets.

XV

Of the Earth, having only one satellite, the Moon, it seems difficult at first to know the proportion by which the force acts that makes the Moon turn around the Earth, since one lacks a comparative point.

HOW M. *NEWTON* ARRIVED AT THE DISCOVERY THAT THE ATTRACTIVE FORCE OF THE EARTH FOLLOWS THE SAME PROPORTION.

M. *Newton* found the means to supply it, and here is how he achieved this.

All of the bodies that fall here on Earth go, according to the progression discovered by *Galileo*, through spaces that are as the squares of the time needed to fall.

We know the mean distance of the Moon to the Earth, which is 60 half-diameters of the Earth in round numbers, and all bodies here on Earth are supposed to be a half-diameter from the center of the Earth. Thus, if the same force causes bodies to fall and the Moon to circle in its orbit, and if this force decreases as the square of the distance, it must act 3,600 times more on bodies placed on the surface of the Earth than on the Moon, since the Moon is 60 times farther than they from the center of the Earth. The orbit of the Moon is known, since we now know the measurements of the Earth, and that the Moon takes 27 days 7 hours and 43' to cover this orbit, consequently, we know the arc it describes in a minute. Now one sees (prop. 4, cor. 9) that the arc described in a given time by a body that turns in a uniform motion and with a given centripetal force in a circle, is the mean proportional between the diameter of this circle and the line on which this body descended toward the center in the same time.[97]

It is true that the Moon does not exactly describe a circle around the Earth, but in this case it can be assumed without observable error, and this assumption made, it turns out that the line expressing the quantity by which the Moon falls by centripetal force toward the Earth in one minute is fifteen *pieds* in round numbers.

Now the Moon, according to the progression of *Galileo*, will describe, in the place where it is, 3,600 times less space in a second than in a minute, and bodies on the surface of the Earth describe, according to the pendulum experiments of M. *Huygens*, around 15 *pieds* in a second, that is to say, 3,600 times more space than the Moon. So the force that makes them fall acts 3,600 times more on them than on the Moon, which is precisely the proportion of the square of their distances.

From this example, it can be seen how useful the measuring of the Earth is; for in order to compare this arrow, which expresses the quantity by which the Moon has moved closer to the Earth, to the contemporaneous space through which gravity makes bodies fall to the surface of the Earth in the same time, we need to know the absolute distance of the Moon to the Earth, reduced to *pieds*, as well as the length of the pendulum, because it is not sufficient in this case to have ratios. Absolute sizes are needed.[98]

XVI

THE MEASURING OF THE EARTH WAS NECESSARY FOR THIS DISCOVERY. ANALOGY LEADS US TO CONCLUDE THAT ATTRACTION ALSO FOLLOWS THE SAME RATIO FOR THE PLANETS THAT DO NOT HAVE SATELLITES.

Thus, Jupiter, Saturn, and our Earth attract bodies in the same proportion as the Sun attracts them, and induction leads us to conclude that gravity follows the same ratios in Mars, Venus, and Mercury. For from everything we

97. Ibid., Book I, proposition 4, corollary 9, 451.
98. Du Châtelet is referring to a geometric representation of this relationship.

know of these three planets, they appear to us as bodies of the same kind as Earth, Jupiter, and Saturn. Thus it can be concluded, with much probability, that they have attractive force, and that this force decreases as the square of the distances.

XVII

BK 3, PROP 5: M. *NEWTON*'S REASONING THAT LED HIM TO CONCLUDE THAT THERE WAS MUTUAL GRAVITATION AMONG ALL CELESTIAL BODIES.[99]
Since it is proved by observation and by induction that all planets have attractive force inversely proportional to the square of the distances, and that by the second law of motion the action is always equal to the reaction, it must be concluded, with M. *Newton*, that all planets gravitate toward one another, and that, just as the Sun attracts the planets, it is reciprocally attracted by them. For since the Earth, Jupiter, and Saturn act on their satellites in inverse proportion to the squares of the distances, there is no reason to believe that this action does not operate at all distances in the same proportion. So the planets must mutually attract, and the effects of this mutual attraction can be observed in the conjunction of Jupiter and Saturn.

XVIII

Analogy leads us to believe that the secondary planets are, in every respect, bodies of the same sort as their principal planets. It is very probable that they also have attractive force, and that, consequently, they attract their principal planet in the same way that they are attracted to it, and that they attract each other, which is confirmed again by the attraction of the Moon on the Earth, the effects of which become perceivable in the tides and in the precession of the equinoxes, as will be seen in what follows.

One can then conclude that the attractive force belongs to all celestial bodies, and that it acts throughout our planetary system according to the inverse proportion of the square of the distances.

XIX

WHY ONE BODY TURNS AROUND ANOTHER, RATHER THAN THE REVERSE. But what causes a body to turn around another? Why, for example, if the Moon and the Earth reciprocally attract each other in inverse proportion to the square of their distances, does the Earth not turn around the Moon, instead of the Moon turning around the Earth? It must certainly be that the law attraction follows not only depends on the distance, and that some other

99. See Cohen translation, *Principia*, Book III, proposition 5, 805–6.

element enters into it that could explain this determination, for here the distance is insufficient, since it is the same for both.

THIS CAUSE APPEARS TO BE THE MASS OF THE CENTRAL BODY.

It is easy, in examining the bodies that compose our planetary system, to suspect that this law is that of the masses. The Sun, around which all the celestial bodies turn, appears much larger than any of them; Saturn and Jupiter are much larger than their satellites; and our Earth is larger than the Moon that turns around it.

BUT TO REACH CERTAINTY, THE MASSES OF THE DIFFERENT PLANETS MUST BE KNOWN.

Now, as size and mass are two different things, to be certain that the gravity of celestial bodies follows the law of masses, these masses have to be known.

But how to know the mass of the different planets, this is what M. *Newton's* theory teaches us.

XX

PATH THAT M. *NEWTON* FOLLOWED TO ARRIVE AT THIS DISCOVERY.

This is the path he followed to arrive at this discovery.

Since attraction of all celestial bodies on the bodies that surround them follows the inverse ratio of the square of the distances, it is very probable that the parts of which they are composed attract in the same proportion.

The overall attractive force of a planet is composed of the attractive force of its parts, for if one knew that many little planets unite to form a big one, the force of this large planet would be composed of the force of all these little planets. And M. *Newton* has proved (prop. 74, 75 and 76) that if the particles of which a sphere is composed attract one another in inverse proportion to the square of the distances, these spheres in their entirety will attract bodies exterior to them, however distant, in this same inverse proportion of the square of their distances.[100] And of all the laws of attraction examined by M. *Newton*, he only found this one in inverse proportion to the square of the distances, and the one that followed the proportion of the simple distance, in which spheres in their entirety attract bodies that are exterior to them in the same ratio that their parts attract one another.

This shows the force of the reasoning that led M. *Newton* to conclude (prop. 74, cor. 3) that since it is proved on the one hand by theory, that when the particles of a sphere reciprocally attract in inverse proportion to

100. Ibid., Book I, propositions 74, 75, 76, 593–97.

the square of the distances, the sphere as a whole attracts exterior bodies in the same proportion, and that on the other hand, observations demonstrate that celestial bodies attract in the same proportion the bodies that are exterior to them, it is very simple to conclude that the particles of which celestial bodies are composed attract reciprocally in the same proportion.[101]

HE BEGAN BY FINDING THE WEIGHTS OF THE SAME BODY ON DIFFERENT PLANETS, AT EQUAL DISTANCE.

In Bk 3, prop 8, M. *Newton* seeks what the same body would weigh on different planets, and he finds the answer by using prop 4, cor 2, in which he has demonstrated that the weights of equal bodies that move in circles are as the diameters of the circles directly, and as the inverse square of their periodic times. Thus, knowing the periodic times of Venus around the Sun, of the satellites of Jupiter around this planet, of the moons of Saturn around Saturn, and of the Moon around the Earth, and the distance of these bodies to the centers around which they turn, and supposing that these bodies describe circles in their revolution, which may be supposed in the circumstances, one finds the weight of the same body transported successively the same distance from the centers of the Sun, of Jupiter, of Saturn, and of the Earth.[102]

AND HE NEXT DEMONSTRATED THAT THE QUANTITY OF MATTER IS PROPORTIONAL TO THE WEIGHT OF THE SAME BODY ON THE DIFFERENT PLANETS, AT AN EQUAL DISTANCE FROM THE CENTER.

The weight of the same body on different planets, at equal distance from their center, being known, M. *Newton* concludes from that the quantity of matter that each contains. For attraction depending on mass and distance, at equal distances the attractive forces are as the quantities of matter of the bodies that attract. Thus, the masses of the different planets are as the weights of the same body supposed to be at equal distance from their centers.

XXI

WHENCE HE DEDUCED THEIR DENSITY.

By the same means, we can know the density of the Sun and the planets that have satellites, that is to say, the proportion between their diameter and the quantity of matter they contain, because M. *Newton* (Bk I, prop 72) proved that the weight of equal bodies placed on the surfaces of homogeneous and unequal spheres are as the diameters of these spheres.[103] Thus, if

101. Ibid., proposition 74, corollary 3, 594.

102. Ibid., Book III, proposition 8, 811–15; Book I, proposition 4, corollary 2.

103. See Cohen translation, *Principia*, Book I, proposition 72, 592–93.

those spheres were heterogeneous and equal, the weight of the bodies on their surface would be as their density, supposing that in the law of attraction only the distance and the mass of the body attracting pertain. Thus, on the surface of heterogeneous and unequal spheres, the weights of equal bodies will be in a compound proportion of the density of these spheres and of their diameter; so the densities will be as the weights of the bodies divided by their diameters.

XXII

THE SMALLEST PLANETS AND THE DENSEST ARE THE CLOSEST TO THE SUN.
We learn from this that the smallest planets are the densest, and that they are the closest to the Sun. For, we saw in chapter 1, where all the proportions of our system were given, that the Earth, which is smaller and closer to the Sun than Jupiter and Saturn, is denser than those planets.

XXIII

THE REASON FOR THIS ACCORDING TO M. *NEWTON*.
M. *Newton* derives from this the reason for the arrangement of the celestial bodies in our planetary system, which is such as required by the density of their matter, so that each is more or less heated by the Sun in proportion to its density and its distance. For it is well known that the denser a body, the more difficult it is to heat, from which M. *Newton* concludes that the matter of Mercury would have to be seven times denser than that of Earth for vegetation to occur there. We know, all things being equal, that the illumination to which heat is proportional is as the square of the distances. Now one knows the proportion of the distance of Mercury and of the Earth to the Sun, and from this proportion one knows that Mercury is illuminated seven times more brightly and, consequently, is seven times warmer than the Earth. And M. *Newton* says that he found by his experiments that the heat of our summer, increased sevenfold, makes water boil; so, if the Earth were placed where Mercury is, all our water would evaporate. If it were where Saturn is, it would be always frozen. In both cases all vegetation would cease, and all animal kind would die.

XXIV

WE CAN ONLY KNOW ALL THE PROPORTIONS FOR THE PLANETS THAT HAVE SATELLITES. WITH THE EXCEPTION OF THE MOON.
From this it is clear that we can only know the mass and the density of planets with satellites, since to arrive at this result we must compare the times of the revolutions of the bodies that turn around these planets, except for the Moon, which I will discuss later.

XXV

The mass of the planets being known, one sees that the bodies that have the
least mass turn around those that have more, and that the greater the mass
of a body, the more attractive force it has, all things being equal. Thus all
the planets turn around the Sun because the Sun has much more mass than
any planet, for the mass of the Sun is to that of Jupiter and Saturn, nearly
as 1 to 1,100, and 3,000 respectively. These two planets being those of our
system with the most mass, it follows that the Sun must be the center of the
motion of our system.

XXVI

THE ALTERATIONS THAT SATURN AND JUPITER MUTUALLY CAUSE IN
THEIR COURSES HAPPEN BY REASON OF THEIR MASSES.

If attraction is proportional to mass, the alteration caused by the action of
Jupiter on the orbit of Saturn in their conjunction must be much greater
than that which is caused in the orbit of Jupiter by the action of Saturn,
since Jupiter has a much greater mass than Saturn. And this is indeed what
happens; the alteration in Jupiter's orbit in its conjunction with Saturn, although perceptible, is, however, much less than what one observes in the
orbit of Saturn.

XXVII

BK 3, PROP. 6.

But if the effect of attraction, or the path made by the attracted body, depends on the mass of the attracting body, why will it not also depend on the
mass of the body being attracted? This assuredly merits examination.

   We know that all bodies here on Earth fall equally fast toward the Earth,
air resistance being discounted; for once the air has been pumped from *Boyle's*
machine, gold and feathers reach the bottom at the same time.[104]

BK 2, PROP 24; BK 3, PROP. 6.[105]

M. *Newton* confirmed this experiment by another where the smallest differences become perceptible, despite the coarseness of our organs. He re-

---

104. Du Châtelet is referring to the celebrated experiments by Boyle with his vacuum pump,
or *machine de vuide*. She and Voltaire certainly experimented with one at Cirey.

105. See Cohen translation, *Principia*, Book II, proposition 24, 700–701; Book III, proposition
6, 806–10.

ports that he made several pendulums in very different materials, such as water, wood, gold, glass, etc., and that having suspended them by strings of equal length, they produced perceivably isochronal oscillations for a very long time.

XXVIII

ATTRACTION IS PROPORTIONAL TO MASS WITHOUT REGARD TO THE
FORM OF THE TYPE OF BODY ATTRACTED.

Undoubtedly, therefore, the attractive force of our Earth is proportional to the mass of the bodies it attracts, and at the same distance it only depends on their mass, that is to say, on their quantity of matter. Thus, supposing that the bodies on Earth have been transported to the orbit of the Moon, since it has been proved above that the same force acts on the Moon and on these bodies and that it decreases as the square of the distances, the distances then being equal, it follows that, supposing the Moon lost its projectile motion, these bodies and the globe of the Moon would reach the surface of the Earth simultaneously and would traverse the same spaces, supposing [as well] air resistance to be nonexistent.

XXIX

The same thing has been proved for the planets with satellites such as Jupiter and Saturn. Supposing the satellites of Jupiter, for example, to be all placed at the same distance from the center of this planet, and all deprived of their projectile motion, they would all fall toward Jupiter and would reach the surface simultaneously. This proposition follows from the proportion between the distances of the satellites and the times of their revolutions.

XXX

In the same way it is proved, by the proportion between the periodic times and the distances of the principal planets to the Sun, that this star acts on each in proportion to its mass, because at equal distances their periodic times would be equal, and if in this supposition the planets lost all their projectile force, they would all reach the Sun at the same time. Thus the Sun attracts each planet in direct proportion to its mass.

XXXI

The regularity of the orbits of Jupiter's satellites around this planet is another proof of this truth. For M. *Newton* proved (prop. 65, cor. 3) that when a system of bodies moves in circles or in regular ellipses, it means that these bodies must experience perceptible action only because of the attractive force that makes them describe their curves.[106] Now Jupiter's satellites describe circular orbits around this planet, orbits perceptibly regular and concentric to this planet. Therefore the distances of Jupiter's satellites, and that of Jupiter itself, to the Sun must be regarded as equal, considering the small proportion between the differences of their distances and the total distance. Thus, if one of Jupiter's satellites, or Jupiter itself, were more attracted by the Sun than another satellite because of its mass, this stronger attraction of the Sun would disturb the orbit of this satellite; and M. *Newton* says (Bk 3, prop. 6) that if this action of the Sun on one of Jupiter's satellites was more or less great in proportion to the mass of this satellite than that which it had on Jupiter because of its mass, if only by a thousandth of its total gravity, the distance from the center of the orbit of this satellite to the Sun would be lesser or greater than the distance from the center of Jupiter to the Sun by 1/2000 of its total distance, that is to say, by the fifth part of the distance of the satellite to Jupiter of its most remote satellite, which would make its orbit perceptibly eccentric.[107] Thus, since these orbits are perceptibly concentric to Jupiter, the accelerating effect of the gravity of the Sun on Jupiter and on its satellites is as their quantity of matter.

The same reasoning applies to Saturn and its satellites, the orbits of which are perceptibly concentric to Saturn.

ATTRACTION IS ALWAYS RECIPROCAL.

Experiments and observations thus lead us to conclude that the attraction of celestial bodies is proportional to mass, both in the attracting bodies and in the bodies attracted; that mass determines that one body turns around another; that all bodies can be regarded as attracting and attracted. Finally, it can be concluded that attraction is always reciprocal between two bodies, and that it is the proportion between their masses that decides whether this double attraction can be perceived.

106. Ibid., Book I, proposition 65, corollary 3, 570. Note that Newton began his Book III and the separate *System of the World* (as it was translated from Latin into English) with this discussion of Jupiter's satellites, rather than in the more conventional way Du Châtelet chose.

107. See Cohen translation, *Principia*, Book III, proposition 6, 806–10.

XXXII

ATTRACTION ACTS UNIFORMLY AND CONTINUALLY, AND PRODUCES
EQUAL ACCELERATIONS IN EQUAL TIMES, WHETHER THE BODIES ON
WHICH IT ACTS ARE MOVING OR AT REST.

Attraction has yet another property, which is to act equally on moving
bodies and on bodies at rest, and to produce equal accelerations in equal
times, from which it follows that its action is continuous and uniform. This
is proved by the manner in which gravity accelerates bodies that fall here
on Earth, and it follows from the motion of the planets that are, as we have
demonstrated, only bigger projectiles, always subject to the same laws.

XXXIII

EFFECT OF THE ATTRACTION OF THE PLANETS ON THE SUN.

Since the proportion between the masses of attracting bodies determines the
path that one makes toward another, it is clear that the Sun, having much
more mass than the planets, the attraction they exert on it cannot be perceptible. Nevertheless, the attraction of the planets on the Sun, although too
small to be perceptible, is not nonexistent. And in considering it, one sees
that the center around which each planet turns is not the center of the Sun
but the point that constitutes the common center of gravity of the Sun and
of the celestial body whose revolution is being considered. We saw in chapter one, §.19, that the Sun's matter is to that of Jupiter, for example, as 1 to
1/1067, and that the distance of Jupiter to the Sun is to the half-diameter
of the Sun in a ratio a little greater; thus it follows that the common center
of gravity of Jupiter and the Sun falls at a point very close to the surface of
the Sun.

By the same reasoning we find that the common center of gravity of
Saturn and the Sun falls on the Sun's surface. And making the same calculations for all the planets, M. *Newton* says that if the Earth and all the other
planets were placed on the same side, the common center of gravity of the
Sun and all the planets would be scarcely farther from the center of the Sun
than one of its diameters. For although the mass of Venus, Mercury, or Mars
is not known, nonetheless, as these planets are much smaller than Saturn and
Jupiter, which have themselves infinitely less mass than the Sun, it can be
concluded that their mass does not upset this proportion.

XXXIV

THIS EFFECT CONSISTS IN MAKING THE SUN OSCILLATE AROUND THE
COMMON CENTER OF GRAVITY OF OUR PLANETARY SYSTEM.

The planets turn around this common center of gravity, and the Sun itself
oscillates around this common center of gravity according to the proportions of the planets' attraction on it. Thus, it is wrong when one considers
the movement of two bodies, one of which turns around the other, to regard the central body as fixed. The two bodies, that is to say, the central
body and that which turns around it, both turn around their common center
of gravity, but the path they describe around this center of gravity being in
reciprocal proportion to their mass, the curve described by the body that
has much more mass is almost imperceptible. This is why only the curve described by the body whose revolution is perceptible is taken into consideration, and the little movement of the central body that one regards as fixed
is overlooked.

XXXV

So the Earth and the Moon turn around their common center of gravity, and
this center itself turns around the center of gravity of the Earth and the Sun.
It is the same for Jupiter and its moons, for Saturn and its satellites, and finally for the Sun and all the planets. Thus the Sun, according to the different positions of the planets, must move successively to all directions around
the common center of gravity of our planetary system.

XXXVI

THIS COMMON CENTER OF GRAVITY IS AT REST.

This common center of gravity is at rest, for the different parts of this system
always correspond to the same fixed stars. Given so many years of observation, if this center were not at rest, and if it moved uniformly in a straight
line, changes in the relationships of the different parts of our planetary system to the fixed stars would have been noticed; but, as no change can be
seen, it must be concluded that our planetary system's common center of
gravity is at rest.

This center is the point at which all the bodies that compose our planetary system would converge if they lost their projectile motion.

THUS THIS CENTER CANNOT BE THE CENTER OF THE SUN WHICH MOVES
PERPETUALLY.
Our planetary system's center of gravity being at rest, the center of the Sun
cannot be this common center of gravity, since it has just been seen that it
moves according to the different positions of the planets, although it never
perceptibly changes its place, because of the very small distance between
the common center of gravity of our planetary world, and the center of
the Sun.

XXXVII

ATTRACTION PERTAINS TO EVERY PARTICLE OF MATTER.
Since attraction is proportional to the mass of the attracting body, and to
that of the body attracted to it, it must be concluded that attraction pertains
to every part of matter and that all the parts of which a body is composed
must mutually attract. For if attraction did not pertain to every part of matter, it would not follow the proportion of the masses.

XXXVIII

ANSWER TO THE OBJECTION THAT SPRINGS FROM THE ATTRACTION OF
BODIES ON EARTH NOT BEING PERCEPTIBLE.
This property of attraction, of being proportional to masses, furnishes a response to the objection often made against the mutual attraction of bodies.
If all bodies have this property of mutual attraction, why, it is asked, is the
attraction they exert on one another on Earth not perceived? But it is easily
understood that attraction being proportional to the masses of the bodies
attracting each other, the attraction that the Earth exerts on bodies here on
Earth is very much stronger than that which they exert on each other, and
that, consequently, these partial attractions are absorbed and rendered imperceptible by that of the Earth.

XXXIX

IT DEVIATES IN CERTAIN CASES, AS IN THE DEVIATION OF A PLUMB LINE
AT THE FOOT OF CHIMBORAZO.[108]
The Academicians who measured a degree of the meridian in Peru believed
they perceived that the attraction of the mountain of Chimborazo, the highest known, caused a perceptible deviation in the plumb line; and it is certain,
from the theory, that the attraction of this mountain must affect the line and

108. Chimborazo is an inactive volcano, the highest summit in Ecuador, and part of the An-
dean chain of mountains visited by La Condamine's expedition.

all bodies: but it remains to be seen if the amount of the observed deviation is what must result from the size of the mountain. For apart from the fact that these observations do not provide an exact amount for the deviation, because of the inevitable practical errors, there is also another problem, that the theory gives no way of assessing the exact amount of this deviation because the total shape of the mountain, its density, etc. are not known.

XL

The same reason that prevents our perceiving the attraction of bodies here on Earth causes the mutual attraction of celestial bodies to be only very rarely perceptible. For the Sun's attraction, which operates on them much more strongly, prevents this mutual attraction from being visible. However, there are cases in which it can be perceived, as in the conjunction of Saturn and Jupiter that then reciprocally disrupts their orbits in a perceptible manner, because the attraction of these two planets is too strong to be absorbed by that of the Sun.

XLI

THE ATTRACTIONS OF MAGNETS AND OF ELECTRICITY HAVE DIFFERENT CAUSES, AND DO NOT FOLLOW THE SAME PROPORTIONS AS THE UNIVERSAL ATTRACTION OF BODIES.

In regard to the perceivable attractions of some bodies on Earth, such as those of the magnet and of electricity, they follow other laws, and probably have other causes than the universal attraction of matter being discussed here.

M. *Newton* proved (prop. 66) that the mutual attraction of two bodies that turn around a third disturbs the regularity of their motions less when the body around which they turn is impelled by their attractions than if it were at rest; thus the little modification noticed in the motion of the planets is yet another proof of the mutuality of attraction.[109]

XLII

BK 3, PROP 14 AND BK 1, PROP 1 AND 2. THE APHELIA OF PLANETS ARE AT REST. EXCEPTIONS TO THIS RULE CAUSED BY THE MUTUAL ACTIONS OF THE PLANETS.[110]

The aphelia of the planets, as well as their nodes, and the planes in which they move are at rest, disregarding the action of the planets on one another.

109. See Cohen translation, *Principia*, Book I, proposition 66, 570–85.
110. Ibid., Book III, proposition 14, 819; Book I, propositions 1 and 2, 444–46.

Mars, Venus, Mercury, and the Earth, being very small planets, do not cause any perceptible alteration in their respective motions; so their aphelia and their nodes can only be disturbed by the action of Jupiter and Saturn. M. *Newton* concludes from his theory that for this reason and as a consequence, the aphelia of these four planets move very little in relation to the fixed stars, and he asserts that these motions follow the sesquiplicate [one and a half] proportion to the distances of these planets to the Sun; from which he derives, Bk 3, prop 14. . . .

[Du Châtelet continues to explain this phenomenon with references to Book 3, prop. 14, and to Newton's *De Systemate mundi*, 1731 edition, p. 36. In XLIII she notes that one "overlooks these alterations and that many astronomers do not acknowledge them at all." This perceptible "rest" of the aphelia offers a "new proof that attraction acts in proportion to the inverse square of distances."]

XLIV

The planets have another motion that I have not discussed in this chapter, because it does not seem to depend on their gravity, it is their rotation on their axes.

ONE DOES NOT KNOW THE CAUSE OR THE RATIO OF THE ROTATING
MOTION OF THE PLANETS.
We saw in chapter one that this rotation is a certainty only as regards the Sun, the Earth, Mars, Jupiter, and Venus, and that astronomers do not yet agree on the time taken by a revolution of this last planet, although they all agree that it turns. But although observations have not yet shown that Mercury, Saturn, and the satellites of Jupiter and of Saturn turn on their center, from the uniformity observed by nature in its operations, it is very probable that these planets also rotate around their axes and that all celestial bodies in our system experience this revolution.

This motion of the planets around their axes is the only uniform celestial motion; this motion, as I have said, does not appear to depend on their gravity, and its cause is not yet known.

XLV

THE MUTUAL GRAVITY OF THE PARTS THAT COMPOSE THE PLANETS
PREVENTS THEM FROM DISPERSING AS AN EFFECT OF THEIR ROTATION.
The mutual gravity of the parts of which the planets are composed prevents them from dispersing as an effect of this rotation. For we know that any body moving in a circle acquires centrifugal force by which it tends to move

away from the center of its revolution; thus, without the mutual gravity of the parts of matter, the rotation of the planets would disperse their parts. For if the gravity of one part whatever of the surface of a rotating body were destroyed, this part, instead of turning with the body would escape by the tangent. So if gravity did not oppose the centrifugal force that the parts of celestial bodies acquire upon rotating on their axes, this force would separate their parts.

XLVI

THE ROTATION MOTION MUST BROADEN THE PLANETS AT
THEIR EQUATOR.

If this tendency of the parts of celestial bodies toward one another opposes the effect of centrifugal force, it does not destroy it, and the effect produced by this force is to make unequal the diameter of supposedly fluid revolving bodies. For the planets are composed of matter, parts of which tend equally toward their center at an equal distance, they would be spherical if they were at rest. But the rotating motion causes their parts to tend by their centrifugal force to move away from their center with all the more force, as they are placed closer to the equator of the revolving sphere. For we know by the theory of centrifugal force that this force, supposing equal times, increases in the same proportion as the radius of the circle the body describes; thus, supposing fluid to be the material of which the celestial bodies are composed, the rotation will increase the diameter of their equator and consequently diminish that of their poles.

XLVII

M. *NEWTON* DEDUCED FROM THESE PRINCIPLES THE PROPORTION OF THE
AXES OF THE EARTH.

By means of telescopes, this difference has been observed in the diameters in Jupiter, and the measurement of degrees has made it possible to calculate it for the Earth.

In the next chapter we shall see how M. *Newton* went about deducing the shape of the Earth from his theory, and what observations have shown on this matter.

CHAPTER THREE:
ON THE DETERMINATION OF THE SHAPE OF THE EARTH,
ACCORDING TO THE PRINCIPLES OF M. NEWTON

I

CENTRIFUGAL FORCE BROADENS THE EQUATORIAL REGIONS IN THE DIURNAL ROTATION.

Since the centrifugal force of bodies that move in a circle increases the size of the circle described when the time of the revolution is the same, the rotational motion must create protuberance at the equatorial regions. For, supposing that the Earth had been spherical and composed of homogeneous, fluid matter before acquiring the rotational motion, in order for the matter of which it is composed to conserve its equilibrium in this rotation and for the Earth's shape to remain constant, it is necessary that the column, the gravity of which is diminished by centrifugal force, be longer than that in which centrifugal force has not altered the gravity. Thus the Earth's axis, which goes through its equator, must be greater than that which passes through the poles.

II

M. *NEWTON'S* METHOD FOR FINDING THE SHAPE OF THE EARTH.

M. *Newton*, in Bk 3, prop. 19, determined the quantity by which the column at the equator must be longer than that of the axis, supposing, as in all the rest of his work, that the gravity bodies experience here on Earth is nothing but the result of the attractions of all the particles of which the Earth is composed, in that he views the Earth as homogeneous. He uses as givens in this problem: (1) the size of the Earth's radius initially taken as spherical, and determined by M. *Picard* as 19,615,800; (2) the length of the pendulum that beats seconds at the latitude of Paris, which is 3 *pieds* 8 5/9 *lignes*.[111]

---

111. See Cohen translation, *Principia*, Book III, proposition 19. *Ligne* is an eighteenth-century French term of measurement; twelve *lignes* made up one *pouce* (or 2.256 mm). In the following discussion Du Châtelet presents the two views of the shape of the Earth. Her mentor, Maupertuis, led the fight in the Academy of Sciences against the view that the Earth was elongated at the poles. He and La Condamine led the Academy-sponsored expeditions that measured the oscillations of pendulums at the Arctic Circle (in Lapland) and at the equator. Their results proved Newton's hypothesis that the Earth was flattened at the poles, oblate in shape. As Newton based this hypothesis on his theory that gravity acted in the universe as "attraction," this offered a proof of his theory of attraction.

It is proved by the theory of oscillation and by this measurement of the pendulum beating the second that a body at the latitude of Paris traverses 2174 *lignes* in a second, making the necessary correction for air resistance.

A body that makes its circular revolution at a distance of 19,615,800 feet from the center, which is the half-diameter of the Earth, in 23h56'4", and which is the exact time of the diurnal revolution, traverses in a second, supposing its motion to be uniform, an arc of 1433.46 *pieds*, of which the versed sine is 0.0523656 *pieds*, or 7.54064 *lignes*.[112] So the force that makes heavy bodies fall at the latitude of Paris is to the centrifugal force that the bodies acquire at the equator from the Earth's rotation as 2174 to 7.54064. Adding then to the gravitational force that makes the heavy bodies fall at the latitude of Paris the amount that centrifugal force takes away from this force at this latitude, in order to obtain the total force that carries heavy bodies toward the center of the Earth at the latitude of Paris, M. *Newton* proves that this total force is to the centrifugal force below the equator as 289 to 1. So that below the equator centrifugal force diminishes centripetal force by 1/289.

In prop. 91, cor. 2, M. *Newton* gave the proportion between attraction exerted by a spheroid on a corpuscle placed on the prolongation of its axis and that which would be exerted on the same corpuscle by a sphere whose diameter was the minor axis of the spheroid.[113] Thus, using this proportion, and supposing the Earth to be homogeneous and without any motion, he finds (Bk 3, prop. 19) that if its form is that of a spheroid the minor axis of which was to the major axis as 100 to 101, the gravity at the pole of this spheroid must be to the gravity of the pole of a sphere described on the minor axis of the spheroid as 126 to 125.[114]

For the same reason, imagining a spheroid of which the equatorial radius was to be the axis of revolution, the gravity at the equator, which then would be the pole of this new spheroid, would be to the gravity of the sphere at the same point, supposing the sphere to have the same axis of revolution, as 125 to 126.

M. *Newton* next supposes that the proportional mean between these two gravities expresses the gravity of the parts of the Earth at the same place, that is to say, at the equator, and that thus the gravity of the parts of the

---

112. *Versed sine*, also known as *versine* (Latin: *sinus versus* [flipped sine]), is now little used.
113. See Cohen translation, *Principia*, Book III, proposition 91, corollary 2, 616–17.
114. Ibid., proposition 19, 821–26.

Earth at the equator is to the gravity of the parts of the sphere with the same axis of revolution as 125½ to 126. And using what he demonstrated in prop. 72, namely, that the homogenous spheres attract to their surface in direct proportion to their radii, he concludes that the attractions exerted by the Earth at the pole and at the equator, given the assumption of the preceding spheroid, are in a compound ratio of 126 to 125, 126 to 125½, and 100 to 101, that is to say, as 501 to 500.[115]

But he had demonstrated (prop. 91, cor. 3) that if one supposes the corpuscle to be placed in the inside of the spheroid, then it would be attracted in proportion to the simple distance to the center.[116] So, the gravities in the two columns corresponding to the equator and the pole will be as the distances to the center of the bodies in which they are placed. Then, supposing the communicating columns or canals to be divided by transverse planes that pass at proportional distances to these canals, the weight of each of these parts in one of these canals will be to the weight of each of the parts in the other canal as the heights of these canals. Consequently, these weights will be to each other as each of these parts and as their joint accelerating gravity, that is to say, as 101 to 100, and as 500 to 501, that is to say, as 505 to 501. So if the centrifugal force of any part whatever, in the canal that passes through the equator, is to the absolute weight of the same part as 4 to 505, that is to say, if the centrifugal force subtracts 4/505 parts from the weight of any part whatever, in the column that passes through the equator, the weight of each part of one and the other canal will become equal, and the fluid will be in equilibrium. But we have just seen that the centrifugal force of any part whatever below the Earth's equator is to its weight as 1 to 289, not as 4 to 505. So, for the axes a ratio other than 100 to 101 must be taken, and such a one that the centrifugal force below the equator is only the 289th part of gravity.

WHENCE HE CONCLUDED THAT THE RATIO OF THE AXES OF THE EARTH IS AS 229 TO 230.

Now this is the immediate result of a rule of three: for if the ratio of 100 to 101 in the axes gives a ratio of 4 to 505 for the proportion of centrifugal force to gravity, it is clear that the ratio will have to be 229 to 230 to give the relationship of 1 to 289 for centrifugal force to gravity.[117]

115. Ibid., Book I, proposition 72, 592–93.
116. Ibid., Book III, proposition 91, corollary 3, 617–18.
117. See I. Bernard Cohen, "Guide," 8:11, 231–38.

III

THE FLATTENING OF THE EARTH MUST ALWAYS RESULT FROM THE
THEORIES OF CENTRIFUGAL FORCES AND OF FLUIDS, WHATEVER
HYPOTHESES ABOUT GRAVITY ONE MIGHT HOLD.

This conclusion of M. *Newton*, namely, the amount of flattening that he determined, is founded on the principle of the mutual gravity of the parts of
matter; but the flattening would in any case result from the theory of fluids
and that of centrifugal forces, even if one did not accept the discoveries of
M. *Newton* on gravity, unless one were making very improbable hypotheses
on gravity.

IV

HOWEVER, MEASUREMENTS TAKEN IN FRANCE CAST DOUBT ON THE
SHAPE OF THE EARTH.

Despite M. *Newton*'s authority, and even though M. *Huygens* reached the same
conclusion about the flattening, even though he held another hypothesis
about gravity than M. *Newton*'s; even though, moreover, the experiments
made with pendulums in different regions of the Earth had all given the
diminution of gravity toward the equator, consequently favoring flattening at the poles; it is well known that measurements taken in France, which
gave lesser degrees going toward the north, cast doubt on the shape of the
Earth. Hypotheses were made about gravity that gave the Earth, supposed
at rest, a shape, the alteration of which agreed with the theory of centrifugal forces, and with the elongated shape toward the poles that resulted from
actual measurements.

For this great question of the shape of the Earth depends on the law
about gravity. It is certain, for example, that if this force depended on a cause
that attracted it now to one side and now to another, which increased and
diminished irregularly, neither the theory nor the practice could ever determine this shape.

V

THE MEASUREMENTS TAKEN BY THE FRENCH ACADEMICIANS
AT THE POLAR CIRCLE AND IN PERU HAVE CONFIRMED THE
FLATTENED SHAPE.

In the end, it was necessary to measure a degree at the equator and another
at the polar circle to decide this matter. We had erred, but we corrected our
mistake, and the measurements of the French Academicians vindicated M.

Newton's theory of the Earth's shape, so that flattening toward the poles is at present generally recognized.[118]

The measurements carried out in Lapland and Peru give a greater flattening than that demonstrated as a result of M. *Newton's* theory, because these measurements give the ratio of the axes as 173 to 174.[119]

VI

TWO SUPPOSITIONS M. *NEWTON* MAKES IN DETERMINING THE
FLATTENING OF THE EARTH. M. *CLAIRAUT* VERIFIED THE FIRST OF THESE
SUPPOSITIONS, WHICH M. *NEWTON* HAD NEGLECTED TO DO.

In determining the ratio of the Earth's axes, M. *Newton*, in addition to the mutual gravity of the parts of matter, also supposed that the Earth was an elliptical spheroid, and further, that its matter was homogeneous. M. *Clairaut*, in his book on the shape of the Earth, demonstrated that the first supposition was legitimate, a demonstration that M. *Newton* neglected to do, even though this was so important for insuring that the true ratio of the Earth's axes was known.

IT IS VERY POSSIBLE THAT THE OTHER SUPPOSITION MIGHT BE FALSE.

It is not the same for the second supposition on the homogeneity of the Earth's matter; it is quite possible (M. *Newton* himself suspected it in Bk 3, prop. 20) that the matter of which the Earth is composed is all the denser as one nears the center.[120] Now the different densities of the layers of matter that compose the Earth must change the law according to which the parts making up the Earth gravitate, and consequently, must alter the ratio of its axes.

118. With the publication of books by Maupertuis and the mathematician Alexis-Claude Clairaut (1716–1765), Maupertuis' companion on the expedition to the polar circle, most European physicists and mathematicians accepted the confirmation of the Earth's shape, even though not all accepted attraction as the cause of the flattening of the Earth at the poles. Some, such as Huygens, continued to believe in Descartes' theory of moving particles of matter as the motive force in the universe, while accepting Newton's system of the solar system.

119. This is one of the places in which Du Châtelet corrects Newton's *Principia* by providing the most recent studies by Continental mathematicians and physicists. Note that she did not change his words in the translation, but instead uses the Commentary to bring knowledge of the universe up to date. For a complete study of Du Châtelet's pattern of corrections, see Antoinette Emch-Dériaz and Gérard G. Emch, "On Newton's French Translator: How Faithful Was Mme Du Châtelet?" in *Emilie Du Châtelet: Rewriting Enlightenment Philosophy and Science*, ed. Judith P. Zinsser and Julie Candler Hayes (Oxford: Oxford University Press, 2006), 226–51.

120. Newton does indeed suggest differing densities in the layers of the Earth's matter. See Cohen's translation, *Principia*, Book III, proposition 20, 826–32.

VII

M. *CLAIRAUT* PROVED THAT THE RATIO OF THE AXES MUST DIMINISH IN PROPORTION AS THE GRAVITY AT THE POLE INCREASES.

M. *Clairaut* demonstrated in his theory of the shape of the Earth I just mentioned, that in all the most probable hypotheses about the density of the interior parts of the Earth there is always, supposing attraction, such a relationship between the fraction expressing the difference of the axes and that expressing the diminution of gravity from the pole to the equator, that if one of these two fractions exceeds 1/230, the other must be less, precisely by the same amount. Supposing, for example, the excess of the equator to the axis to be 1/173, which agrees well enough with actual measurements, one would have 1/173–1/230 or 1/698 for the amount that must be deducted from 1/230 in order to have the total shortening of the pendulum, going from the pole to the equator, that is to say, this shortening or, which is the same thing, the total diminution of gravity, will be 1/230–1/698, that is to say, about 1/343.

Now all the experiments with the pendulum have demonstrated that the diminution of gravity from the pole to the equator, far from being smaller than 1/230 as would be required to be in agreement with this theory, is on the contrary greater. It follows that current measurements are not in agreement with M. *Newton* on this point with the theory.

VIII

M. *NEWTON* HAD REACHED A VERY DIFFERENT CONCLUSION.

It cannot be concealed that M. *Newton* had drawn a very different conclusion from the supposition that the parts of the Earth were denser as one nears the center; he believed that, in this case, the ratio of the axes must increase.

M. *NEWTON*'S WORDS ON THIS SUBJECT IN THE SECOND EDITION OF THE PRINCIPLES.

Here is what he says on p. 386 of the second edition of the Principles: *This delay of the pendulum at the equator proves the diminution of gravity at this place; and there, the lighter matter is, the higher it will have to be in equilibrium with that of the pole.*[121]

WHERE HE ERRED

M. *Newton* believed that with the density increasing toward the center, gravity increased from the equator to the pole in a greater ratio than in the case

---

121. Ibid., proposition 20, 827 n[ote] bb, 828 n[ote] cc.

of homogeneity, which is true. But he thought that gravity at each point of the spheroid was in inverse proportion to the distances at the center of the spheroid, whether the spheroid was homogeneous or not, or its density varied in any manner whatever. Whence he had concluded that, in the case of increased density from the circumference to the center, gravity increasing in a greater ratio than in homogeneity, the flattening would be greater, which is false; being based on a supposition that only applies in a homogeneous spheroid.

IX

It follows from M. *Clairaut's* theory that, admitting the suppositions he makes about the interior of the Earth to be the most natural of those that spring to mind, the flattening can never be greater than 229 to 230, since this ratio is found in the supposition of the Earth's homogeneity. And as a result of this theory, in all other cases gravity increasing, the flattening must be less.

X

THE WEIGHT OF A BODY IN THE DIFFERENT REGIONS OF THE EARTH.
Having determined the ratio of the Earth's axes, homogeneity being assumed (Bk 3, prop. 20), M. *Newton* seeks in the following manner to determine the weight of bodies in the different regions of the Earth. Since one has seen that the columns of matter that correspond to the pole and the equator were in equilibrium when their lengths were to each other in proportion as 229 to 230, and that the weights of the equal parts, also placed in the same way in these two columns, must be in reciprocal proportion to these columns, or as 230 to 229, it can be seen, by similar reasoning, that in all the columns of matter that compose the spheroid, the weight of bodies must be in inverse proportion to these columns, that is to say, to their distance to the center. Thus, supposing that the distance of any place from the Earth's surface to the center is known, one will be able to calculate the gravity in this place and consequently the amount by which gravity increases or diminishes going toward the pole or toward the equator. Now, as the distance of any place whatever to the center roughly decreases almost as the square of the straight sine of the latitude, as can be shown by calculation, we can understand how M. *Newton* made the table of Bk 3, prop. 20, where he gave the diminution of gravity from the pole to the equator.[122]

---

122. Ibid., proposition 20, 826–32. The table appears on 828.

XI

THEY ARE IN PROPORTION TO THE LENGTH OF THE PENDULUMS.
Gravity being the only cause of the oscillations of pendulums, the slowing down of these oscillations proves the diminution of gravity, and their acceleration proves that gravity acts more strongly. For, we know that the speed of the oscillations of the pendulums is in inverse relation to the length of the string by which they are suspended. So if, in order to make the vibrations of a pendulum in a region isochronous to its vibrations in another, one must be shortened or lengthened, it must be concluded that the gravity is lesser or greater in this region than in the other. Since M. *Huygens'* time the ratio between the quantity by which one lengthens or shortens the pendulum has been known, as well as the diminution or increase of gravity. Thus, this quantity being proportional to the increases and decreases in weight, M. *Newton* gave the lengths of pendulums rather than the weights in a table.

XII

THE DEGREES OF LATITUDE ARE IN THE SAME PROPORTION.
The degrees of latitude diminishing in M. *Newton's* spheroid in the same proportion as the weights, the same table gives the size of the degrees of latitude beginning at the equator where the latitude is 0° to the pole where it is 90°.

XIII

EXPERIMENTS SHOW THAT GRAVITY IS A LITTLE LESS TOWARD THE EQUATOR THAN GIVEN IN M. *NEWTON'S* TABLE.
M. *Newton's* table gives a slightly smaller decrease for gravity toward the equator than that which results from actual measurements, but this table was only made for the homogeneous case. And he warns, at the end of the proposition in which he gives it, that in the case when the density of the parts of the Earth increases from the circumference to the center, one must also increase the decrement in gravity from the pole to the equator.

XIV

HE ATTRIBUTES THE DIFFERENCE TO THE HEAT OF THE EQUATORIAL REGIONS THAT LENGTHENS THE PENDULUM IN THESE REGIONS. BUT THE LATEST EXPERIMENTS HAVE DEMONSTRATED THAT THE LENGTHENING PRODUCED BY THE HEAT OF THESE REGIONS CANNOT CAUSE THESE DIFFERENCES.
Although M. *Newton* seems inclined to believe from the observations he reports in this same prop. 20 on the lengthening of the pendulum caused by

the heat in the equatorial regions that these differences come from the different temperatures of the places where the observations were made, the care taken to conserve the same degree of heat by means of a thermometer in the experiments that have been done since M. *Newton* on the length of pendulums in the different regions of the Earth, prove that these differences must not be attributed to this cause and that there really is a decrease of gravity from the pole to the equator greater than the one M. *Newton* gave in his table.

XV

METHOD GIVEN BY M. *NEWTON* TO FIND THE AXIS OF ANY PLANET.

At the end of Bk 3, prop. 19, M. *Newton* explains how to find the ratio of the axes of any planet for which one knows the density and the time of the diurnal revolution, using the ratio found between the axes of the Earth serving as a term of comparison. For whether a planet be greater or smaller than the Earth, if its density were the same and the time of its diurnal revolution equal to that of Earth, there would be the same proportion between centrifugal force and its gravity, and consequently between its diameters, as one found for those on Earth; but if its diurnal motion is more or less faster than the Earth's by some ratio, then the centrifugal force, and consequently the difference of the diameters will be more or less great in squared ratio to this speed, which follows the theory of centrifugal forces.[123] And if the density of this planet is greater or smaller than that of the Earth by some ratio, gravity on this planet will increase or decrease in the same ratio, and the difference of the diameters will increase in proportion to the diminished gravity, and will decrease in proportion to the increased gravity, which follows from the theory of attraction that M. *Newton* supposes in this matter.

XVI

DETERMINATION OF THE AXES OF JUPITER BY THIS METHOD.

Thus, for example, the difference in Jupiter's diameters, the diurnal revolution and the density of which are known, will be, at its small diameter in a compound ratio of the squares of the times of the diurnal revolution of the Earth and Jupiter, of the densities of Jupiter and the Earth, and of the difference of the diameters of the Earth compared with the minor axis of the Earth, that is to say, very nearly as $29/5 \times 400/94\frac{1}{2} \times 1/229$ to 1, that is to say, very nearly as 1 to $9\frac{1}{3}$; so the diameter of Jupiter from orient to occident is to its diameter between its poles as $10\frac{1}{3}$ to roughly $9\frac{1}{3}$. M. *Newton* adds that he supposed in this determination that the matter of which Jupiter

123. Ibid., proposition 19, 821–26; esp. 824-25.

was composed was of a uniform density, but that, as it is quite possible that because of the heat of the Sun it is denser toward the equatorial regions than toward the polar regions, its diameters can be in a relationship as 12 to 11, 13 to 12, or even 14 to 13, and that thus his theory agrees with observations, since observations teach that Jupiter is flattened, and that this flattening is less than 10⅓ to 9⅓, and between 11 to 12, and 13 to 14.

XVII

IMPROBABLE REASON GIVEN BY M. *NEWTON* FOR THE FLATTENING OF JUPITER BEING LESS THAN WHAT RESULTS FROM THE THEORY.

This means used by M. *Newton* to explain a flattening less than that given by homogeneity seems quite improbable, and it is surprising that in explaining the flattening of Jupiter he had recourse to a cause whose effect would be more perceptible on Earth than on Jupiter, since the Earth is much closer to the Sun than Jupiter.

Had he known M. *Clairaut's* proposition, namely, that the density increasing at the center diminishes the flattening, he would have found a very natural cause for the phenomenon he wanted to explain. Supposing Jupiter denser in the center than at its surface is a hypothesis that agrees with all the laws of mechanics.

XVIII

IN THE FIRST EDITION OF THE PRINCIPLES, M. *NEWTON* HAD GIVEN JUPITER MUCH LESS FLATTENING AND WHY.

In the first edition of the Principles, M. *Newton* had not taken density into account in the proportion of the diameters of Jupiter, and had concluded that the ratio of the axes was as 40 to 39, only considering the diurnal revolution, and the ratio of the axis to the Earth.

XIX

WHY ONLY THE PROPORTION OF THE AXES OF JUPITER, THE EARTH, AND THE SUN CAN BE KNOWN. THE PROPORTION OF THE SUN'S AXES IS TOO SLIGHT TO BE PERCEPTIBLE.

As it is only for the Earth, Jupiter, and the Sun that both elements necessary to determine the axes are known, namely, the diurnal revolution and the density, only the ratio of the axes of these three celestial bodies can be found. We just discussed the ratio of the axes of the Earth and of Jupiter; the ratio of the axes of the Sun could be found by taking the compound ratio of the square of 27½ to 1, of the density of the Earth to that of the Sun, and of 229 to 230: this would give, for the ratio of the axes of the Sun, an amount too small to be observed.

CHAPTER FOUR:
HOW M. NEWTON EXPLAINED
THE PRECESSION OF THE EQUINOXES

[Du Châtelet gives Newton's explanation for the observed movement of the Earth's axes, known as the precession of the equinoxes.[124] Using the principle of gravitation, Newton created a geometrical model demonstrating "the combined attraction of the Sun and the Moon on the Earth's protuberance at the equator." This phenomenon, Newton asserted, proved that the Earth must be broader at the equator and flattened at the poles.]

CHAPTER FIVE:
OF THE EBB AND FLOW OF THE SEA

THE EXPLANATION OF THE EBB AND FLOW OF THE SEA, LIKE THAT OF THE PRECESSION OF THE EQUINOXES, IS DERIVED FROM PROP. 66 OF THE FIRST BOOK OF THE PRINCIPLES AND ITS COROLLARIES.

I

The relationship between the ebb and flow of the sea and the precession of the equinoxes is easy to see. M. *Newton* deduces his explanation of the tides from the same cor. of prop. 66, from which, as we saw he derived his explanation of the precession of the equinoxes; these two phenomena are both necessary results of the attraction of the Moon and the Sun on the parts that compose the Earth.

II

GALILEO'S ERROR ON THE CAUSES OF THE EBB AND FLOW.
Galileo thought that the phenomenon of the tides could be explained by the rotation of the Earth and by its motion of translation around the Sun. But if this great man had paid more attention to the circumstances that accompany the ebb and flow, he would have seen that because of the diurnal movement, the waters must, in truth, rise toward the equator, which must give the Earth the shape of a spheroid depressed at the poles, but that this rotating mo-

---

124. This chapter has been summarized because of its complexity and in the interests of brevity for the volume as a whole. Overall, it demonstrates Du Châtelet's mastery of Newton's ideas and his mathematical methods. She refers to various sections of the *Principia*. See Cohen's translation, *Principia*, Book III, Lemmas, 885–88; Book III, proposition 39, 885–88 Book I, proposition 66, corollary 21, 583–84, and see also corollaries 17–19, 582–88; Book I, proposition 66, corollary 22, 584–85.

tion could never cause any reciprocal motion in the waters of the sea, as M. *Newton* demonstrated (prop. 66, cor. 19). M. *Newton* also demonstrates in this same corollary, using what he had demonstrated in corollaries 5 and 6 of the laws of motion, that the translation of the Earth in its great orbit can change nothing in the motions performed on its surface, and that consequently, the translative motion of the Earth around the Sun cannot cause the movement of the ebb and flow of the waters of the sea.[125]

III

THE EBB AND FLOW ARE A RESULT OF THE SUN'S AND THE MOON'S ACTION ON THE WATERS OF THE SEA. M. *NEWTON* DEMONSTRATED THAT IT IS BY THEIR ATTRACTION THAT THE SUN AND THE MOON ACT ON THE SEA.

By paying attention to the circumstances that accompany the ebb and flow, it was easy to perceive that these phenomena depend on the position of the Earth in relation to the Sun and the Moon, but it was easy not to understand the manner in which these two celestial bodies produce them, and the quantity each contributes. Only by the effects in which these actions are so intermingled can they be seen, so that without M. *Newton's* principles they could not have been separated one from the other, nor the quantity of each assigned. It was reserved for this great man to find the true causes of the ebb and flow and to subject these causes to calculation. Here is the path that he followed to arrive at them.

IV

THE PATH THAT HE FOLLOWED TO SUCCEED IN ASSIGNING THE QUANTITY EACH OF THESE CELESTIAL BODIES CONTRIBUTES TO THESE PHENOMENA.

In prop. 66 he begins by examining the principal phenomena that must result from the motion of three mutually attracting bodies in reciprocal relationship to the square of the distances, the smaller ones turning around the biggest.

After having seen in the first 17 corollaries of this proposition what are, in such a system, the disruptions the biggest body must cause in the motion of the smallest, which turns around the third, and given by this means the foundations of the theory of the Moon, he considers in corollary 18 many fluid bodies that turn around the third, and next supposes that these fluid

---

125. See Cohen translation, *Principia*, Book I, proposition 66, corollary 19, 582; Laws of Motion, corollaries 5–6, 423.

bodies become contiguous and form a ring that turns around the body, serving as its center.[126] And he demonstrates that this ring's motion must experience, through the action of the biggest body, the same disruptions as the single body of which he supposes this ring has taken the place. Finally, in corollary 19 he assumes that the body around which this ring revolves extends out to it, that this solid body contains the water of this ring in a canal crossed around it and that it turns on its axis in a uniform motion. And he demonstrates that then the motion of the water in this canal will be accelerated and retarded in turn by the action of the greatest body, and that this motion will be quicker in the syzygies of this water and slower in its quadratures, and, finally, that this water will experience an ebb and flow like our sea.

In Bk 3, prop. 24, M. *Newton* applies this proposition 66 and its corollary to the phenomena of the sea, and he demonstrates that they are a result of the combined attraction of the Sun and the Moon on the parts that compose the Earth.[127]

V–VII

[In sections V–VII Du Châtelet considers the action of the Sun and the proportions *Newton* discovered to describe its contribution to the tides. He concluded that these effects must be a combination of the Sun's attraction, centrifugal force, and the periodic times of the Earth around the Sun, and that it means that the Sun raises or lowers the water by approximately two Paris *pieds*.[128]]

VIII

[In this section Du Châtelet considers the action of the Moon and the variations depending upon the Moon's proximity to the Earth in combination with the Sun's effects. Using the example of variation in the tides in Bristol in the spring and the autumn, she explains:

"For, the first height is produced by the united forces of the Sun and

126. See Cohen translation, *Principia*, Book I, proposition 66, corollaries 1–17, 573–81; corollary 18, 581–82.

127. Ibid., Book III, proposition 24, 835–39. *Syzygy* signifies the point of an orbit at which a planet is in conjunction or opposition to another body; *quadrature* is the midpoint between the syzygies.

128. In this section, Du Châtelet makes reference to Book I, proposition 66, 570–85 and especially to corollary 14, 579–80; Book III, proposition 25, 839–40; see also proposition 30, 874–75 (pagination is from the Cohen translation of the *Principia*). This chapter has been cut and sections summarized in the interests of brevity.

the Moon, and the last by their differences; thus the sum of the forces of the Sun and the Moon on the sea, when these two bodies are at the equator and at their average distance to the Earth, is to their difference as 45 to 25, or as 9 to 5."[129]

Du Châtelet then describes how *Newton* separates out the force of the Moon from that of the Sun. He concluded that the force of the Moon raised the seas by about 9 feet, so that the force of the two together was about 10½ feet and increased to 12 when the Moon was in its perigee.]

IX

M. *BERNOULLI* BELIEVES THAT THESE FORCES ARE VERY MUCH GREATER THAN THOSE REPRESENTED BY M. *NEWTON*.

In his dissertation on the seas, which won the prize from the Academy for the year 1738, M. *Daniel Bernoulli* states that the absolute forces of the Sun and the Moon that cause the tides are very much greater than M. *Newton* supposes them to be.[130] And instead of considering, as M. *Newton* does, the Earth as composed of homogeneous parts, he imagines the density of the Earth's layers increasing from the circumference to the center, which is very probable for many physical reasons, and he asserts that with this supposition one can increase the forces of the Sun and the Moon on the sea as much as the phenomena will require.

X

WHAT LED M. *BERNOULLI* TO DISTANCE HIMSELF FROM THE SENTIMENT OF M. *NEWTON*.

[Du Châtelet, in her summary of Bernoulli's memoir, describes how Newton's supposition of a homogeneous Earth did not generate enough force to cause the tides observed in the Moon's quadratures; that Bernoulli also noted all the "accidental circumstances" that caused the greatest and least tides to vary from port to port, a sort of oscillation. She then continues her explanation.]

. . . from which M. *Bernoulli* concludes that it would be surer to evaluate the respective forces of the Sun and the Moon on the tides by their dura-

---

129. She refers to Book III, proposition 28, 844–46; proposition 37, 875–80 (pagination is from the Cohen translation of the *Principia*).

130. Daniel Bernoulli (1700-1782), one of the sons of Johann Bernoulli, was a mathematician and experimental philosopher, known for his work on astronomy and nautical phenomena, such as the tides.

tion and their intervals than by their heights, and in using this method, he finds that the force of the Moon is in a lesser proportion to that of the Sun than M. *Newton* found.

HOW IT CAN HAPPEN THAT THE MOON'S ATTRACTION HAS SO MUCH INFLUENCE ON THE WATERS OF THE SEA, AND DISTURBS THE EARTH'S MOTION SO LITTLE.

First, it is astonishing that the force of the Sun's attraction on the Earth is strong enough to force it to turn around it, while that of the Moon scarcely causes perceptible alterations of the Earth's orbit. Nevertheless, the Moon has much more influence than the Sun on the movements of the sea. But if one notes that the sea's movements happen because its parts are attracted differently from those of the rest of the globe, because their fluidity means that they yield much more easily to causes that act on them, one will see that the Sun's action, which is very strong on the whole Earth, attracts all its parts almost equally because of its great distance from the Earth, whereas the Moon, being much closer to the Earth, must act more unequally on the different parts of our globe, and that this inequality must be much more perceptible.

Having demonstrated that the combined attraction of the Sun and the Moon on the waters of the sea is the cause of the tides, and having determined the quantity that each of these two celestial bodies contributes, M. *Newton* embarks on the explanation of the circumstances that accompany the phenomena of the sea.

THREE SORTS OF VARIATION CAN BE DISTINGUISHED IN THE MOTION OF THE SEA.

Three types of motion in the sea have always been known: its daily movement, which makes it rise and fall twice a day; the regular alterations in this movement each month, which follow the positions of the Moon in relation to the Earth; and finally, those that take place each year and are caused by the greater proximity of the Earth to the Sun at certain times of the year.

[Du Châtelet explains that the diurnal variations in lowering and raising the sea twice a day correspond to the passage of the Moon to the meridian.[131]]

---

131. She refers in this section to Book I, proposition 66, corollaries 19 and 20, 582–83 (pagination refers to the Cohen translation of the *Principia*).

XIII[132]

[Du Châtelet describes the effect of the seabed and the water's inertia in the twice-daily movement of the sea that can delay the highest levels for two to three hours after the Moon passes through the meridian. "So that the sea's movement is perpetually accelerated in the six hours that precede the passage of the celestial body to the meridian." The same factors delay the perceivable withdrawal of the sea for about three hours after the culmination of the Moon.][133]

M. *Culler*, from whose dissertation I have taken many things in this chapter, says that if only the vertical motion of the water was considered, its greatest elevation should take place simultaneously with the passage of the Moon to the meridian, and even increase because of the action of the Sun. He attributes the greatest part of the delay in the rising of the water to its horizontal movement, by which it rubs against the bed in which it runs.[134]

In the regions where the sea does not connect with the ocean, the tides come later, sometimes by as much as twelve hours. And in other places it is often said that the tide precedes the passage of the Moon to the meridian; at the Port of Le Havre, for example, where the tide is nine hours later, it is believed that it precedes by three hours the passage of the Moon to the meridian; but the truth is that this tide is the effect of the preceding culmination.

XIV–XXIII

[In these sections Du Châtelet explains all of the variations in tides—daily, monthly, yearly—and their causes: the Moon's position in relation to the Earth; the combined effect of the Sun and the Moon; when their combined force is greatest and least; the significance of the declination of the Sun and Moon; how to calculate the effect of a place's latitude. Finally, she explains how the position of the waters in relation to land or to an ocean can be such that there is no ebb or flow. For example, she states that there is none in the Baltic, Black or Caspian Seas; that it is "hardly perceivable" in the Mediter-

---

132. The sections have been misnumbered; there is no XI or XII.

133. The rise of the Moon to its highest point.

134. Du Châtelet may mean Samuel Colepresse, whose observations at Plymouth were reported in the Royal Society's *Philosophical Transactions* for 1668. I am grateful to Ronald K. Smeltzer for suggesting a solution to this mystery.

ranean; and that the greater depth of the Adriatic led the Venetians to be the first to notice the ebb of the Mediterranean.[135]]

XXIV

THERE ENTER INTO THE PHENOMENA OF THE EBB AND FLOW MANY
CAUSES THAT ARE NOT ASSIGNABLE.

Thus, apart from the assignable causes by which one can account for the phenomena of the sea, there are still several that cause inequalities in its motions which are not reducible to any law, because they depend on elements that change with each place; such are the beds over which the waters pass, straits, the different depths of the seas, their width, the opening of mouths of rivers, the winds, etc., all causes that can alter the quantity of motion of the water and, consequently, delay the flow, augment it, or diminish it, and that cannot be subjected to calculation. This is why in some places the flow arrives three hours after the culmination of the Moon, and others where it only arrives twelve hours later. In general, the bigger the tides, the later they arrive, and it must be thus, since the causes that delay them act for a correspondingly longer time.

If the flow was infinitely small, it would take place in the very moment of the culmination, because the obstacles that delay the flow would act infinitely little; it is partly for this reason that the greatest tides, which happen toward the new and the full Moon, follow the passage of the Moon at the meridian much later than those that happen toward the quadratures; for the latter tides are the smallest.

XXV

[Du Châtelet gives the results of Leonhard Euler's calculations for the speed of the water in the tides at St. Malo, Dunkirk, and Ostend, sites known for long, rapidly moving, extreme tides.][136]

XXVI

THE TIDES ARE ALWAYS GREATER TOWARD THE COASTS, AND WHY.

Tides are always greater toward the coasts than on the open sea for several reasons: first, the seas lash against the shores, and the impact adds to

---

135. Du Châtelet refers in this section to the *Principia*, Book III, proposition 37, 875–80; Book III, proposition 24, 835–39 (pagination is from the Cohen translation of the *Principia*). Newton's *De Systemate mundi*, 58 [her page reference].

136. Leonhard Euler (1707–1783) was a noted mathematician and physicist with whom Du Châtelet corresponded.

the height; second, it arrives with the speed that it had in the ocean where the depth is very great, and it arrives in great quantity and because it is repelled by the shores, the tide rises even more; finally, when it passes through straits, its height increases greatly, because, being repelled by the shores, it comes with the force it acquired by the effort it made to flood them. This is why at *Bristol* the tide rises to a very great height toward the syzygies; on this coast the shore is full of sinuosities and sand banks against which the water lashes with great force and which it cannot ebb from as swiftly as if the shore were even.

XXVII–XXVIII

[Du Châtelet gives Newton's explanations in his *De Systemate mundi* for the many tidal variations in ports around the world.]

XXIX

AT THE MOUTHS OF RIVERS THE EBB LASTS LONGER THAN THE FLOW,
AND WHY.

At the mouths of rivers, the ebb and flow are different yet again, for the current of the river going into the sea resists the motion of the flow of the sea and aids the motion of the ebb, and this cause must consequently make the ebb last longer than the flow, which is indeed what happens. *Sturmnius* reports that above *Bristol*, at the mouth of the river Oundale, the flow lasts five hours and the ebb seven. Moreover, this is also why, all things being equal, the greatest flows happen later at the mouths of rivers than elsewhere.[137]

XXX–XXXII

[Du Châtelet describes tidal variations in different climate zones caused by the relation of the Moon to the equator. For example, there are no daily tides at the poles because the Moon is at the same elevation on the horizon all twenty-four hours. She notes that, except for the tides, the effects of the Sun's and Moon's forces are imperceptible on Earth.]

XXXIII

[Du Châtelet suggests that perhaps the changes in Jupiter's spots result in part from the action of its satellites on its waters. She notes that "if one observed that these changes could be analogous with the aspects of these sat-

---

137. Samuel Sturmy (1633–1699) reported his observations in the *Philosophical Transactions* of the Royal Society in 1668. I am grateful to Ronald K. Smeltzer for this reference.

ellites that follow from this theory of the tides, one would have a proof that this was the true cause."]

[CHAPTER SIX]:
HOW M. NEWTON EXPLAINS THE PHENOMENA OF THE SECONDARY PLANETS, AND PRINCIPALLY THOSE OF THE MOON[138]

I

The first phenomenon the secondary planets displayed to physicists is their tendency toward their principal planet, following the same law as the principal planets toward the Sun. We have sufficiently established this tendency in the second chapter in reference to the principal planets, disregarding, as is done at first to simplify the question, all the inequalities the planets produce between them or that they can receive from the Sun. But it is now appropriate to examine these inequalities, to see in a more satisfactory manner the universality of the principle of attraction and the harmony of the system built upon it. The Moon is, of all these planets, the one whose variations are best known and whose course is most amenable to the theory.

For a complete examination of the other secondary planets we lack one element that seems impossible to supply, knowledge of their masses, which is necessary to measure their reciprocal actions, and the resulting disruptions of their orbits. And even if, abandoning the hope of calculating by theory alone the movements of these celestial bodies, we were to propose only to demonstrate a posteriori that the phenomena have nothing contrary to the principle of attraction, we would be none the wiser, because the phenomena themselves, considered astronomically, are not well enough ascertained. The theory of these planets is thus reduced to having seen the forces with which they act on one another, or those with which the Sun acts on them to disturb their orbits, which forces are very small in comparison with the attraction that they experience toward their principal planets, and is reduced

---

138. Du Châtelet omitted the chapter heading here, perhaps overlooked in the rush of correcting the proofs in June, July, and August of 1749. This may well have been the last chapter she wrote as it is the one describing Clairaut's work and the most recent ideas on this, the three-body problem. His handwriting appears on the draft of her longer discussion of this phenomenon at a few points in the margins.

to knowing that this attraction is like all the others in inverse proportion to the squares of the distances.[139]

The different sorts of motion of the Moon that had been observed long ago and the laws of these motions found by famous astronomers furnished M. *Newton* the means to apply his theory to this planet with success. This great man, who had already made so many discoveries in the other parts of the system of the world, wanted to perfect it further; and although the method he followed on this occasion was less clear and less satisfying than the one he used for other phenomena, one must be very grateful to him for exerting himself here.

We will give some slight idea of the method he followed in this investigation.

II

MANNER OF TAKING INTO ACCOUNT THE INEQUALITY OF THE FORCE OF THE SUN ON THE EARTH AND ON THE MOON. BK 1, PROP. 66.[140]

It is easy to see that if the Sun was at a distance from the Earth and from the Moon that was infinite in relation to that between these two planets, it would in no way affect the motions of the Moon around the Earth; since equal forces, whose directions are parallel, that act on any two bodies could not alter their relative motions. But as the angle made by the lines drawn from the Moon and from the Earth to the Sun, though very small, cannot be regarded as zero, it must be taken into account, and from it must be deduced the inequality of the Sun's action on the two bodies in question.

THE FORCE OF THE SUN SPLITS INTO TWO OTHERS. ONE PUSHES THE MOON TOWARD THE EARTH. THE OTHER ACTS FOLLOWING THE LINE DRAWN FROM THE EARTH TO THE SUN.

Taking then, like M. *Newton*, on the line drawn from the Moon to the Sun a straight line to represent the force with which the Sun attracts it, let us consider this line as the diagonal of a parallelogram one side of which is the line drawn from the Moon to the Earth and the other a parallel leading from the Moon to the line that connects the Sun and the Earth. It is clear that these two sides of the same parallelogram will represent two forces that one can

---

139. This is a classic eighteenth-century rhetorical practice used in science in which the author argues from the observations to formulate the hypothesis.

140. See Cohen translation, *Principia*, Book I, proposition 66, 570–85. This chapter goes through what was identified in Du Châtelet's time as the three-body problem: how to formulate geometrical abstractions and then to calculate the effect of attraction when three bodies interact—typically, the Sun, the Earth, and the Moon.

substitute for the force of the Sun on the Moon, and that the first of these two forces, that which pushes the Moon toward the Earth, will not disturb in any way the observance of *Kepler*'s rule about the areas proportional to the times, but that it will only change the law determining the force with which the Moon will tend toward the Earth, and will in consequence alter the form of its orbit. As for the second force, that which acts following the parallel to the radius of the orbit of the Earth, if it were equal to the force with which the Sun acts on the Earth, it is easy to see that it would not disturb in any way the orbit of the Moon; but this equality can only happen at the points where the Moon is at a distance from the Sun equal to that where the Earth is at the same time, which happens near the quadratures. At any other point, these two quantities being unequal, it is their difference that expresses the perturbative force of the Sun on the Moon's orbit, enough both to disturb the description of the areas in equal times and to prevent the Moon from always moving in the same plane.

III–VI

[Du Châtelet explains Newton's method for measuring this perturbative force of the Sun on the Moon and his use of mathematics and the laws of mechanics to calculate the variations in the Moon's orbit.[141]]

VII–XII

[Du Châtelet explains how Newton developed the laws governing the changes in the position of nodes and the inclination of the Moon at any given time.[142]]

XIII–XXI

Du Châtelet explains: ". . . In the examination of the first inequalities, although the reader might not be very satisfied with some suppositions and abstractions used to make the problem easier, he has at least this advantage, that he sees the route taken by the author, and that he acquires new principles which he can be sure will help him to go much further. But as regards the motion of the apogee and the variation of the eccentricity, and all the other inequalities in the Moon's motion, M. *Newton* is content with results

141. She credits Tycho Brahe for first observing and recording the variations in the Moon's orbit. She refers to Book III, proposition 25, 839–40; Book III, proposition 3, 802–3; Book III, propositions 26–29, 840–48 (pagination is from Cohen's translation of the *Principia*).

142. She refers to Book III, propositions 30, 31, 33, 35, 848–60, 867–69 (pagination is from Cohen's translation of the *Principia*).

that seem adequate for astronomers to construct tables of the Moon's motion, and he maintains that his theory of gravity led him to these results."

[Du Châtelet then goes on to cite information from such tables and seven corrective equations to give the true place of the Moon in its orbit.[143]]

XXII

To retrace the path that may have led M. *Newton* to all these equations, we can think only of some corollaries of proposition 66 of the first book, where he gives the way of estimating the perturbative forces of the Sun that I set forth in this chapter. In truth, we readily admit that one of the two forces that act in the direction of the radius of the orbit of the Moon, joined with the force of the Earth, changes the inverse proportion of the square of the distances, and must change both the curvature of the orbit and the time the Moon takes to traverse it. But how did M. *Newton* use these alterations by the central force, what principles did he follow to avoid or to conquer the extreme complexity and the difficulties of the calculation this investigation presents? This has not yet been understood, at least not in a satisfactory manner.

Admittedly, in the first book of the principles can be found a proposition on the motion of the apsides in general, that at first promises to be very useful for the theory of the apsides of the Moon, but when one comes to apply it, it is soon clear that it does not lead very far in this investigation.[144]

The proposition I mean teaches that, if to a force that acts inversely as the square of the distances is added one inversely proportional to the cube, this new force will not change the nature of the curve described by the first force, but will give a circular motion to the plane on which it is described. I mean to say that the addition of this new force that follows the inverse proportion of the cube causes the body, instead of describing an ellipse around the center of the forces on an immobile plane, as it would have described under the influence of the mere force in inverse proportion to the square, to describe the curve traced by a point moving in an ellipse, while the plane of this ellipse itself turns around the center of the forces. In corollaries of this

---

143. She refers to Book III, proposition 35, 867–79. See also the Scholium to proposition 35, 869–74 (pagination is from Cohen's translation of the *Principia*).

144. Du Châtelet refers probably to Book I, proposition 45, problem 31, 539–45 (pagination is from the Cohen translation of the *Principia*). Note that this is an occasion on which Du Châtelet explicitly points out weaknesses in Newton's theorizing.

impulsion, gives an orbit that is always a conic section, with the Sun as its focus. To confirm this theory, comets needed to have no other motion than those that one can relate to these curves, and the areas traversed by them around the Sun had to be proportional to the times it took for their course or description.

V

HE DETERMINED THE ORBIT OF ANY COMET BY MEANS OF THREE OBSERVATIONS.

Calculation and observations, the faithful guides of this great man, easily verified this conjecture. He resolved this beautiful astronomic-geometric problem: three locations for a comet, assumed to move in a parabolic orbit, describing around the Sun areas proportional to the time, being given with locations on the Earth at the same time, how to find the position of the axis, of the apex and the parameter of the parabola, or, what amounts to the same, how to find the orbit of the comet?

This problem, difficult as it is with respect to the parabolic orbit, would have been so perplexing in the case of the ellipse and the hyperbola that it was appropriate to reduce it to this lesser degree of difficulty. Besides, the hypothesis of the ellipse, the only probable one practically speaking, amounted to almost the same as that of the parabola, because comets, having only a small part of their orbits within range of our observations, must follow very elongated ellipses, and such curves can, as we know, in the part closest to their focus, be taken without significant error for parabolas.[148]

VI

HE VERIFIED HIS CALCULATION BY OBSERVATIONS OF A GREAT NUMBER OF COMETS.

So, M. *Newton,* having resolved the problem we have just discussed, applied it to all observed comets, and from this deduced the complete confirmation of his conjecture. For all positions determined by calculation based on three longitudes and latitudes of the celestial body were found to be so close to the positions found directly by observations that their agreement, considering how difficult it is to make accurate observations of this kind, is surprising.

148. This reasoning comes from the *Principia,* Book III, proposition 40; see the Cohen translation of the *Principia,* 895.

VII

THE DURATION OF THEIR PERIOD CAN ONLY BY FOUND BY LOOKING AT
THE APPEARANCES OF THE COMETS IN THE SAME CIRCUMSTANCES AND
AT EQUAL INTERVALS RECORDED IN THE PAST.

As for the length of the periods of the comets, it cannot be deduced from
the same calculation, because, as we just said, their orbits being so elongated
that one can take them without significant error as parabolas, excessive differences in their duration would hardly produce any change in their appearances, in the arc of their orbit that we know. But it is nonetheless satisfactory for M. *Newton's* theory to see at the points at which they are visible that
they observe exactly M. *Kepler's* law concerning the areas proportional to the
times, and the law according to which the Sun attracts them as all other celestial bodies in inverse proportion to the square of their distance.

VIII

M. *HALLEY* USED THE PERIOD OF THAT OF 1680 TO CORRECT THE ORBIT
OF THIS COMET.

M. *Halley*, to whom all parts of astronomy owe so much, and who carried
the doctrine of comets so far, made on the occasion of the famous comet
of 1680 an investigation that brought M. *Newton* much satisfaction. Finding
that three observations for comets mentioned in historical accounts agreed
with that one in a number of remarkable circumstances, and that they had
reappeared at a distance of 575 years from one to the other, he supposed that
this could be one and the same comet making its revolution around the Sun
in this time period. So he supposed the parabola changed to an ellipse that
took 575 years for the comet to describe, and that the curvature conformed
with the parabola in the instances of its orbit near the Sun.

Having then calculated the positions of the comet in this elliptical orbit, he found that they corresponded so well with those where the comet
was observed that the variations did not exceed the difference found between the calculated positions of the planets and those that can be observed,
although the motion of the latter has been the object of astronomers' inves-
tigations for thousands of years.

IX

THE COMET OF 1682 MUST REAPPEAR IN 1758.

The comet of 1680 having so considerable a time of duration, its return that
is to take place toward the year 2255, is of little interest. But there is another

comet whose return is so near that it promises a very agreeable spectacle for the astronomers of our time. It is the comet that appeared in 1682, in circumstances so similar to those of the comet that appeared in 1607 that it is difficult not to believe it is not one and the same planet, making its revolution in seventy-five years around the Sun. If this conjecture is found verified, the same comet will reappear in 1758, and this will be a very pleasing moment for the partisans of M. *Newton*.[149] This comet seems to be among those that move away least in our system. For when most remote from the Sun, it is not farther away from us than four times Saturn's distance from us. If it is visible when it passes again in the inferior part of its orbit in 1758, one will not hesitate to number it among the planets.[150]

X

DIFFERENT OPINIONS ON THE TAILS OF COMETS. M. *NEWTON* ASSERTS THAT THEY ARE ONLY A SMOKE THE BODY OF THE COMET EXHALES.

The tails of comets that caused appearances of these celestial bodies to be considered ill omens are now placed among ordinary phenomena, which excite the attention only of philosophers. A few insisted that the rays of the Sun, passing across the body of the comet that they supposed to be transparent, produced the appearance of their tails, in the same way as we perceive the space traversed by the Sun's rays, passing through the eye of a camera obscura.[151] Others imagined that the tails were the light of the comet, refracted when it reached us and producing an elongated image, such as the Sun produces by the refraction of a prism. M. *Newton*, having stated these two opinions and refuted them, reports a third that he accepted himself. It consists in regarding the tail of the comet as a vapor continually rising from the body of the comet toward the parts opposed to the Sun, for the same reason that vapors or smoke rise in the Earth's atmosphere and in the void of the vacuum pump. Because of the motion of the comet's body, the tail is

149. In fact, the conjecture was proved correct. Clairaut labored for two years from 1757 to 1759 with the assistance of two other mathematicians, Joseph Jérôme Lefrançois de Lalande and Mme Nicole-Reine Etable de la Brière Lapaute (1723–1788), to make the calculations predicting this comet's return. Their prediction was accurate to within a few days. The excitement over the return of this comet probably explains his interest in the publication of Du Châtelet's translation in an incomplete form in 1756, and then in its current form in 1759.

150. Note that Du Châtelet uses some astronomical terms indiscriminately, for example, *planet* seems to have been a term for any clearly perceptible body that orbited the Sun.

151. Du Châtelet is drawing on her own experiments and observations and on the *Principia*, Book III, proposition 41, 919–28 (pagination is from the Cohen translation of the *Principia*).

slightly curved toward the place where the core or head passed, a little like smoke rising from a burning coal that is being moved around.[152]

XI

WHAT CONFIRMS THIS OPINION.

What further confirms this opinion is that the tails are always greatest when the comet leaves its perihelion, that is to say, the place where it is at its least distance from the Sun, where it receives the most heat and where the atmosphere of the Sun is at its greatest density. The head appears after this to be darkened by the thick vapor that rises from it abundantly, but at the center can be discovered a part very much more luminous than the rest, which is called the core.

XII

PURPOSE OF THESE TAILS ACCORDING TO M. *NEWTON*.

By this rarefaction a great part of the tails of comets must be shed into the solar system, some, by its gravity, can fall toward the planets to mix with their atmosphere and to replace the fluids consumed in the operations of nature.

XIII

THE COMETS MIGHT UNDERGO GREAT ALTERATIONS IN THE EXTREMITIES OF THEIR ORBITS.

If one considers all that can affect comets in the most distant parts of their orbits, where the Sun's force on them becomes extremely weak, and where they may be in the vicinity of other celestial bodies, one sees that the permanence of their period is not as necessary as for other planets. So if it happened that some of the comets that we wait for did not reappear, this would be much less prejudicial to the *Newtonian* system than the support drawn for this system by their constancy in following *Kepler's* first rule, that of the spaces proportional to the times.[153]

---

152. I. Bernard Cohen, Newton's English translator, uses the word *head*, Du Châtelet, the word *noyau* or *core*.

153. Newton's supporters took four key hypotheses as the way to prove the existence of universal attraction acting in the universe; his explanation for comets was one of these. Here Du Châtelet, in a sense, prepares her audience for the fact that predictions of their return may not be the only way in which comets prove Newton's theory, the fact of their orbit being elliptical like that of the planets increased the probability of its truth.

XIV

SOME OF THE COMETS MIGHT WELL FALL INTO THE SUN.

The resistance that comets encounter in traversing the Sun's atmosphere when they are in the inferior part of their orbits can also affect their movements, slow them down from revolution to revolution, and make them approach closer and closer to the Sun, until they finally are swallowed up in this immense globe of fire.

The comet of 1680 passed at a distance from the Sun's surface that did not exceed the sixth part of the diameter of this globe; it is probable that it will approach even closer in the next revolution and that it will finally fall completely into the Sun.

XV

CONJECTURES M. *NEWTON* MADE ON THE CONSIDERABLE CHANGES THAT HAPPENED TO FIXED STARS.

M. *Newton* suspects that the stars from which the light sometimes appeared to weaken considerably and which next appeared brilliant may have owed their new brightness to the fall of some comet that served as a nutrient for their fire.

VI

DISCOURSE ON HAPPINESS

VOLUME EDITOR'S INTRODUCTION

The last selection presented here is the complete *Discourse on Happiness*, the first of Du Châtelet's writings to receive a critical edition and to be issued as a paperback.[1] Du Châtelet must have written the essay in stages, from the end of the 1730s when she and Voltaire were at Cirey into the1740s when she was an active courtier at Louis XV's Versailles, and then at King Stanislas's Lunéville. She probably put it down as other projects took precedence, and finished it in the spring of 1748 as a gift for her new lover, Jean-François de Saint-Lambert. One can easily imagine Du Châtelet formulating some of her ideas on this subject during her extended time at Cirey from 1735 to 1738, when she and Voltaire were discussing Mandeville and Locke, reading the English poet Alexander Pope, and thinking about human nature and why men and women act as they do.[2] To write an essay on happiness was a common amusement for men of the Republic of Letters, but there are a number of ways in which hers differs from others of its type.[3]

Initially, Du Châtelet followed the pattern established by learned men, and without indicating her sex, discussed what she believed to be the sources

---

1. The critical edition was edited by Robert Mauzi; the popular paperback had no annotation, only a preface by the French writer Elizabeth Badinter. With Du Châtelet's knowledge it circulated in multiple manuscript copies, three are known. Her son prevented publication until 1779. It appeared again with an early selection of her letters in 1806.

2. The echoes of her translation of Mandeville are particularly evident. She followed his sentiments ascribing most "good" behavior to pride and self-esteem, to the desire to be thought well of by others, and to the fear of being publicly shamed and ridiculed. She speaks of the pleasures of gambling in both writings, and the uncertainty it excites.

3. For example, both Voltaire and their younger friend, Claude Adrien Helvétius (1715–1771) wrote on this theme. See the study by Robert Mauzi, *L'Idée du bonheur dans la littérature et la pensée françaises au XVIIIe siècle* (Paris: Librairie Armand, 1969).

of happiness. In every other way, she made her essay more personal and more daring than those of male authors. About half way through the *Discourse*, she reveals herself to be female and uses numerous anecdotes from her own life rather than making long references to classical and contemporary authors to illustrate her points. She applauds the passions as a key source of happiness and rejects any sense of sin. She specifically advises her readers not to spend time repenting their excesses. Instead she believes that reason, individual conscience, and one's fear of public shame will act as the needed restraints—by implication rejecting the role of the established Church in maintaining public propriety. She assumes that people of her station in life will be ambitious and describes "study" as the only avenue to glory and fame available to women, and the only source of happiness in which one is not dependent upon others. Despite her belief in reason, she condones "illusion" as a means to happiness, even in love, the passion she identifies as the ultimate source of happiness.

In the last section of the *Discourse*, Du Châtelet abandons any pretense of generality and speaks of specific events and in her own voice, as a woman past thirty, a confident, ambitious member of elite society. She makes veiled, but nonetheless obvious, references to Voltaire, the end of the sexual aspect of their relationship, and his abrupt betrayal in 1745 when he began his affair with his niece, Mme Marie-Louise Denis. The final pages echo her letters to Saint-Lambert in 1748 and her effort to make reasonable her surrender to her love for the younger soldier-poet.

The first two letters that precede the *Discourse* show the depth and the pleasures of her feelings for him and the nature of her last days before the birth of their daughter on 4 September 1749. The third letter, to the official of the Royal Library in Paris is from that same week. Saint-Lambert was with her for the birth and at her side the night she died. The inventory of her estate contains the receipt from Abbé Sallier for her manuscripts. As she had feared, when she first realized she was pregnant, the translation and commentary were her last projects.

The letters to Saint-Lambert are from the collection edited by Anne Soprani; the one to the royal librarian is from the Besterman collection.

☞

RELATED LETTERS

To Saint-Lambert[4]

[Commercy? Summer 1749?] At 9 thirty, Tuesday

Yesterday you said things so tender and so touching that you have penetrated my heart, so love me always in this way. Believe that, when you love me, I adore you. I spent the most agreeable night that I could spend without you, you never left my thoughts. . . . You wish me to tell you about what I will do today. What I want to do every day of my life: I will see you, I will love you, I will say it to you. But I want to read that in the charming eyes that I adore.

To Saint-Lambert[5]

[Lunéville] Saturday evening [end of August 1749]

You know me very little, you do little justice to the eagerness of my heart if you believe that I can be two days without having your letters, when it may be otherwise. Your confidence in the possibility of going on guard as soon as you arrive scarcely agrees with the impatience with which I endure your absence. In a word, you have business and duties in Haroué; it is better than having pleasures there.[6]

When I am with you, I endure my state with patience, I often do not notice it. But when I have lost you, I see only the dark side of things. Yesterday I again walked to my little house, and my stomach is so terribly low, I feel such pain in my back.[7] I am so sad this evening that I would not be surprised if I gave birth this night. But I would be sorry, although I know that you would be pleased. I will endure the pains of childbirth more patiently when I know you are in the same place as I. I wrote you eight pages yester-

4. No. 95 in the Soprani edition, 236; no. 407 in the Besterman collection, vol. 2, 207–8.

5. No. 99 in the Soprani edition, 242–43; no. 485 in the Besterman collection, vol. 2, 305–6.

6. Haroué was the Beauvau-Craon family château and estates. The Prince de Beauvau-Craon was Saint-Lambert's patron.

7. Although Du Châtelet uses the word *stomach*, she is referring to her uterus and the way in which it "falls" as the baby changes position in preparation for the birth. The "little house" she refers to was one of the summer pavilions that King Stanislas arranged to have prepared for his favorite courtiers.

day, you will only receive them on Monday. You did not state clearly if you will return on Tuesday, and if you will be able to avoid going to Nancy in the month of September thanks to this guard duty. Do not leave me in uncertainty. I would be frightened of how sad and despondent I feel if I believed in presentiments. All I wish is to see you again. It is a long time from now to Tuesday! The prince will be very fortunate to have you with him! He will not value your presence as much as I do! I finally received a letter from his sister. The princess comes tomorrow to have supper with me. If you do not reassure my heart, if you do not write tenderly to me, I will be greatly to be pitied.

At the least, say to the prince that you will not go to Haroué before my lying-in; I would not allow it. I will have myself bled only on your return. I was hoping to work during your absence, I have not been able to yet. I am too unwell, I have an insupportable backache, and am prey to discouragement in the mind and in my whole person, from which my heart alone is spared.

My letter, which is at Nancy, will please you more than this one. I did not love you better, but had more strength to tell you, there was less time since you had left. I stop because I cannot write any more.

To Abbé Sallier, Bibliothèque du roi[8]

Around the 1st September 1749

I use the liberty that you gave me, monsieur, to put into your hands manuscripts that I very much wish will remain after me. I very much hope that I will be able to thank you again for this service and that my lying-in, which I am expecting at any moment, will not be fatal, as I fear. I earnestly beg you to put a number on these manuscripts and to have them registered, so that they will not be lost. M. de Voltaire, who is here with me, sends his warmest compliments to you, and, I reiterate, monsieur, the assurances of the sentiments with which I will never cease to be your very humble and very obedient servant.

Breteuil Du Châtelet

ॐ

8. No. 486 in the Besterman collection, vol. 2, 306–7; D4004 in the *Oeuvres complètes*. Abbé Claude Sallier was in charge of the King's Library, now the Bibliothèque nationale.

DISCOURSE ON HAPPINESS

It is commonly believed that it is difficult to be happy, and there is much reason for such a belief; but it would be much easier for men to be happy if reflecting on and planning conduct preceded action. One is carried along by circumstances and indulges in hopes that never yield half of what one expects. Finally, one clearly perceives the means to be happy only when age and self-imposed fetters put obstacles in one's way.

Let us anticipate the reflections that we make too late: those who will read these pages will find what age and the circumstances of their life would provide too slowly. Let us prevent readers from losing a part of the precious short time that all of us have to feel and to think; and from giving their time to caulking their ship, time which they should devote to securing the pleasures that they can enjoy on their voyage.

In order to be happy, one must have freed oneself of prejudices, one must be virtuous, healthy, have tastes and passions, and be susceptible to illusions; for we owe most of our pleasures to illusions, and unhappy is the one who has lost them. Far then, from seeking to make them disappear by the torch of reason, let us try to thicken the varnish that illusion lays on the majority of objects. It is even more necessary to them than are care and finery to our body.

One must begin by saying to oneself, and by convincing oneself, that we have nothing to do in the world but to obtain for ourselves some agreeable sensations and feelings. The moralists who say to men, curb your passions and master your desires if you want to be happy, do not know the route to happiness.[9] One is only happy because of satisfied tastes and passions; I say tastes because one is not always happy enough to have passions, and lacking passions, one must be content with tastes.[10] It is passions then that one should ask of God, if one dared to ask him for something, and Le Nôtre was quite right to ask the Pope for temptations rather than indulgences.[11]

But, some will object, do not the passions cause more unhappiness than happiness? I do not have the instrument necessary to weigh in general the

---

9. Du Châtelet is referring to the Greek moralists, the Stoics of the third century BCE, and to their Roman followers such as Seneca, Epictetus, and Marcus Aurelius in the first and second centuries CE.

10. *Goûts*, the French word used here, has been translated as *tastes*, and Du Châtelet sees these as less emotionally intense than *passions*.

11. André Le Nôtre (1613–1700) was Louis XIV's landscape architect for Versailles.

good and the bad that they have done to men;[12] but one must remark that the unhappy are known because they have need of others, that they love to tell their misfortunes, that they seek in the telling some remedy and physical relief. Happy men and women seek nothing and do not notify others of their happiness; the unhappy are interesting, the happy are unknown.

This is why when two lovers are reconciled, when their jealousy is gone, when the obstacles that separated them have been surmounted, they are no longer proper drama. The play is over for the spectators, and the scene of Rinaldo and Armida would not interest us as much as it does if the spectator did not expect that the love of Rinaldo is the effect of an enchantment that must be dispelled, and that the passion displayed by Armida in this scene will make her unhappiness more interesting.[13] The same motives move our soul at the theater and in the events of life. So one knows more of love by the unhappiness it causes than by the often obscure happiness it produces in men's lives. But let us suppose for a moment that the passions cause more unhappiness than happiness, I say they are still to be more desired, because they are a necessary condition for the enjoyment of great pleasures. Now, the only point of living is to experience agreeable sensations and feelings; and the stronger the agreeable feelings are, the happier one is. So it is desirable to be susceptible to the passions, and let me say it again: passions do not come for the asking.

It is for us to make them serve our happiness, and that often depends on us. Whoever knows how to make the most of his station in life and the circumstances in which fortune has placed him so well—that he succeeds in putting his mind and his heart in an untroubled state, that he is susceptible to all the feelings, to all the agreeable sensations his situation carries—is surely an excellent philosopher and should thank nature.

I say his station in life and the circumstances where fortune placed him because I believe that one of the things that contribute the most to happiness is to content oneself with one's situation and to think of ways to make it happy rather than to change it.

My goal is not to write for all sorts of social orders and all sorts of people; all ranks are not susceptible to the same kind of happiness. I write only for those who are called people of quality, that is to say, for those who

12. This reference is both to the scientific instrument and to the Christian idea that the souls of the dead would be weighed in the balance on the Day of Judgment with a feather on the other side of the scales. Those heavy with sin went to purgatory or hell, those light with virtue to heaven.

13. Du Châtelet is referring to a famous rendering of the medieval story of two lovers, Rinaldo and Armida, by the Italian poet Torquato Tasso (1544–1595), in *Jerusalem Delivered*.

are born with a fortune already made, more or less distinguished, more or less opulent, but such that they can maintain their station without being ashamed, and they are perhaps not the easiest to make happy.

But in order to have passions, to be able to satisfy them, one must certainly be healthy; this is the first good. Now, this good is not as independent of us as one may think. As we are all born healthy (in general that is) and our bodies made to last a certain time, there is no doubt that if we did not destroy our health by overeating, by late nights, in short, by excesses, we would all live approximately to what one calls full adulthood. I exclude from this the violent deaths that one cannot predict, and with which consequently it is useless to concern oneself.

But some will tell me, if your passion is eating fine foods[14] you will be very unhappy because if you want to be healthy you will have to restrain yourself perpetually. To this I answer that happiness being your goal, in satisfying your passions, nothing must divert you. If the stomachache or the gout that the excesses of the table give you causes pain more acute than the pleasure you find in satisfying your love of fine food, you calculate badly. If you value the enjoyment of the sensual pleasure at the expense of your happiness, you stray from your goal and your unhappiness is your own fault.

Do not complain that you like your food so much, as this passion is a source of continual pleasure; but you should know how to make it serve your happiness. This will be easy if you stay home and order your servants to bring you only what you want to eat. Occasionally go on a diet; if you wait for your stomach to be truly hungry, all that will be presented will give you as much pleasure as more sought-after dishes, which you will not dream of because they will not be right in front of your eyes. This abstemiousness that you impose on yourself will enhance the pleasure when next enjoyed. I do not recommend this to put an end to your desire for fine food, but to prepare you for a more delicious pleasure. As for people who are ill, those who are old and frail and bothered by everything, they have other kinds of happiness. To be warm enough, to digest their chicken well, to use the commode[15] is a true delight for them. Such a happiness, if it is one, is too insipid to bother about attaining. It seems that such people live in a world where

14. The word *la gourmandise*, here translated as *eating fine foods*, carries in French both the meaning of enjoyment of fine food, and of *gluttony*, one of the seven capital sins.

15. "Aller a la garde-robe," the phrase used here, meant to go to the small closetlike room that served as a modern-day toilet. There would be a "commode," a chair with a rounded hole in the seat and a chamber pot placed under it that a servant would empty periodically.

happiness, physical enjoyment, and agreeable feelings are out of reach. Those people are to be pitied; but nothing can be done for them.

When one has been persuaded that without health one cannot enjoy any pleasure and any good thing, one finds it easy to make some sacrifices to preserve one's own. I may say that I am a good example of this. I have a very good constitution; but I am not at all robust, and there are some things that would be sure to destroy my health. Such is wine, for example, and all liqueurs; I have forbidden myself these from early youth, I have an abundance of fire in my nature, I spend all morning drenching myself with liquids.[16] Last, I too often give myself up to the enjoyment of fine food, a taste with which God has endowed me, and I counter these excesses by rigorous diets that I impose on myself at the first sign of discomfort and that have always prevented illnesses for me. These diets do not cost me anything, because at such times I always stay home at mealtimes. And, as nature is wise enough not to make us feel the pangs of hunger when we have glutted ourselves with nourishment, and my delight in food not being excited by the presence of dishes, I do not deny myself anything by not eating, and thus I restore my health without depriving myself of anything.

Another source of happiness is to be free from prejudices; and the decision rests with us to rid ourselves of them. We all have a sufficient share of intelligence to examine things that others want to oblige us to believe; to know, for example, if two and two make four, or five; besides, in this century, there are a great many ways to gain instruction. I know that there are other prejudices than those of religion, and I believe that it is good to shake them off, though no prejudices influence our happiness and our unhappiness so much as those of religion.[17] Prejudice is an opinion that one has accepted without examination, because it would be indefensible otherwise. Error can never be a good, and it is surely a great evil in the things on which the conduct of life depends.

Prejudices must not be confused with the proprieties. There is no truth

16. The phrase she uses is "tempérament de feu," which refers to the Greek/medical concept of the four humors that controlled the body: women were usually considered to be wet and cool, men hot and dry. Here she takes the masculine humors for her nature and suggests that she needs to counter their influence with liquids to place the humors in equilibrium again.

17. Here and at the end of this sentence, in the two other extant versions of the *Discourse on Happiness*, Du Châtelet used the word *superstition* instead of *religion*. In those versions she was following Voltaire, who often referred to the religious dogma of the Catholic and Protestant churches as *superstition*. Du Châtelet's equation of religion with prejudice is even more controversial.

in prejudices, and they can only be useful to malformed souls; for there are corrupted souls just as there are deformed bodies. Those have fallen out of line,[18] and I have nothing to say to them. The proprieties have a conventional truth, and that is enough to convince all good people never to deviate from them. No book teaches the proprieties; nevertheless, no one can in good faith claim not to be aware of them. They vary according to rank in society, age, and circumstances. Whoever wishes for happiness must never deviate from them. Exact conformity to the proprieties is a virtue, and I have said that in order to be happy one must be virtuous. I know that preachers in their pulpits, and Juvenal himself,[19] say that one must love virtue for its own sake, for its own beauty; but one must try to understand the meaning of these words, and if one does, one sees that they come down to this: one must be virtuous, because one cannot be immoral and happy at the same time. By *virtue* I understand all that contributes to the happiness of society, and consequently to ours, since we are members of that society.

I say that one cannot be happy and immoral, and the demonstration of this axiom lies in the depths of the hearts of all men. I put it to them, even to the most villainous, that there is not one of them to whom the reproaches of his conscience—that is to say, of his innermost feeling, the scorn that he feels he deserves and that he experiences, as soon as he is aware of it—there is not one to whom these are not a kind of torture. By villains I do not mean thieves, assassins, poisoners; they do not belong in the category of those for whom I write. Villains are the false and perfidious, the slanderers, the informers, the ungrateful. In a word, all those who have vices the laws do not curb, but against which custom and society have brought formal judgments.[20] These formal judgments are all the more terrible, as they are always carried out.

I maintain then that there is no one on earth who can feel that he is despised and not feel despair. This public disdain, this turning away of people of good will, is a torture more cruel than all those that the public executioner could inflict, because it lasts much longer, and because hope never accompanies it.

---

18. The French phrase is *hors de rang*, a military term used to indicate that an individual had fallen out of the line when marching.

19. Juvenal (60–140 CE) wrote satiric poems (sixteen extant in whole or part). The beginning of Satire 8 asserts that nobility lies in individual virtue rather than in bloodlines. In Satire 10 (lines 357ff.), he expresses the view that we should pray for a sound mind in a sound body (*mens sana in corpore sano*) and that the only path to happiness is through virtue (line 364).

20. The French term *arrêt*, translated here as *formal judgment*, had a legal meaning, and was also synonymous with being arrested by the authorities.

So one must never be immoral if one does not want to be unhappy. But it is not enough for us not to be unhappy; life would not be worth the effort of living if the absence of suffering was our only goal; nothingness would be better; for assuredly, that is the state of least suffering. One must, then, try to be happy. One must be at ease with oneself for the same reason that one must be comfortable in one's own home. We would hope in vain for enjoyment of this satisfaction if we were not virtuous:

> Mortals' eyes are easily dazzled;
> But one cannot deceive the vigilant eye of the gods,

as one of our best poets has said;[21] but it is the ever vigilant eye of one's own conscience that one can never deceive.

One is an exacting judge of oneself, and the more one can bear witness to oneself that one has fulfilled one's duties, done all the good that one could do, that in short, one is virtuous, the more one tastes this interior satisfaction that one can call the health of the soul.[22] I doubt that there is a more delicious feeling than what one experiences after doing a virtuous action, an action that merits the esteem of honorable men. To the inner satisfaction caused by virtuous actions can be added the pleasure of enjoying universal esteem, but even though rogues cannot refuse their esteem to integrity, only the esteem of honorable men is truly worthwhile. Finally, I say that to be happy one must be susceptible to illusion, and this scarcely needs to be proved; but, you will object, you have said that error is always harmful: is illusion not an error? No: although it is true, that illusion does not make us see objects entirely as they must be in order for them to give us agreeable feelings, it only adjusts them to our nature. Such are optical illusions: now optics does not deceive us, although it does not allow us to see objects as they are, because it makes us see them in the manner necessary for them to be useful to us. Why do I laugh more than anyone else at the puppets, if not because I allow myself to be more susceptible than anyone else to illusion, and that after a quarter of an hour I believe that it is Polichinelle, the puppet, who speaks?[23] Would we have a moment of pleasure at the theater if we did not lend ourselves to the illusion that makes us see famous individuals that

21. These are lines from Voltaire's play, *Sémiramis*, Act I, scene 3, first performed at the Comédie française 29 August 1748.
22. Here again Du Châtelet refers to the Christian idea of the soul being judged after death.
23. Polichinelle is the French name for "Punch," the humpbacked fool famous for defying all authority with his loud voice and trusty bat, and a favorite of the eighteenth-century Parisian marionette and hand-puppet stage.

we know have been dead for a long time, speaking in Alexandrine verse?[24] Truly, what pleasure would one have at any other spectacle where all is illusion if one was not able to abandon oneself to it? Surely there would be much to lose, and those at the opera who only have the pleasure of the music and the dances have a very meager pleasure, one well below that which this enchanting spectacle viewed as a whole provides. I have cited spectacles, because illusion is easier to perceive there. It is, however, involved in all the pleasures of our life, and provides the polish, the gloss of life. Some will perhaps say that illusion does not depend on us, and that is only too true, up to a point. We cannot give ourselves illusions any more than we can give ourselves tastes, or passions; but we can keep the illusions that we have; we can seek not to destroy them. We can choose not to go behind the set, to see the wheels that make flight, and the other machines of theatrical productions.[25] Such is the artifice that we can use, and that artifice is neither useless nor unproductive.

These are the great machines of happiness, so to speak; but there are yet other, lesser skills that can contribute to our happiness.

The first is to be resolute about what one wants to be and about what one wants to do. This is lacking in almost all men; it is, however, the prerequisite without which there is no happiness at all. Without it, one swims forever in a sea of uncertainties, one destroys in the morning what one made in the evening;[26] life is spent doing stupid things, putting them right, repenting of them.

This feeling of repentance is one of the most useless and most disagreeable that our soul can experience.[27] One of the great secrets is to know how to guard against it. As no two things in life are alike, it is almost always useless to see one's errors, or at least to pause a long time to consider them and to reproach oneself with them. In so doing we cover ourselves with confusion in our own eyes for no gain. One must start from where one is, use all one's sagacity to make amends and to find the means to make amends, but

24. Alexandrine is the classical meter used by French playwrights from the sixteenth to the nineteenth century. Each line had twelve syllables, each pair of lines rhymed.

25. The French word *coulisses*, translated here as *set*, literally means *wings* and is probably more equivalent to the English stage term *hanging flats*, scenes painted on canvas raised and lowered for different acts and tableaux in a theatrical production.

26. This is a reference to Homer's *Odyssey* and the story of Penelope, Ulysses' wife. To fend off numerous suitors, she told them she could not remarry until she had completed her weaving. Each night she undid what she had woven during the day (*Odyssey*, 2.93ff.). Du Châtelet reversed the sequence in her reference to the story.

27. The act of repenting, of *Penance*, is a sacrament of the Catholic Church, thus this section on *repenting* would have had many connotations for Du Châtelet's French readers.

there is no point in looking back, and one must always brush from one's mind the memory of one's errors. The ability to benefit from an initial examination, dismiss sad ideas and substitute agreeable ideas, is one of the mainsprings of happiness, and we have this in our power, at least up to a point. I know that a violent passion that makes us unhappy proves that it does not depend entirely on us to banish from our mind the ideas that distress us; but we are not always in such violent situations, all illnesses are not malign fevers, and the trifling misfortunes, those sensations that are disagreeable, though weak, should be avoided. Death, for example, is an idea that always distresses us whether we foresee our own, or think of that of the people we love. So we must avoid with care all that can remind us of this idea. I very much disagree with Montaigne, who congratulated himself on having so accustomed himself to death that he was sure he would see it approach without being afraid.[28] It may be seen by the complacency with which he reports this victory that it was a costly effort for him. And in this the wise Montaigne had miscalculated, because surely it is a folly to poison with this sad and humiliating idea part of the little time we have to live, all this in order to endure more patiently a moment that bodily sufferings always make very bitter, in spite of our philosophy. Moreover, who knows if the weakening of our mind, caused by illness or old age, will allow us to reap the benefit of our reflections; perhaps our efforts will have been all in vain, as so often happens in this life? When the idea of death recurs, let us always have this line of Gresset in mind: "Suffering is a century, and death a moment."[29]

Let us turn the mind away from all disagreeable ideas; they are the source of all metaphysical anxieties, and it is above all those anxieties that it is almost always in our power to avoid.

Wisdom must always have counters in her hand to play with;[30] *wise* and

28. Michel de Montaigne (1533–1592) was the author of numerous essays, published between 1580 and 1588. Du Châtelet read *Les Essais* [The Essays] with great pleasure, and often cited them. This is one of her rare disagreements with Montaigne; she refers here to the sentiments he expressed in "That to philosophize is to learn to die," written in the early 1570s. Note that Montaigne's later essays take a view similar to hers, suggesting that she had not yet read these. See, for example, "Of physiognomy," written in the late 1580s. I am grateful to Robert Zaretsky for identifying this change in Montaigne's views.

29. Jean-Baptiste-Louis Gresset (1709–1777), originally trained for the Church, became famous for his verse and his plays. These lines are from his "Epître à ma soeur sur ma convalescence [Epistle to my sister on my convalescence]."

30. *Counters*, or *jetons* in French, has two meanings: counters used for keeping track of sums as in a shop, similar to an abacus, and the chips used in gambling. Gambling at cards and at *cavagnole*, a board game, was a source of great pleasure for Du Châtelet throughout her life. See also her additions on gambling to Mandeville's "Remarks," in *Studies on Voltaire*, ed. Ira O. Wade (Princeton NJ: Princeton University Press, 1947), 168–72.

happy mean the same, at least in my dictionary. One must have passions to be happy; but they must be made to serve our happiness, and there are some that must absolutely be prevented from entering our soul. I am not speaking here of the passions that are vices, like hatred, vengeance, rage; but ambition, for example, is a passion that I believe one must defend one's soul against, if one wants to be happy. This is not because it does not give enjoyment, for I believe that this passion can provide that; it is not because ambition can never be satisfied—that is surely a great good. Rather, it is because ambition, of all the passions, makes our happiness dependent on others. Now the less our happiness depends on others the easier it is for us to be happy. Let us not be afraid to reduce our dependence on others too much, our happiness will always depend on others quite enough. If we value independence, the love of study is, of all the passions, the one that contributes most to our happiness. This love of study holds within it a passion from which a superior soul is never entirely exempt, that of glory. For half the world, glory can only be obtained in this manner, and it is precisely this half whose education made glory inaccessible and made a taste for it impossible.[31]

Undeniably, the love of study is much less necessary to the happiness of men than it is to that of women. Men have infinite resources for their happiness that women lack. They have many means to attain glory, and it is quite certain that the ambition to make their talents useful to their country and to serve their fellow citizens, perhaps by their competency in the art of war, or by their talents for government, or negotiation, is superior to that which one can gain for oneself by study.[32] But women are excluded, by definition, from every kind of glory, and when, by chance, one is born with a rather superior soul, only study remains to console her for all the exclusions and all the dependencies to which she finds herself condemned by her place in society.[33]

The love of glory that is the source of so many pleasures of the soul and of so many efforts of all sorts that contribute to the happiness, the instruction, and the perfection of society, is entirely founded on illusion. Nothing is so easy as to make the phantom after which all superior souls run disap-

31. In her "Translator's Preface" to her version of *The Fable of the Bees*, Du Châtelet also refers to the obstacles encountered by women who wished to go beyond what she considered to be appropriate education for girls, even in her privileged world.

32. In Du Châtelet's era the word *competency* carried the connotation of *the right to* a particular occupation because of birth as well as skill. She is certainly thinking of the men in her family; her husband and son, her father, uncles, and cousins succeeded in careers in the army, government, and as representatives to foreign courts.

33. Du Châtelet makes the Cartesian distinction between the *soul* or *mind* and the body.

pear; but there would be much to lose for them and for others! I know there is some substance in the love of glory that one can enjoy in one's lifetime; but there are scarcely any heroes, of whatever kind, who would want to close themselves off entirely from the plaudits of posterity, from which one expects more justice than from one's contemporaries. One does not always acknowledge the enjoyment of the ill-defined desire to be spoken of after one has passed out of existence; but it always stays deep in our heart. Philosophy would have us feel the vanity of it; but the feeling prevails, and this pleasure is not an illusion; for it proves to us the very real benefit of enjoying our future reputation. If our only source of good feeling were in the present, our pleasures would be even more limited than they are. We are made happy in the present moment not only by our actual delights but also by our hopes, our reminiscences. The present is enriched by the past and the future. Would we work for our children, for the greatness of our lineage, if we did not enjoy the future? Whatever we do, self-esteem is always the more or less hidden driving force of our actions; it is the wind that fills the sails, without which the boat would not move at all.

I have said that the love of study is the passion most necessary to our happiness. It is an unfailing resource against misfortunes, it is an inexhaustible source of pleasures, and Cicero is right to say: *The pleasures of the senses and those of the heart are, without doubt, above those of study; study is not necessary for happiness: but we may need to feel that we have within us this resource and this support.*[34] One may love study and spend whole years, perhaps one's whole life, without studying. Happy is he who spends it thus: for only more lively pleasures cause one to sacrifice a pleasure that one is sure to find and that can be made lively enough to compensate for the loss of others.

One of the great secrets of happiness is to moderate one's desires and to love the things already in one's possession. Nature, whose goal is always our happiness (and by nature, I understand all that is instinctive and without reasoning), nature, I say, only gives us the desires appropriate to our rank and circumstances. We only naturally desire things by degrees and within our purview: an infantry captain wishes to be a colonel, and he is not unhappy not to command the armies, whatever talent he feels he has. It is for our mind and our reflections to strengthen the wise moderation of nature; only fulfilled desires can make us happy. So one must allow oneself to desire only the things that can be obtained without too much care and effort, and this is a place where we can do much for our own happiness. To love what one pos-

---

34. This is probably a paraphrase of Cicero (106–43 BCE) from his *Tusculan Disputations*, a favorite of Du Châtelet's and Voltaire's.

sesses, to know how to enjoy it, to savor the advantages of one's situation in life, not to look too much at those who seem happier than we, to apply one-self to perfect one's own happiness and to make the most of it, this is what is rightly termed happiness. And I believe I define it well in saying that the happiest man is he who least desires to change his rank and circumstances. To enjoy this happiness, one must cure or prevent a sickness of another sort that is entirely opposed to it, but is only too common: restlessness. This state of mind is incompatible with any enjoyment, and, consequently, any kind of happiness. Good philosophy, that is to say, the firm belief that all we have to do in this world is to be happy, is a sure remedy against this sickness of which lively minds—those who are capable of reasoning on first causes and consequences—are almost always exempt.

There is a passion, very unreasonable in the eyes of philosophers and of reason, the motive of which, however disguised it may be, is even humili-ating, and should be enough to cure one of it and which, nevertheless can make one happy: it is the passion of gambling. It is a good passion to have, if it can be moderated and kept for the time in our life when this resource will be necessary to us, and this time is old age. There is no doubt that the love of gambling has its source in the love of money; there is no individual for whom playing for high stakes is not an interesting activity (and I call high stakes gambling that can make a difference to our fortune).

Our soul wants to be moved by the passions of hope or fear; it is made happy only by things that cause it to feel alive. Now gambling places us per-petually in the grip of these two passions, and consequently holds our soul in an emotion that is one of the great principles of happiness to be found in us. The pleasure that gambling has given me has often served to console me for not being rich. I believe I have a good enough mind to be happy with what would seem to others a mediocre fortune; and in that case—if I were rich—gambling would become dull for me. At least I was afraid that it would, and this fear of boredom convinced me that I owe the pleasure of gambling to my limited fortune, and that consoled me for not being rich.

There is no doubt that physical needs are the source of the pleasures of the senses, and I am convinced that there is more pleasure in a mediocre fortune than in great abundance. A new snuffbox, a new piece of furniture or of china, is a true delight to me; but if I owned thirty snuffboxes, I would be less appreciative of the thirty-first.[35] Our tastes are easily blunted by satia-

---

35. In other manuscript versions Du Châtelet chose the numbers 300 and 301. Snuffboxes were her particular weakness, she probably had well over thirty at the time of her death. They also seem to have been the preferred gift to exchange at court.

tion, and one must give thanks to God for giving us the necessary privations to preserve them. This is what causes a king to be so often bored, and why it is impossible for him to be happy, unless Heaven has given him a soul magnanimous enough to be susceptible to the pleasures of his position, that is to say, the pleasure of making a great number of men happy. Then this position becomes the first above all for the happiness it brings, as it is by its power.

I have said that the more our happiness depends on us, the more assured it is; yet the passion that can give us the greatest pleasures and make us happiest, places our happiness entirely in the hands of others. You have already gathered that I am speaking of love.

This passion is perhaps the only one that can make us wish to live, and bring us to thank the author of nature, whoever he is, for giving us life.[36] My Lord Rochester is right to say that the gods have put this heavenly drop in the chalice of life to give us the courage to bear it:

> One must love, it is that which sustains us:
> Because without love, it is sad to be man.[37]

If this mutual taste, which is a sixth sense, and the most refined, the most delicate, the most precious of all, brings together two souls equally sensitive to happiness, to pleasure, all is said, one need not do anything more to be happy, everything else is inconsequential except for health. All the faculties of one's soul must be used to enjoy this happiness. One must give up life when one loses that happiness, and acknowledge that to have attained Nestor's age is nothing when balanced with a quarter hour of such bliss.[38] It is appropriate that such a happiness should be rare; if it were common, one would choose to be a man rather than a god, at least such as we can conceive of God. The best thing we can do is to persuade ourselves that this happiness is not impossible. However, I do not know if love has ever brought together two people who are so made for each other that they have never known the satiety of delight, nor the cooling of passion caused by a sense of security, nor the indolence and the tedium that arise from the ease and the

---

36. Not to use the word *God* is significant; the skepticism is obvious in the phrase, "whoever he is."

37. Voltaire was introduced to the verses of the notorious libertine and cross dresser, John Wilmot, Lord Rochester (1647–1680) by his English friends during his exile in England from 1726 to 1728. Voltaire, in turn, gave them to Du Châtelet, who has here translated lines from Rochester's "A Satire against Man."

38. Nestor, Odysseus's ancient tutor in Homer's *Odyssey*, was symbolic of extreme age and wisdom.

continuity of a relationship, and whose power of illusion never wanes (for where is illusion more important than in love?); and, last, whose ardor remains the same whether in the enjoyment or in the deprivation of the other's presence, and equally tolerates both unhappiness and pleasure.[39]

The creation of a heart capable of such a love, a soul so tender, and so steadfast appears to exhaust the power of the deity; only one is born in a century. It seems that to produce two such hearts would be beyond the deity's powers, or if he has produced them, and if they could meet, he would be jealous of their pleasures. But love can make us happy at less cost: a tender and sensitive soul is made happy by the sheer pleasure it finds in loving. I do not mean that unrequited love could make one perfectly happy; but I say that, although our ideas of happiness are not entirely satisfied by the love given us, the pleasure we feel in giving ourselves up to our feelings of tenderness can suffice to make us happy. And if this soul still has the good fortune to be susceptible to illusions, it is not impossible that it should not believe itself more loved perhaps than it is in fact. This soul must love so much that it loves for two, and the warmth of its heart supplies what is, in fact, lacking in its happiness. A feeling character, keen and susceptible to the passions, must pay the price of the inconveniences attached to these qualities, and I do not know if I must say they are good or bad; but I believe that however composed one's character, one would still want to have them. A first passion carries a soul tempered in this way so much beyond itself that it is inaccessible to any reflection and to any moderate ideas; this soul can probably look forward to great sorrows; but the greatest inconvenience attached to such sensibility is that someone who loves to this excess cannot possibly be loved, for there is scarcely a man whose amorous inclination does not diminish with the experience of such a passion. That must appear quite strange to him who does not yet know enough about the human heart; but however little one may have reflected about what experience offers us, one will feel that to hold the heart of one's beloved for a long time, hope and fear must always operate on him. Now, a passion, such as I have just depicted, produces an abandon that makes one incapable of any art. Love bursts out on all sides; you are initially adored, it can only be so; but soon the certainty of being loved, and the tedium of having one's wishes always anticipated, the misfortune of having nothing to fear, dulls one's inclination. Such is the human heart, and no one should think that I speak out of resentment. I have been endowed by God, it is true, with one of these loving and steadfast souls

39. In this and subsequent paragraphs Du Châtelet's contemporaries would have assumed that she was describing herself and giving the history of her relationship with Voltaire.

that know neither how to disguise nor how to moderate its passions, that know neither their diminution nor disgust with them, and whose tenacity can resist everything, even the certainty of being no longer loved. But I was happy for ten years because of the love of the man who had completely seduced my soul; and these ten years I spent tête-à-tête with him without a single moment of distaste or hint of melancholy.[40] When age, illness, as well as perhaps the ease of pleasure made his inclination less, for a long time I did not perceive it; I was loving for two, I spent all my time with him, and my heart, free from suspicion, delighted in the pleasure of loving and in the illusion of believing myself loved. True, I have lost this happy state, and this has cost me many tears. Terrible shocks are needed to break such chains. The wound to my heart bled for a long time; I had grounds to complain, and I have pardoned all.[41] I was fair enough to accept that in the whole world, perhaps only my heart possessed the steadfastness that annihilates the power of time; that if age and illness had not entirely extinguished his desire, it would perhaps still have been for me, and that love would have restored him to me; lastly, that his heart, incapable of love, felt for me the most tender affection, and caused him to dedicate his life to me. The certainty that a return of his inclination and his passion was impossible—I know well that such a return is not in nature—imperceptibly led my heart to the peaceful feeling of deep affection; and this sentiment, together with the passion for study, made me happy enough.

But can such a tender heart be satisfied by a sentiment as peaceful and as weak as that of close friendship? I do not know if one must hope, or even wish, to cling forever to this sensibility, once one has reached the kind of apathy to which it is so difficult to lead such a soul. Only lively and agreeable feelings make one happy; why then forbid oneself love, the most lively and most agreeable of all? But what one has experienced, the reflections that one has had to make to lead one's heart to this apathy, the very pain that caused one to bring it to this state—all this must make one fear to leave a situation that at least is not unhappy, in order to venture out to meet with the misfortunes which one's age and the loss of one's beauty would make pointless.[42]

40. *Langueur*, the word Du Châtelet uses in French, was considered a physical state in her era, just short of *melancholy*, the depression particularly associated with love.

41. This is a reference to Du Châtelet's discovery that Voltaire had begun a sexual relationship with his niece, Mme Marie-Louise Denis, after having told her a few years before that he was no longer capable of a sexual relationship. See Judith P. Zinsser, *Emilie Du Châtelet: Daring Genius of the Enlightenment* (originally published as *La Dame d'Esprit: A Biography of the Marquise Du Châtelet*) (New York: Penguin, 2006), 244–49.

42. At this point in the *Discourse* Du Châtelet turns to her reflections on beginning a new affair and her increasing passion for Jean-François de Saint-Lambert.

Fine reflections, you will say, and very useful ones! You will see whether they are useful or not, if you ever feel inclination for someone who falls in love with you; but I think that it is wrong to believe these reflections useless. The passions, after thirty years of age, no longer carry us off with the same impetuosity. Please believe that one could resist one's inclination, if one was really intent on it, and was quite certain that giving in to it would make one unhappy. One only gives in because one is not quite convinced of the reliability of the maxims presented here, and still hopes to be happy, and one is right to persuade oneself that this is possible. Why forbid oneself the hope of experiencing happiness, and of the most intense kind? But even if we should not forbid ourselves this hope, we must not deceive ourselves about the means to happiness; experience must at the least teach us to rely on ourselves and to make our passions serve our happiness. One can keep control of oneself up to a point. No doubt complete self-control is out of reach, but a measure of it is not; and I suggest, without fear of being wrong, that there is no passion that one might not overcome once one is fully convinced that it can only lead to unhappiness. What misleads us on this point in our early youth is that we are incapable of reflecting, that we have no experience at all, and imagine that we will recapture the good that we have lost if we run after it long enough; but experience and knowledge of the human heart teaches us that the more we run after it, the more it flees from us.[43] It is a deceptive prospect that disappears when we believe we have reached it. We cannot will, we cannot persuade, ourselves into feelings of attraction for someone, and attraction can hardly ever be rekindled. What is your goal when you give in to the attraction you feel for someone? Is it not in order to be happy because of the pleasure of loving and being loved? So it would be ridiculous to refuse oneself this pleasure for fear of an unhappiness that perhaps will only be experienced after having known great happiness. Thus, there would be an overall balance, and you must only think of recovering from this malady and of not repenting, in the same way a reasonable person would blush if she did not take her happiness into her own hands, and if she put it entirely in those of another.

The great secret for preventing love from making us unhappy is to try never to appear in the wrong with your lover, never to display eagerness when his love is cooling, and always to be a degree cooler than he. This will not bring him back, but nothing could bring him back; there is nothing for us to do then but to forget someone who ceases to love us. If he still loves

---

43. This passage could be taken as a reference to Du Châtelet's youthful affair with Count Louis Vincent de Goesbriand. It ended in ridicule when she apparently threatened suicide if he were to leave her. See Zinsser, *Emilie Du Châtelet,* 88–89.

you, nothing can revive his love and make it as fiery as it was at first, except the fear of losing you and of being less loved. I know that for the susceptible and sincere this secret is difficult to put into practice; however, no effort will be too great, all the more so as it is much more necessary for the susceptible and sincere than for others. Nothing degrades as much as the steps one takes to regain a cold or inconstant heart. This demeans us in the eyes of the one we seek to keep, and in those of other men who might take an interest in us. But, and this is even worse, it makes us unhappy and uselessly torments us. So we must follow this maxim with unwavering courage and never surrender to our own heart on this point. We must attempt, before surrendering to our inclination, to become acquainted with the character of the person to whom we are becoming attached. Reason must be heard when we take counsel with ourselves; not the reason that condemns all types of commitment as contrary to happiness, but that which, in agreeing that one cannot be very happy without loving, wants one to love only in order to be happy, and to conquer an attraction by which it is obvious that one would only suffer unhappiness. But if and when this inclination has prevailed, when, as happens only too often, it has triumphed over reason, one must not pride oneself on a constancy that would be as ridiculous as it would be misplaced. This is a case in point for putting into practice the proverb *The shortest follies are the best;* above all the shortest follies cause the shortest unhappiness. For there are follies that would make us very happy if they lasted all our lives. One should not blush to have been mistaken; one must cure oneself, at whatever cost, and above all avoid the presence of the object that can only excite one and make one lose the fruit of one's reflections. For with men, flirtation survives love. They want to lose neither their conquest nor their victory, and by a thousand coquetries they know how to rekindle a fire imperfectly extinguished and to hold you in a state of uncertainty as ridiculous as it is intolerable. Drastic action must be taken, one must break off once and for all; friendship must be carefully unstitched and love torn up, says M. de Richelieu.[44] Lastly, it is for reason to make our happiness. In childhood, our senses alone attend to this task; in youth, the heart and the mind become involved, with the proviso that the heart makes all the decisions; but in middle age reason must take part in the decision, it is for reason to make us feel that we must be happy, whatever it costs. Every age has its own pleasures; those of old age are the most difficult to obtain: *gambling, studying,* if one is still ca-

---

44. The Duke de Richelieu was a relative by marriage and also a long-time friend of Du Châtelet's; he gave her counsel when she was contemplating moving to Cirey to join Voltaire. He was one of their principal allies and patrons at Louis XV's court.

pable of it, the *enjoyment of fine foods, respect,* those are the mainsprings of old age. No doubt these are only consolations. Thank goodness, it is up to us to choose the time of our death, if it is too slow in coming;[45] but as long as we prefer to endure life, we must open ourselves to pleasure by all the doors leading to our soul; we have no other business.

So let us try to be healthy, to have no prejudices, to have passions, to make them serve our happiness, to replace our passions with inclinations, to cherish our illusions, to be virtuous, never to repent, to keep away sad ideas, and never to allow our heart to sustain a spark of inclination for someone whose inclination for us diminishes and who ceases to love us. We must leave love behind one day, if we do indeed age, and that day must be the one when love ceases to make us happy. Lastly, let us think of fostering a taste for study, a taste which makes our happiness depend only on ourselves. Let us preserve ourselves from ambition, and, above all, let us be certain of what we want to be; let us choose for ourselves our path in life, and let us try to strew that path with flowers.[46]

45. This is an unusual sentiment for this era; Du Châtelet is suggesting suicide as a possibility, an act considered a sin by the Catholic Church.

46. Voltaire uses similar phrasing at the conclusion of his tale *Candide,* published in 1759. Note that this also echoes the idea of her "Translator's Preface" to Mandeville's *The Fable of the Bees,* that of accepting what one has, even when contrasted to what others have who have more.

SERIES EDITORS'
BIBLIOGRAPHY

PRIMARY SOURCES

Agnesi, Maria Gaetana, Giuseppa Eleonora Barbapiccola, Diamante Medaglia Faini, Aretafila Savini de' Rossi, and the Accademia de' Ricovrati. *The Contest for Knowledge*. Ed. and trans. Rebecca Messbarger and Paula Findlen, introd. Rebecca Messbarger. The Other Voice in Early Modern Europe. Chicago: University of Chicago Press, 2005.

Agrippa, Henricus Cornelius. *Declamation on the Nobility and Preeminence of the Female Sex*. Ed. and trans. Albert Rabil Jr. The Other Voice in Early Modern Europe. Chicago: University of Chicago Press, 1996.

Alberti, Leon Battista. *The Family in Renaissance Florence*. Trans. Renée Neu Watkins. Columbia: University of South Carolina Press, 1969.

D'Aragona, Tullia. *Dialogue on the Infinity of Love*. Ed. and trans. Rinaldina Russell and Bruce Merry, introd. Rinaldina Russell. The Other Voice in Early Modern Europe. Chicago: University of Chicago Press, 1997.

Arenal, Electa, and Stacey Schlau, eds. *Untold Sisters: Hispanic Nuns in Their Own Words*. Trans. Amanda Powell. Albuquerque: University of New Mexico Press, 1989.

Argula von Grumbach: A Woman's Voice in the Reformation. Ed. and trans. Peter Matheson. Edinburgh: T. & T. Clark, 1995.

Askew, Anne. *The Examinations of Anne Askew*. Ed. Elaine V. Beilin. Women Writers in English, 1350–1850. Oxford: Oxford University Press, 1996.

Astell, Mary. *The First English Feminist: Reflections on Marriage and Other Writings*. Ed. and introd. Bridget Hill. New York: St. Martin's Press, 1986.

Astell, Mary, and John Norris. *Letters concerning the Love of God*. Ed. E. Derek Taylor and Melvyn New. The Early Modern Englishwoman, 1500–1750: Contemporary Editions. Burlington, VT: Ashgate, 2005.

Atherton, Margaret, ed. *Women Philosophers of the Early Modern Period*. Indianapolis, IN: Hackett, 1994.

de l'Aubespine, Madeleine. *Selected Poems and Translations*. A Bilingual Edition. Ed. and trans. Anna Kłosowska. The Other Voice in Early Modern Europe. Chicago: University of Chicago Press, 2007.

Aughterson, Kate, ed. *Renaissance Woman: Constructions of Femininity in England; A Source Book*. New York: Routledge, 1995.

Autobiographical Writings by Early Quaker Women. Ed. David Booy. Burlington, VT: Ashgate, 2004.

Barbaro, Francesco. *On Wifely Duties.* Trans. Benjamin Kohl. In *The Earthly Republic,* edited by B. Kohl and R. G. Witt, 179–228. Philadelphia: University of Pennsylvania Press, 1978. Translation of the Preface and Book 2.

Battiferra degli Ammannati, Laura. *Laura Battiferra and Her Literary Circle.* Ed. and trans. Victoria Kirkham. The Other Voice in Early Modern Europe. Chicago: University of Chicago Press, 2006.

Behn, Aphra. *Love Letters between a Nobleman and His Sister.* Ed. Janet Todd. New York: Penguin, 1996.

———. *Oroonoko.* Ed. Joanna Lipking. Norton Critical Edition. New York: W. W. Norton, 1997.

———. *The Rover.* Ed. Anne Russell. Peterborough, ON: Broadview Press, 1994; 2nd ed, 1999.

———. *The Rover, The Feigned Courtesans, The Lucky Chance, and The Emperor of the Moon.* Ed. and introd. Jane Spencer. The World's Classics. Oxford: Oxford University Press, 1995.

———. *The Works of Aphra Behn.* 7 vols. Ed. Janet Todd. Columbus: Ohio State University Press, 1992–96.

Bigolina, Giulia. *Urania: A Romance.* Ed. and trans. Valeria Finucci. The Other Voice in Early Modern Europe. Chicago: University of Chicago Press, 2005.

Blamires, Alcuin, ed. *Woman Defamed and Woman Defended: An Anthology of Medieval Texts.* Oxford: Clarendon Press, 1992.

Boccaccio, Giovanni (1313–1375). *Corbaccio or the Labyrinth of Love.* Trans. Anthony K. Cassell. 2nd rev. ed. Binghamton, NY: Medieval and Renaissance Texts and Studies, 1993.

———. *Famous Women.* Ed. and trans. Virginia Brown. The I Tatti Renaissance Library. Cambridge, MA: Harvard University Press, 2001.

Booy, David, ed. *Autobiographical Writings by Early Quaker Women.* Burlington, VT: Ashgate, 2004.

Bradstreet, Anne. *The Tenth Muse (1650) and, from the Manuscripts, Meditations Divine and Morall, Together with Letters and Occasional Prose.* Comp. and introd. Josephine K. Piercy. Delmar, NY: Scholars' Facsimiles and Reprints, 1978.

———. *The Works of Anne Bradstreet.* Ed. Jeannine Hensley, forword Adrienne Rich. Cambridge, MA: Belknap Press of Harvard University Press, 1967.

Brown, Sylvia, ed. *Women's Writing in Stuart England: The Mother's Legacies of Dorothy Leigh, Elizabeth Joscelin, and Elizabeth Richardson.* Thrupp, Stroud, Gloucester, UK: Sutton, 1999.

Bruni, Leonardo. "On the Study of Literature to Lady Battista Malatesta of Moltefeltro." In *The Humanism of Leonardo Bruni: Selected Texts.* Trans. and introd. Gordon Griffiths, James Hankins, and David Thompson, 240–51. Binghamton, NY: Medieval and Renaissance Studies and Texts, 1987.

Caminer Turra, Elisabetta. *Selected Writings of an Eighteenth-Century Venetian Woman of Letters.* Ed. and trans. Catherine M. Sama. The Other Voice in Early Modern Europe. Chicago: University of Chicago Press, 2003.

Campiglia, Maddalena. *Flori: A Pastoral Drama.* A Bilingual Edition. Ed., introd., and notes Virginia Cox and Lisa Sampson, trans. Virginia Cox. The Other Voice in Early Modern Europe. Chicago: University of Chicago Press, 2004.

Cary, Elizabeth, Lady Falkland. *The Life and Letters*. Ed. Heather Wolfe. Renaissance Texts from Manuscript. Tempe, AZ: MRTS, 2001.

————. *The Tragedy of Mariam*, 1613. Ed. A. C. Dunstan. Supplement to the introduction Marta Straznicki and Richard Roland. Malone Society Reprints. Oxford: Oxford University Press, 1992.

————. *The Tragedy of Mariam: The Fair Queen of Jewry*. Ed. Stephanie Hodgson-Wright. Peterborough, ON: Broadview Literary Texts, 2000.

————. *The Tragedy of Mariam: The Fair Queen of Jewry. With The Lady Falkland: Her Life, by One of Her Daughters*. Ed. Barry Weller and Margaret W. Ferguson. Berkeley: University of California Press, 1994.

Castiglione, Baldassare. *The Book of the Courtier*. Trans. George Bull. New York: Penguin, 1967.

————. *The Book of the Courtier*. Ed. Daniel Javitch. New York: W. W. Norton, 2002.

Cavendish, Margaret, Duchess of Newcastle. *Bell in Campo and the Sociable Companions*. Ed. Alexandra G. Bennett. Peterborough, ON: Broadview Press, 2002.

————. *The Blazing World and Other Writings*. Ed. Kate Lilley. London: Pickering and Chatto, 1992; repr. London: Penguin, 1994.

————. *The Convent of Pleasure and Other Plays*. Ed. Anne Shaver. Baltimore, MD: Johns Hopkins University Press, 1999.

————. *Observations upon Experimental Philosophy*. Ed. Eileen O'Neill. Cambridge: Cambridge University Press, 2001.

————. *Paper Bodies: A Margaret Cavendish Reader*. Ed. Sylvia Bowerbank and Sara Mendelson. Peterborough, ON: Broadview, 1999.

————. *Poems and Fancies, 1653*. Menston, UK: Scolar Press, 1972.

————. *Political Writings*. Ed. Susan James. Cambridge: Cambridge University Press, 2003.

————. *Sociable Letters, 1964*. Scolar Press Facsimile. Menston, UK: Scolar Press, 1969.

Celeste, Sister Maria. *Galileo's Daughter: A Historical Memoir of Science, Faith, and Love*. Trans. Dava Sobel. New York: Penguin, 1999.

————. *Sister Maria Celeste's Letters to Her Father, Galileo*. Ed. and trans. Rinaldina Russell. New York: Writers Club Press, 2000.

Cerasano, S. P., and Marion Wynne-Davies, eds. *Renaissance Drama by Women: Texts and Documents*. New York: Routledge, 1996.

Cereta, Laura. *Collected Letters of a Renaissance Feminist*. Ed. and trans. Diana Robin. The Other Voice in Early Modern Europe. Chicago: University of Chicago Press, 1997.

Christine de Pizan. *The Book of the City of Ladies*. Trans. Earl Jeffrey Richards, foreword Marina Warner. New York: Persea Books, 1982.

————. *The Book of the City of Ladies*. Trans., introd., and notes Rosalind Brown-Grant. New York: Penguin, 1999.

————. *Epistre au dieu d'Amours*. Ed. and trans. Thelma S. Fenster. In *Poems of Cupid, God of Love*, edited by Thelma S. Fenster and Mary Carpenter Erler. Leiden: E. J. Brill, 1990.

————. *A Medieval Woman's Mirror of Honor: The Treasury of the City of Ladies*. Trans. and introd. Charity Cannon Willard. Ed. and introd. Madeleine P. Cosman. New York: Persea Books, 1989.

————. *The Treasure of the City of Ladies*. Trans. Sarah Lawson. New York: Viking Penguin, 1985.

Clarke, Danielle, ed. *Isabella Whitney, Mary Sidney, and Aemilia Lanyer: Renaissance Women Poets.* New York: Penguin, 2000.

Clifford, Lady Anne. *The Diaries of Lady Anne Clifford.* Ed. D. J. H. Clifford. Phoenix Mill, UK: Alan Sutton, 1990.

———. *Memoir of 1603 and the Diary of 1616–1619.* Ed. Katherine O. Acheson. Peterborough, ON: Broadview, 2007.

de Coignard, Gabrielle. *Spiritual Sonnets.* A Bilingual Edition. Ed. and trans. Melanie E. Gregg. The Other Voice in Early Modern Europe. Chicago: University of Chicago Press, 2004.

Collins, An. *Divine Songs and Meditacions.* Ed. Sidney Gottlieb. Tempe, AZ: MRTS, 1996.

Colonna, Vittoria. *Sonnets for Michelangelo.* A Bilingual Edition. Ed. and trans. Abigail Brundin. The Other Voice in Early Modern Europe. Chicago: University of Chicago Press, 2005.

Convents Confront the Reformation: Catholic and Protestant Nuns in Germany. Ed. and introd. Merry Wiesner-Hanks, trans. Joan Skocir and Merry Wiesner-Hanks. Women of the Reformation. Milwaukee, WI: Marquette University Press, 1996.

Crawford, Patricia, and Laura Gowing, eds. *Women's Worlds in Seventeenth-Century England: A Source Book.* New York: Routledge, 2000.

"Custome Is an Idiot": Jacobean Pamphlet Literature on Women. Ed. Susan Gushee O'Malley, afterword Ann Rosalind Jones. Urbana: University of Illinois Press, 2004.

Daughters, Wives, and Widows: Writings by Men about Women and Marriage in England, 1500–1640. Ed. Joan Larsen Klein. Urbana: University of Illinois Press, 1992.

Davies, Lady Eleanor. *Prophetic Writings of Lady Eleanor Davies.* Ed. Esther S. Cope. Women Writers in English, 1350–1850. New York: Oxford University Press, 1995.

De Erauso, Catalina. *Lieutenant Nun: Memoir of a Basque Transvestite in the New World.* Trans. Michele Stepto and Gabriel Stepto. Foreword Marjorie Garber. Boston: Beacon Press, 1995.

Dentière, Marie. *Epistle to Marguerite de Navarre and Preface to a Sermon by John Calvin.* Ed. and trans. Mary B. McKinley. The Other Voice in Early Modern Europe. Chicago: University of Chicago Press, 2004.

Domestic Politics and Family Absence: The Correspondence (1588–1621) of Robert Sidney, First Earl of Leicester, and Barbara Gamage Sidney, Countess of Leicester. Ed. Margaret P. Hannay, Noel J. Kinnamon, and Michael G. Brennan. The Early Modern Englishwoman, 1500–1750: Contemporary Editions. Burlington, VT: Ashgate, 2005.

DuGard, Lydia. *The Letters of Lydia DuGard, 1665–1672: With a New Edition of The Marriages of Cousin Germans by Samuel DuGard.* Ed. Nancy Taylor. Tempe, AZ: MRTS and Renaissance English Text Society, 2003.

Eighteenth-Century Women: An Anthology. Ed. Bridget Hill. London: George Allen and Unwin, 1984.

Elisabeth of Bohemia, Princess and René Descartes. *The Correspondence between Princess Elisabeth of Bohemia and René Descartes.* Ed. and trans. Lisa Shapiro. The Other Voice in Early Modern Europe. Chicago: University of Chicago Press, 2007.

Elizabeth I: Collected Works. Ed. Leah S. Marcus, Janel Mueller, and Mary Beth Rose. Chicago: University of Chicago Press, 2000.

Elizabeth I. *Elizabeth I: Autograph Compositions and Foreign Language Originals.* Ed. Janel Mueller and Leah S. Marcus. Chicago: University of Chicago Press, 2003.

———. *Elizabeth's Glass: With "The Glass of the Sinful Soul" (1544) by Elizabeth I and "Epistle Dedicatory" & "Conclusion" (1548) by John Bale.* Ed. Marc Shell. Lincoln: University of Nebraska Press, 1993.

———. *The Letters of Queen Elizabeth I.* Ed. G. B. Harrison. New York: Funk and Wagnalls, 1935.

———. *Queen Elizabeth I: Selected Works.* Ed. Steven W. May. New York: Washington Square Press, 2004.

Elyot, Thomas. *Defence of Good Women: The Feminist Controversy of the Renaissance.* Facsimile Reproductions. Ed. Diane Bornstein. New York: Delmar, 1980.

English Women's Voices, 1540–1700. Ed. Charlotte Otten. Miami: Florida International University Press, 1992.

Erasmus, Desiderius (1467–1536). *Erasmus on Women.* Ed. Erika Rummel. Toronto: University of Toronto Press, 1996.

Family Life in Early Modern England: An Anthology of Contemporary Accounts, 1576–1716. Ed. Ralph Houlbrooke. London: Blackwells, 1988.

Fedele, Cassandra. *Letters and Orations.* Ed. and trans. Diana Robin. The Other Voice in Early Modern Europe. Chicago: University of Chicago Press, 2000.

Female and Male Voices in Early Modern England: An Anthology of Renaissance Writing. Ed. Betty S. Travitsky and Anne Lake Prescott. New York: Columbia University Press, 2000.

Female Playwrights of the Restoration: Five Comedies. Ed. Paddy Lyons and Fidelis Morgan. London: Everyman, 1994.

The Female Spectator: English Women Writers before 1800. Ed. Mary R. Mahl and Helene Koon. Bloomington: Indiana University Press, 1977; and Old Westbury, NY: Feminist Press, 1977.

Ferguson, Moira, ed. *First Feminists: British Women Writers, 1578–1799.* Bloomington: Indiana University Press, 1985.

Ferrazzi, Cecilia. *Autobiography of an Aspiring Saint.* Ed. and trans. Anne Jacobson Schutte. The Other Voice in Early Modern Europe. Chicago: University of Chicago Press, 1996.

Fettiplace, Elinor. *Elinor Fettiplace's Receipt Book: Elizabethan Country House Cooking.* Ed. Hilary Spurling. London: Elisabeth Sifton Books, 1986.

The Fifteen Joys of Marriage. Trans. Elizabeth Abbott. New York: Orion Press, 1959.

First Feminists: British Women Writers, 1578–1799. Bloomington: Indiana University Press, 1985.

Fitzmaurice, James, Josephine A. Roberts, Carol L. Barash, Eugine R. Cunnar, and Nancy A. Gutierrez, eds. *Major Women Writers of Seventeenth-Century England.* Ann Arbor: University of Michigan Press, 1997.

Folger Collective on Early Women Critics, eds. *Women Critics, 1660–1820: An Anthology.* Bloomington: Indiana University Press, 1995.

Fonte, Moderata (Modesta Pozzo). *Floridoro: A Chivalric Romance.* Ed. and introd. Valeria Finucci, trans. Julia Kisacky, annot. Valeria Finucci and Julia Kisacky. The Other Voice in Early Modern Europe. Chicago: University of Chicago Press, 2006.

————. *The Worth of Women.* Ed. and trans. Virginia Cox. The Other Voice in Early Modern Europe. Chicago: University of Chicago Press, 1997.

Francisca de los Apóstoles. *The Inquisition of Francisca: A Sixteenth-Century Visionary on Trial.* Ed. and trans. Gillian T. W. Ahlgren. The Other Voice in Early Modern Europe. Chicago: University of Chicago Press, 2005.

Franco, Veronica. *Poems and Selected Letters.* Ed. and trans. Ann Rosalind Jones and Margaret F. Rosenthal. The Other Voice in Early Modern Europe. Chicago: University of Chicago Press, 1998.

Galilei, Maria Celeste. *Sister Maria Celeste's Letters to Her Father, Galileo.* Ed. and trans. Rinaldina Russell. Lincoln, NE and New York: Writers Club Press of Universe. com, 2000; *To Father: The Letters of Sister Maria Celeste to Galileo, 1623–1633.* Trans. Dava Sobel. London: Fourth Estate, 2001.

Gethner, Perry, ed. *The Lunatic Lover and Other Plays by French Women of the 17th and 18th Centuries.* Portsmouth, NH: Heinemann, 1994.

Glückel of Hameln. *The Memoirs of Glückel of Hameln.* Trans. Marvin Lowenthal, new introd. Robert Rosen. New York: Schocken Books, 1977.

de Gournay, Marie le Jars. *Apology for the Woman Writing and Other Works.* Introd. Richard Hillman, ed. and trans. Richard Hillman and Colette Quesnel. The Other Voice in Early Modern Europe. Chicago: University of Chicago Press, 2002.

Graham, Elspeth, Hilary Hinds, Elaine Hobby, and Helen Wilcox, eds. *Her Own Life: Autobiographical Writings by Seventeenth-Century Englishwomen.* New York: Routledge, 1989.

Grimmelshausen, Johann. *The Life of Courage: The Notorious Thief, Whore, and Vagabond.* Trans. and introd. Mike Mitchell. Gardena, CA: SCB Distributors, 2001.

Guasco, Annibal. *Discourse to Lady Lavinia His Daughter.* Ed. and trans. Peggy Osborn. The Other Voice in Early Modern Europe. Chicago: University of Chicago Press, 2003.

de Guevara, María. *Warnings to the Kings and Advice on Restoring Spain.* A Bilingual Edition. The Other Voice in Early Modern Europe. Chicago: University of Chicago Press, 2007.

Harline, Craig, ed. *The Burdens of Sister Margaret: Inside a Seventeenth-Century Convent.* New Haven, CT: Yale University Press, abr. ed., 2000.

Henderson, Katherine Usher, and Barbara F. McManus, eds. *Half Humankind: Contexts and Texts of the Controversy about Women in England, 1540–1640.* Urbana: University of Illinois Press, 1985.

Herbert, Mary Sidney, Countess of Pembroke. *The Collected Works of Mary Sidney Herbert, Countess of Pembroke.* Ed., introd., and notes Margaret P. Hannay, Noel J. Kinnamon, and Michael G. Brennan. New York: Oxford University Press, 1998.

————. *Selected Works.* Ed. Margaret P. Hannay, Noel J. Kinnamon, and Michael G. Brennan. Tempe, AZ: MRTS, 2005.

Herman, Peter C. *Reading Monarch's Writing: The Poetry of Henry VIII, Mary Stuart, Elizabeth I, and James VI/I.* Tempe, AZ: MRTS, 2002.

Hill, Bridget, ed. *Eighteenth-Century Women: An Anthology.* London: George Allen and Unwin, 1984.

Hoby, Lady Margaret. *The Private Life of an Elizabethan Lady: The Diary of Lady Margaret Hoby, 1599–1605.* Ed. Joanna Moody. Phoenix Mill, UK: Sutton Publishing, 1998.

Humanist Educational Treatises. Ed. and trans. Craig W. Kallendorf. The I Tatti Renaissance Library. Cambridge, MA: Harvard University Press, 2002.

Hunter, Lynette, ed. *The Letters of Dorothy Moore, 1612–64.* Burlington, VT: Ashgate, 2004.

Inés, Juana de la Cruz, Sister. *The Answer / La Respuesta: Including a Selection of Poems.* Ed. and trans. Electa Arenal and Amanda Powell. New York: Feminist Press of The City University of New York, 1994.

———. *Poems, Protest, and a Dream.* Trans. and notes Margaret Sayers Peden, introd. Ilan Stavans. New York: Penguin, 1997.

———. *A Sor Juana Anthology.* Trans. Alan Trueblood, foreword Octavio Paz. Cambridge, MA: Harvard University Press, 1988.

Isabella Whitney, Mary Sidney, and Aemilia Lanyer: Renaissance Women Poets. Ed. Danielle Clarke. New York: Penguin, 2000.

Joscelin, Elizabeth. *The Mothers Legacy to Her Unborn Childe.* Ed. Jean LeDrew Metcalfe. Toronto: University of Toronto Press, 2000.

Julian of Norwich. *Revelations of Divine Love.* Trans. Elizabeth Spearing, introd. and notes A. C. Spearing. New York: Penguin, 1998.

de Jussie, Jeanne. *The Short Chronicle.* Ed. and trans. Carrie F. Klaus. The Other Voice in Early Modern Europe. Chicago: University of Chicago Press, 2006.

Kaminsky, Amy Katz, ed. *Water Lilies, Flores del agua: An Anthology of Spanish Women Writers from the Fifteenth through the Nineteenth Century.* Minneapolis: University of Minnesota Press, 1996.

Kempe, Margery. *The Book of Margery Kempe.* Ed. and trans. Lynn Staley. A Norton Critical Edition. New York: W. W. Norton, 2001.

———. *The Book of Margery Kempe.* Trans. B. A. Windeatt. New York: Penguin, 1985.

———. *The Book of Margery Kempe.* Trans. and introd. John Skinner. New York: Doubleday, 1998.

King, Margaret L., and Albert Rabil Jr., eds. *Her Immaculate Hand: Selected Works by and about the Women Humanists of Quattrocento Italy.* Binghamton, NY: Medieval and Renaissance Texts and Studies, 1983; 2nd rev. paperback ed., 1991.

Klein, Joan Larsen, ed. *Daughters, Wives, and Widows: Writings by Men about Women and Marriage in England, 1500–1640.* Urbana: University of Illinois Press, 1992.

Knox, John (1505–1572). *The Political Writings of John Knox: The First Blast of the Trumpet against the Monstrous Regiment of Women and Other Selected Works.* Ed. Marvin A. Breslow. Washington, DC: Folger Shakespeare Library, 1985.

Kors, Alan C., and Edward Peters, eds. *Witchcraft in Europe, 400–1700: A Documentary History.* Philadelphia: University of Pennsylvania Press, 2000.

Kottanner, Helene. *The Memoirs of Helene Kottanner, 1439–1440.* Trans. Maya B. Williamson. Library of Medieval Women. Rochester, NY: Boydell and Brewer, 1998.

Krämer, Heinrich, and Jacob Sprenger. *Malleus Maleficarum* (ca. 1487). Trans. Montague Summers. London: Pushkin Press, 1928; repr. New York: Dover, 1971.

Labé, Louise. *Complete Poetry and Prose.* A Bilingual Edition. Ed. and introd. Deborah Lesko Baker, trans. Annie Finch. The Other Voice in Early Modern Europe. Chicago: University of Chicago Press, 2006.

———. *Sonnets.* Introd. and commentary Peter Sharratt, trans. Graham Dunstan Martin. Edinburgh Bilingual Library. Austin: University of Texas Press, 1972.

de Lafayette, Marie-Madeleine Pioche de La Vergne, Comtesse. *Zayde: A Spanish Romance.* Ed. and trans. Nicholas D. Paige, The Other Voice in Early Modern Europe. Chicago: University of Chicago Press, 2006.

Lanyer, Aemilia. *The Poems of Aemilia Lanyer: Salve Deus Rex Judæorum.* Ed. Susanne Woods. Women Writers in English, 1350–1850. New York: Oxford University Press. 1993.

Larsen, Anne R., and Colette H. Winn, eds. *Writings by Pre-Revolutionary French Women: From Marie de France to Elizabeth Vigée-Le Brun.* New York: Garland, 2000.

Lay by Your Needles Ladies, Take the Pen: Writing Women in England, 1500–1700. Ed. Susanne Trill, Kate Chedgzoy, and Melanie Osborne. New York: Arnold, 1997.

Lock, Anne Vaughan. *The Collected Works of Anne Vaughan Lock.* Ed. Susan M. Felch. Renaissance English Text Society. Tempe, AZ: MRTS, 1999.

de Lorris, William, and Jean de Meun. *The Romance of the Rose.* Trans. Charles Dahlbert. Princeton, NJ: Princeton University Press, 1971; repr. University Press of New England, 1983.

Lyons, Paddy, and Fidelis Morgan, eds. *Female Playwrights of the Restoration: Five Comedies.* London: Everyman, 1994.

Mahl, Mary R., and Helene Koon, eds. *The Female Spectator: English Women Writers before 1800.* Bloomington: Indiana University Press and Old Westbury, NY: Feminist Press, 1977.

de Maintenon, Madame. *Dialogues and Addresses.* Ed. and trans. John J. Conley, S. J. The Other Voice in Early Modern Europe. Chicago: University of Chicago Press, 2004.

Major Women Writers of Seventeenth-Century England. Ed. James Fitzmaurice, Josephine A. Roberts, Carol L. Barash, Eugine R. Cunnar, and Nancy A. Gutierrez. Ann Arbor: University of Michigan Press, 1997.

Makin, Bathsua. *Woman of Learning.* Ed. Frances Teague. Lewisburg, PA: Bucknell University Press, 1998.

Marcus, Leah S., Janel Mueller, and Mary Beth Rose, eds. *Elizabeth I: Collected Works.* Chicago: University of Chicago Press, 2000.

Marguerite d'Angoulême, Queen of Navarre. *The Heptameron.* Trans. P. A. Chilton. New York: Viking Penguin, 1984.

Marinella, Lucrezia. *The Nobility and Excellence of Women and the Defects and Vices of Men.* Ed. and trans. Anne Dunhill, introd. Letizia Panizza. The Other Voice in Early Modern Europe. Chicago: University of Chicago Press, 1999.

Markham, Gervase. *The English Housewife: Containing the inward and outward virtues which ought to be in a complete woman; as her skill in physic, cookery, banqueting-stuff, distillation, perfumes, wool, hemp, flax, dairies, brewing, baking, and all other things belonging to a household.* Ed. Michael R. Best. Montreal: McGill-Queen's University Press, 1986.

Mary of Agreda. *The Divine Life of the Most Holy Virgin.* Abr. of *The Mystical City of God.* Abr. Fr. Bonaventure Amedeo de Caesarea, M.C. Trans. from French Abbé Joseph A. Boullan. Rockford, IL: Tan Books, 1997.

Mary, Queen of Scots. *Bittersweet within My Heart: The Collected Poems of Mary, Queen of Scots.* Trans. and ed. Robin Bell. London: Pavilion Books, 1992.

Matraini, Chiara. *Selected Poetry and Prose.* A Bilingual Edition. Ed. and trans. Elaine Maclachlan, introd. Giovanna Rabitti. The Other Voice in Early Modern Europe. Chicago: University of Chicago Press, 2007.

de' Medici, Lucrezia Tornabuoni. *Sacred Narratives*. Ed. and trans. Jane Tylus. The Other Voice in Early Modern Europe. Chicago: University of Chicago Press, 2001.

Medieval Women Writers. Ed. Katharina M. Wilson. Athens: University of Georgia Press, 1984.

The Meridian Anthology of Restoration and Eighteenth-Century Plays by Women. Ed. Katharine M. Rogers. New York: Penguin, 1994.

Millman, Jill Seal, and Gillian Wright, eds. *Early Modern Women's Manuscript Poetry*. Introd. Elizabeth Clarke and Jonathon Gibson. Manchester: Manchester University Press, 2005.

de Montpensier, Anne-Marie-Louise d'Orléans, Duchesse. *Against Marriage: The Correspondence of La Grande Mademoiselle*. Ed. and trans. Joan DeJean. The Other Voice in Early Modern Europe. Chicago: University of Chicago Press, 2002.

Moore, Dorothy. *The Letters of Dorothy Moore, 1612–64: The Friendships, Marriage, and Intellectual Life of a Seventeenth-Century Woman*. Ed. Lynette Hunter. The Early Modern Englishwoman, 1500–1750: Contemporary Editions. Burlington, VT: Ashgate, 2004.

Morata, Olympia. *The Complete Writings of an Italian Heretic*. Ed. and trans. Holt N. Parker. The Other Voice in Early Modern Europe. Chicago: University of Chicago Press, 2003.

Moulsworth, Martha. *"My Name Was Martha": A Renaissance Woman's Autobiographical Poem*. Ed. and commentary Robert C. Evans and Barbara Wiedemann. West Cornwall, CT: Locust Hill Press, 1993.

Mullan, David George. *Women's Life Writing in Early Modern Scotland: Writing the Evangelical Self, c. 1670–c. 1730*. Burlington, VT: Ashgate, 2003.

Myers, Kathleen A., and Amanda Powell, eds. *A Wild Country Out in the Garden: The Spiritual Journals of a Colonial Mexican Nun*. Bloomington: Indiana University Press, 1999.

Nogarola, Isotta. *Complete Writings: Letterbook, Dialogue on Adam and Eve, Orations*. Ed. and trans. Margaret L. King and Diana Robin. The Other Voice in Early Modern Europe. Chicago: University of Chicago Press, 2004.

Osborne, Dorothy. *Letters to Sir William Temple*. Ed., introd., and notes Kenneth Parker. New York: Penguin, 1987.

Ostovich, Helen, and Elizabeth Sauer, eds. *Reading Early Modern Women: An Anthology of Texts in Manuscript and Print, 1550–1700*. New York: Routledge, 2004.

Otten, Charlotte F, ed. *English Women's Voices, 1540–1700*. Miami: Florida International University Press, 1992.

Ozment, Steven. *Magdalena and Balthasar: An Intimate Portrait of Life in 16th-Century Europe Revealed in the Letters of a Nuremberg Husband and Wife*. New York: Simon and Schuster, 1986; paperback New Haven, CT: Yale University Press, 1989.

Parr, Katherine. *Prayers or Medytacions and The Lamentation of a Synner*. Ed. Janel Mueller. The Early Modern Englishwoman: A Facsimile Library of Essential Works. Part 1: Printed Writings, 1500–1640. Burlington, VT: Ashgate, 1996.

Pascal, Jacqueline. *A Rule for Children and Other Writings*. Ed. and trans. John J. Conley, S.J. The Other Voice in Early Modern Europe. Chicago: University of Chicago Press, 2003.

Petersen, Johanna Eleonora. *The Life of Lady Johanna Eleonora Petersen, Written by Herself*. Ed. and trans. Barbara Becker-Cantarino. The Other Voice in Early Modern Europe. Chicago: University of Chicago Press, 2005.

Poullain de la Barre, François. *Three Cartesian Feminist Treatises.* Introd. and notes Marcelle Maistre Welch, trans. Vivien Bosley. The Other Voice in Early Modern Europe. Chicago: University of Chicago Press, 2002.

Pulci, Antonia. *Florentine Drama for Convent and Festival.* Ed. and trans. James Wyatt Cook. The Other Voice in Early Modern Europe. Chicago: University of Chicago Press, 1996.

Reading Early Modern Women: An Anthology of Texts in Manuscript and Print, 1550–1700. Ed. Helen Ostovich and Elizabeth Sauer. New York: Routledge, 2004.

Reading Monarch's Writing: The Poetry of Henry VIII, Mary Stuart, Elizabeth I, and James VI/I. Ed. Peter C. Herman. Tempe, AZ: MRTS, 2002.

Renaissance Drama by Women: Texts and Documents. Ed. S. P. Cerasano and Marion Wynne-Davies. New York: Routledge, 1996.

The Renaissance Englishwoman in Print: Counterbalancing the Canon. Ed. Anne M. Haselkorn and Betty S. Travitsky. Amherst: University of Massachusetts Press, 1990.

Renaissance Woman: Constructions of Femininity in England; A Source Book. Ed. Kate Aughterson. New York: Routledge, 1995.

Riccoboni, Sister Bartolomea. *Life and Death in a Venetian Convent: The Chronicle and Necrology of Corpus Domini, 1395–1436.* Ed. and trans. Daniel Bornstein. The Other Voice in Early Modern Europe. Chicago: University of Chicago Press, 2000.

des Roches, Madeleine and Catherine. *From Mother and Daughter.* Ed. and trans. Anne R. Larsen. The Other Voice in Early Modern Europe. Chicago: University of Chicago Press, 2006.

Rogers, Katharine M., ed. *The Meridian Anthology of Restoration and Eighteenth-Century Plays by Women.* New York: Penguin, 1994.

Rosen, Barbara, ed. *Witchcraft in England, 1558–1618.* 1969; repr. with a new preface. Amherst: University of Massachusetts Press, 1991.

Russian Women, 1698–1917, Experience & Expression: An Anthology of Sources. Compiled, ed., annot., and introd. Robin Bisha, Jehanne M. Gheith, Christine Holden, and William G. Wagner. Bloomington: Indiana University Press, 2002.

Salazar, María de San José. *Book for the Hour of Recreation.* Introd. and notes Alison Weber, trans. Amanda Powell. The Other Voice in Early Modern Europe. Chicago: University of Chicago Press, 2002.

Salter, Thomas. *A Critical Edition of Thomas Salter's The Mirrhor of Modestie.* Ed. Janis Butler Holm. The Renaissance Imagination. New York: Garland, 1987.

Sarrocchi, Margherita. *Scanderbeide: The Heroic Deeds of George Scanderbeg, King of Epirus.* Ed. and trans. Rinaldina Russell. The Other Voice in Early Modern Europe. Chicago: University of Chicago Press, 2006.

van Schurman, Anna Maria. *Whether a Christian Woman Should Be Educated and Other Writings from Her Intellectual Circle.* Ed. and trans. Joyce L. Irwin. The Other Voice in Early Modern Europe. Chicago: University of Chicago Press, 1998.

Schütz Zell, Katharina. *Church Mother: The Writings of a Protestant Reformer in Sixteenth-Century Germany.* Ed. and trans. Elsie McKee. The Other Voice in Early Modern Europe. Chicago: University of Chicago Press, 2006.

de Scudéry, Madeleine. *Selected Letters, Orations, and Rhetorical Dialogues.* Ed. and trans. Jane Donawerth and Julie Strongson. The Other Voice in Early Modern Europe. Chicago: University of Chicago Press, 2004.

————. *The Story of Sappho.* Ed. and trans. Karen Newman. The Other Voice in Early Modern Europe. Chicago: University of Chicago Press, 2003.

Shepherd, Simon, ed. *The Woman's Sharp Revenge: Five Women's Pamphlets from the Renaissance.* New York: St. Martin's Press, 1985.

Siegemund, Justine. *The Court Midwife.* Ed. and trans. Lynne Tatlock. The Other Voice in Early Modern Europe. Chicago: University of Chicago Press, 2005.

The Southwell-Sibthorpe Commonplace Book, Folger MS. V.b.198. Ed. Jean Klene, C.S.C. Renaissance English Text Society. Tempe, AZ: MRTS, 1997.

Speght, Rachel. *The Polemics and Poems of Rachel Speght.* Ed. Barbara Kiefer Lewalski. Women Writers in English, 1350–1850. New York: Oxford University Press, 1996.

Stampa, Gaspara. *Selected Poems.* Ed. and trans. Laura Anna Stortoni and Mary Prentice Lillie. New York: Italica Press, 1994.

Stortoni, Laura Anna, ed. *Women Poets of the Italian Renaissance: Courtly Ladies and Courtesans.* Trans. Stortoni and Mary Prentice Lillie. New York: Italica Press, 1997.

Stuart, Lady Arbella. *The Letters of Lady Arbella Stuart.* Ed. Sara Jayne Steen. Women Writers in English, 1350–1850. New York: Oxford University Press, 1994.

Tarabotti, Arcangela. *Paternal Tyranny.* Ed. and trans. Letizia Panizza. The Other Voice in Early Modern Europe. Chicago: University of Chicago Press, 2004.

Teresa of Avila, Saint. *The Collected Letters of St. Teresa of Avila. Volume One: 1546–1577.* Trans. Kieran Kavanaugh. Washington, DC: Institute of Carmelite Studies, 2001. Volume Two is forthcoming.

————. *The Life of Saint Teresa of Avila by Herself.* Trans. J. M. Cohen. New York: Viking Penguin, 1957.

Tilney, Edmund. *The Flower of Friendship: A Renaissance Dialogue Contesting Marriage.* Ed. and introd. Valerie Wayne. Ithaca, NY: Cornell University Press, 1993.

Trapnel, Anna. *The Cry of a Stone.* Ed. and introd. Hilary Hinds. Tempe, AZ: MRTS, 2000.

Travitsky, Betty, ed. *The Paradise of Women: Writings by Englishwomen of the Renaissance.* Westport, CT: Greenwood Press, 1981.

The Trial of Joan of Arc. Trans. and introd. Daniel Hobbins. Cambridge, MA: Harvard University Press, 2005.

Trill, Susanne, Kate Chedgzoy, and Melanie Osborne, eds. *Lay by Your Needles Ladies, Take the Pen: Writing Women in England, 1500–1700.* New York: Arnold, 1997.

de Villedieu, Madame. *Memoirs of the Life of Henriette-Sylvie de Molière: A Novel.* Ed. and trans. Donna Kuizenga. The Other Voice in Early Modern Europe. Chicago: University of Chicago Press, 2004.

Vives, Juan Luis. *The Education of a Christian Woman: A Sixteenth-Century Manual.* Ed. and trans. Charles Fantazzi. The Other Voice in Early Modern Europe. Chicago: University of Chicago Press, 2000.

————. *The Instruction of a Christen Woman.* Ed. Virginia Walcott Beauchamp, Elizabeth H. Hageman, and Margaret Mikesell. Urbana: University of Illinois Press, 2002.

Weamys, Anna. *A Continuation of Sir Philip Sidney's Arcadia.* Ed. Patrick Colborn Cullen. New York: Oxford University Press, 1994.

Weston, Elizabeth Jane. *Collected Writings.* Ed. and trans. Donald Cheney and Brenda M. Hosington. Toronto: University of Toronto Press, 2000.

Weyer, Johann. *Witches, Devils, and Doctors in the Renaissance: Johann Weyer, De praestigiis daemonum.* Ed. George Mora with Benjamin G. Kohl, Erik Midelfort, and Helen Bacon; trans. John Shea. Binghamton, NY: Medieval and Renaissance Texts and Studies, 1991.

Wilson, Katharina M., ed. *Medieval Women Writers.* Athens: University of Georgia Press, 1984.

Wilson, Katharina M., ed. *Women Writers of the Renaissance and Reformation.* Athens: University of Georgia Press, 1987.

Wilson, Katharina M., and Frank J. Warnke, eds. *Women Writers of the Seventeenth Century.* Athens: University of Georgia Press, 1989.

Witchcraft in England, 1558–1619. Ed. Barbara Rosen. 1969; repr. with a new preface Amherst: University of Massachusetts Press, 1991.

Witchcraft in Europe, 1100–1700: A Documentary History. Ed. Alan C. Kors and Edward Peters. Philadelphia: University of Pennsylvania Press, 1972.

Wollstonecraft, Mary. *A Vindication of the Rights of Men and a Vindication of the Rights of Women.* Ed. Sylvana Tomaselli. Cambridge: Cambridge University Press, 1995.

————. *The Vindications of the Rights of Men, The Rights of Women.* Ed. D. L. Macdonald and Kathleen Scherf. Peterborough, ON: Broadview Press, 1997.

Woman Defamed and Woman Defended: An Anthology of Medieval Texts. Ed. Alcuin Blamires. Oxford: Clarendon Press, 1992.

The Woman's Sharp Revenge: Five Women's Pamphlets from the Renaissance. Ed. Simon Shepherd. New York: St. Martin's Press, 1985.

Women Critics, 1660–1820: An Anthology. Ed. the Folger Collective on Early Women Critics. Bloomington: Indiana University Press, 1995.

Women Philosophers of the Early Modern Period. Ed. Margaret Atherton. Indianapolis, IN: Hackett Publishing, 1994.

Women Poets of the Italian Renaissance: Courtly Ladies and Courtesans. Ed. Laura Anna Stortoni, trans. Stortoni and Mary Prentice Lillie. New York: Italica Press, 1997.

Women's Life Writing in Early Modern Scotland: Writing the Evangelical Self, c. 1670–c. 1730. Ed. David G. Mullan. The Early Modern Englishwoman, 1500–1750: Contemporary Editions. Burlington, VT: Ashgate, 2003.

Women's Worlds in Seventeenth-Century England: A Source Book. Ed. Patricia Crawford and Laura Gowing. New York: Routledge, 2000.

Women Writers of the Renaissance and Reformation. Ed. Katharina M. Wilson. Althens: University of Georgia Press, 1987.

Women Writers of the Seventeenth Century. Ed. Katharina M. Wilson and Frank J. Warnke. Athens: University of Georgia Press, 1989.

Women's Writing in Stuart England: The Mother's Legacies of Dorothy Leigh, Elizabeth Joscelin, and Elizabeth Richardson. Ed. Sylvia Brown. Thrupp, Stroud, Gloucester, UK: Sutton, 1999.

Wroth, Lady Mary. *The First Part of the Countess of Montgomery's Urania.* Ed. Josephine A. Roberts. Renaissance English Text Society. Tempe, AZ: MRTS, 1995.

————. *The Second Part of the Countess of Montgomery's Urania.* Ed. Josephine R. Roberts, Suzanne Gossett, and Janel Mueller. Renaissance English Text Society. Tempe AZ: MRTS, 1999.

————. *Lady Mary Wroth's "Love's Victory": The Penshurst Manuscript.* Ed. Michael G. Brennan. London: The Roxburghe Club, 1988.

————. *Pamphilia to Amphilanthus.* Ed. G. F. Waller. Elizabethan and Renaissance Studies. Salzburg, Austria: Institut für Englische Sprache und Literatur Universität Salzburg, 1977.

————. *The Poems of Lady Mary Wroth.* Ed. Josephine A. Roberts. Baton Rouge: Louisiana State University Press, 1983.

de Zayas, Maria. *The Disenchantments of Love.* Trans. H. Patsy Boyer. Albany: State University of New York Press, 1997.

————. *The Enchantments of Love: Amorous and Exemplary Novels.* Trans. H. Patsy Boyer. Berkeley: University of California Press, 1990.

SECONDARY SOURCES

Abate, Corinne S., ed. *Privacy, Domesticity, and Women in Early Modern England.* Burlington, VT: Ashgate, 2003.

Aemilia Lanyer: Gender, Genre, and the Canon. Ed. Marshall Grossman. Lexington: University Press of Kentucky, 1998.

Ahlgren, Gillian. *Teresa of Avila and the Politics of Sanctity.* Ithaca, NY: Cornell University Press, 1996.

Åkerman, Susanna. *Queen Christina of Sweden: The Transformation of a Seventeenth-Century Philosophical Libertine.* Leiden: E. J. Brill, 1991.

Akkerman, Tjitske, and Siep Sturman, eds. *Feminist Thought in European History, 1400–2000.* London and New York: Routledge, 1997.

Allen, Sister Prudence, R.S.M. *The Concept of Woman: The Aristotelian Revolution, 750 B.C.–A.D. 1250.* Grand Rapids, MI: William B. Eerdmans Publishing Company, 1997.

————. *The Concept of Woman. Vol. 2, The Early Humanist Reformation, 1250–1500.* Grand Rapids, MI: William B. Eerdmans, 2002.

Altmann, Barbara K., and Deborah L. McGrady, eds. *Christine de Pizan: A Casebook.* New York: Routledge, 2003.

Ambiguous Realities: Women in the Middle Ages and Renaissance. Ed. Carole Levin and Jeanie Watson. Detroit, MI: Wayne State University Press, 1987.

Amussen, Susan D. *An Ordered Society: Gender and Class in Early Modern England.* Oxford and New York: Basil Blackwell, 1988.

Amussen, Susan D., and Adele Seeff, eds. *Attending to Early Modern Women.* Newark: University of Delaware Press, 1998.

Anderson, Bonnie S., and Judith P. Zinsser. *A History of Their Own: Women in Europe from Prehistory to the Present.* 2 vols. 2nd rev. ed. New York: Oxford University Press, 2000.

Anderson, Karen. *Chain Her by One Foot: The Subjugation of Women in Seventeenth-Century New France.* New York: Routledge, 1991.

Andreadis, Harriette. *Sappho in Early Modern England: Female Same-Sex Literary Erotics, 1550–1714.* Chicago: University of Chicago Press, 2001.

Arcangela Tarabotti: A Literary Nun in Baroque Venice. Ed. Elissa B. Weaver. Ravenna: Longo Editore, 2006.

Architecture and the Politics of Gender in Early Modern Europe. Ed. Helen Hills. Burlington, VT: Ashgate, 2003.

Armon, Shifra. *Picking Wedlock: Women and the Courtship Novel in Spain.* New York: Rowman and Littlefield, 2002.

Atkinson, Clarissa W. *Mystic and Pilgrim: The Book and the World of Margery Kempe*. Ithaca, NY: Cornell University Press, 1983.

Attending to Early Modern Women. Ed. Susan D. Amussen and Adele Seeff. Newark: University of Delaware Press, 1998.

Attending to Women in Early Modern England. Ed. Betty S. Travitsky and Adele F. Seef. Newark, DE: University of Delaware Press, 1994.

Backer, Dorothy Anne Liot. *Precious Women*. New York: Basic Books, 1974.

Baernstein, P. Renée. *A Convent Tale: A Century of Sisterhood in Spanish Milan*. New York: Routledge, 2002.

Bainton, Roland H. *Women of the Reformation in France and England*. Minneapolis, MN: Augsburg, 1973.

——. *Women of the Reformation in Germany and Italy*. Minneapolis, MN: Augsburg, 1971.

Ballaster, Rosalind. *Seductive Forms*. New York: Oxford University Press, 1992.

Barash, Carol. *English Women's Poetry, 1649–1714: Politics, Community, and Linguistic Authority*. New York: Oxford University Press, 1996.

Bardsley, Sandy. *Venomous Tongues: Speech and Gender in Late Medieval England*. The Middle Ages Series. Philadelphia: University of Pennsylvania Press, 2006.

Barker, Alele Marie, and Jehanne M. Gheith, eds. *A History of Women's Writing in Russia*. Cambridge: Cambridge University Press, 2002.

Barroll, Leeds. *Anna of Denmark: A Cultural Biography*. Philadelphia: University of Pennsylvania Press, 2001.

Barry, Jonathan, Marianne Hester and Gareth Roberts, eds. *Witchcraft in Early Modern Europe: Studies in Culture and Belief*. New York: Cambridge University Press, 1996.

Barstow, Anne L. *Joan of Arc: Heretic, Mystic, Shaman*. Lewiston, NY: Edwin Mellen Press, 1986.

Battigelli, Anna. *Margaret Cavendish and the Exiles of the Mind*. Lexington: University Press of Kentucky, 1998.

Beasley, Faith. *Revising Memory: Women's Fiction and Memoirs in Seventeenth-Century France*. New Brunswick, NJ: Rutgers University Press, 1990.

——. *Salons, History, and the Creation of 17th-Century France*. Burlington, VT: Ashgate, 2006.

Becker, Lucinda M. *Death and the Early Modern Englishwoman*. Burlington, VT: Ashgate, 2003.

Beilin, Elaine V. *Redeeming Eve: Women Writers of the English Renaissance*. Princeton, NJ: Princeton University Press, 1987.

Bell, Rudolph M. *Holy Anorexia*. Chicago: University of Chicago Press, 1985.

Bennett, Judith M., and Amy M. Froide, eds. *Singlewomen in the European Past, 1250–1800*. Philadelphia: University of Pennsylvania Press, 1999.

Bennett, Lyn. *Women Writing of Divinest Things: Rhetoric and the Poetry of Pembroke, Wroth, and Lanyer*. Pittsburgh: Duquesne University Press, 2004.

Benson, Pamela Joseph. *The Invention of Renaissance Woman: The Challenge of Female Independence in the Literature and Thought of Italy and England*. University Park: Pennsylvania State University Press, 1992.

Benson, Pamela Joseph, and Victoria Kirkham, eds. *Strong Voices, Weak History? Medieval and Renaissance Women in Their Literary Canons: England, France, Italy*. Ann Arbor: University of Michigan Press, 2003.

Berry, Helen. *Gender, Society, and Print Culture in Late-Stuart England*. Burlington, VT: Ashgate, 2003.

Berry, Philippa. *Of Chastity and Power: Elizabethan Literature and the Unmarried Queen.* New York: Routledge, 1989.

Beyond Isabella: Secular Women Patrons of Art in Renaissance Italy. Ed. Sheryl E. Reiss and David G. Wilkins. Kirksville, MO: Truman State University Press, 2001.

Beyond Their Sex: Learned Women of the European Past. Ed. Patricia A. Labalme. New York: New York University Press, 1980.

Bicks, Caroline. *Midwiving Subjects in Shakespeare's England.* Burlington, VT: Ashgate, 2003.

Bilinkoff, Jodi. *The Avila of Saint Teresa: Religious Reform in a Sixteenth-Century City.* Ithaca, NY: Cornell University Press, 1989.

————. *Related Lives: Confessors and Their Female Penitents, 1450–1750.* Ithaca, NY: Cornell University Press, 2005.

Bissell, R. Ward. *Artemisia Gentileschi and the Authority of Art.* University Park: Pennsylvania State University Press, 2000.

Blain, Virginia, Isobel Grundy, and Patricia Clements, eds. *The Feminist Companion to Literature in English: Women Writers from the Middle Ages to the Present.* New Haven, CT: Yale University Press, 1990.

Blamires, Alcuin. *The Case for Women in Medieval Culture.* Oxford: Clarendon Press, 1997.

Bloch, R. Howard. *Medieval Misogyny and the Invention of Western Romantic Love.* Chicago: University of Chicago Press, 1991.

Blumenfeld-Kosinski, Renate. *Not of Woman Born: Representations of Caesarean Birth in Medieval and Renaissance Culture.* Ithaca, NY: Cornell University Press, 1990.

Bogucka, Maria. *Women in Early Modern Polish Society, against the European Background.* Burlington, VT: Ashgate, 2004.

Bornstein, Daniel, and Roberto Rusconi, eds. *Women and Religion in Medieval and Renaissance Italy.* Trans. Margery J. Schneider. Chicago: University of Chicago Press, 1996.

Brant, Clare, and Diane Purkiss, eds. *Women, Texts, and Histories, 1575–1760.* London and New York: Routledge, 1992.

Breisach, Ernst. *Caterina Sforza: A Renaissance Virago.* Chicago: University of Chicago Press, 1967.

Bridenthal, Renate, Claudia Koonz, and Susan M. Stuard. *Becoming Visible: Women in European History.* 3rd ed. Boston: Houghton Mifflin, 1998.

Briggs, Robin. *Witches and Neighbours: The Social and Cultural Context of European Witchcraft.* New York: HarperCollins, 1995; Viking Penguin, 1996.

Brink, Jean R., ed. *Female Scholars: A Tradition of Learned Women before 1800.* Montreal: Eden Press Women's Publications, 1980.

Brink, Jean R., Allison Coudert, and Maryanne Cline Horowitz, eds. *The Politics of Gender in Early Modern Europe.* Sixteenth Century Essays and Studies, vol. 12. Kirksville, MO: Sixteenth Century Journal Publishers, 1989.

Broad, Jacqueline S. *Women Philosophers of the Seventeenth Century.* Cambridge: Cambridge University Press, 2002, paperback 2007.

Broad, Jacqueline S., and Karen Green. *A History of Women's Political Thought in Europe, 1400–1700.* Cambridge: Cambridge University Press, 2008.

Broad, Jacqueline S., and Karen Green, eds. *Virtue, Liberty, and Toleration: Political Ideas of European Women, 1400–1700.* Dordrecht: Springer, 2007.

Brodsky, Vivien. *Mobility and Marriage: The Family and Kinship in Early Modern London.* London: Blackwells, 1988.

Broude, Norma, and Mary D. Garrard, eds. *The Expanding Discourse: Feminism and Art History.* New York: HarperCollins, 1992.

Brown, Judith C. *Immodest Acts: The Life of a Lesbian Nun in Renaissance Italy.* New York: Oxford University Press, 1986.

Brown, Judith C., and Robert C. Davis, eds. *Gender and Society in Renaissance Italy.* London: Addison-Wesley Longman, 1998.

Brown-Grant, Rosalind. *Christine de Pizan and the Moral Defence of Women: Reading Beyond Gender.* Cambridge: Cambridge University Press, 1999.

Brucker, Gene. *Giovanni and Lusanna: Love and Marriage in Renaissance Florence.* Berkeley: University of California Press, 1986.

Burke, Mary E., Jane Donawerth, Linda L. Dove, and Karen Nelson, eds. *Women, Writing, and the Reproduction of Culture in Tudor and Stuart Britain.* Syracuse, NY: Syracuse University Press, 2000.

Burns, Jane E., ed. *Medieval Fabrications: Dress, Textiles, Cloth Work, and Other Cultural Imaginings.* New York: Palgrave Macmillan, 2004.

Bynum, Carolyn Walker. *Fragmentation and Redemption: Essays on Gender and the Human Body in Medieval Religion.* New York: Zone Books, 1992.

———. *Holy Feast and Holy Fast: The Religious Significance of Food to Medieval Women.* Berkeley: University of California Press, 1987.

———. *Jesus as Mother: Studies in the Spirituality of the High Middle Ages.* Berkeley: University of California Press, 1982.

Cahn, Susan. *Industry of Devotion: The Transformation of Women's Work in England, 1500–1660.* New York: Columbia University Press, 1987.

Callaghan, Dympna, ed. *The Impact of Feminism in English Renaissance Studies.* New York: Palgrave Macmillan, 2007.

Campbell, Julie DeLynn. "Renaissance Women Writers: The Beloved Speaks Her Part." PhD diss., Texas A&M University, 1997.

Catling, Jo, ed. *A History of Women's Writing in Germany, Austria, and Switzerland.* Cambridge: Cambridge University Press, 2000.

Cavallo, Sandra, and Lyndan Warner. *Widowhood in Medieval and Early Modern Europe.* New York: Longman, 1999.

Cavanagh, Sheila T. *Cherished Torment: The Emotional Geography of Lady Mary Wroth's Urania.* Pittsburgh: Duquesne University Press, 2001.

Cavendish and Shakespeare: Interconnections. Ed. and introd. Katherine Romack and James Fitzmaurice. Burlington, VT: Ashgate, 2006.

Cerasano, S. P., and Marion Wynne-Davies, eds. *Gloriana's Face: Women, Public and Private, in the English Renaissance.* Detroit, MI: Wayne State University Press, 1992.

———. *Readings in Renaissance Women's Drama: Criticism, History, and Performance, 1594–1998.* New York: Routledge, 1998.

Cervigni, Dino S., ed. "Women Mystic Writers." *Annali d'Italianistica* 13 (1995) (entire issue).

Cervigni, Dino S., and Rebecca West, eds. "Women's Voices in Italian Literature." *Annali d'Italianistica* 7 (1989) (entire issue).

Chambers, Anne. *Granuaile: The Life and Times of Grace O'Malley, c. 1530–1603.* Rev. ed. Dublin, Ireland: Wolfhound Press, 1998.

Charlton, Kenneth. *Women, Religion, and Education in Early Modern England.* New York: Routledge, 1999.

Chedgzoy, Kate, Melanie Hansen, and Suzanne Trill, eds. *Voicing Women: Gender and Sexuality in Early Modern Writing.* Pittsburgh: Duquesne University Press, 1996.

Chojnacka, Monica. *Working Women of Early Modern Venice.* Baltimore: Johns Hopkins University Press, 2001.

Chojnacki, Stanley. *Women and Men in Renaissance Venice: Twelve Essays on Patrician Society.* Baltimore: Johns Hopkins University Press, 2000.

Cholakian, Patricia Francis. *Rape and Writing in the Heptameron of Marguerite de Navarre.* Carbondale and Edwardsville: Southern Illinois University Press, 1991.

———. *Women and the Politics of Self-Representation in Seventeenth-Century France.* Newark: University of Delaware Press, 2000.

Cholakian, Patricia Francis, and Rouben Charles Cholakian. *Marguerite De Navarre: Mother of the Renaissance.* New York: Columbia University Press, 2006.

Christine de Pizan: A Casebook. Ed. Barbara K. Altmann and Deborah L. McGrady. New York: Routledge, 2003.

Clogan, Paul Maruice, ed. *Medievali et Humanistica: Literacy and the Lay Reader.* Lanham, MD: Rowman and Littlefield, 2000.

Clubb, Louise George. *Italian Drama in Shakespeare's Time.* New Haven, CT: Yale University Press, 1989.

Clucas, Stephen, ed. *A Princely Brave Woman: Essays on Margaret Cavendish, Duchess of Newcastle.* Burlington, VT: Ashgate, 2003.

Coakley, John W. *Women, Men, and Spiritual Power: Female Saints and Their Male Collaborators.* New York: Columbia University Press, 2006.

Conley, John J., S.J. *The Suspicion of Virtue: Women Philosophers in Neoclassical France.* Ithaca, NY: Cornell University Press, 2002.

Cook, Ann Jennalie. *Making a Match: Courtship in Shakespeare and His Society.* Princeton, NJ: Princeton University Press, 1991.

Couchman, Jane, and Ann Crabb. *Women's Letters across Europe, 1400–1700: Form and Persuasion.* Women and Gender in the Early Modern World. Burlington, VT: Ashgate, 2005.

Cox, Virginia. *Women's Writing in Italy, 1400–1650.* Baltimore: Johns Hopkins University Press, 2008.

Crabb, Ann. *The Strozzi of Florence: Widowhood and Family Solidarity in the Renaissance.* Ann Arbor: University of Michigan Press, 2000.

The Crannied Wall: Women, Religion, and the Arts in Early Modern Europe. Ed. Craig A. Monson. Ann Arbor: University of Michigan Press, 1992.

Crawford, Patricia. *Women and Religion in England, 1500–1750.* New York: Routledge, 1993.

Creative Women in Medieval and Early Modern Italy. Ed. E. Ann Matter and John Coakley. Philadelphia: University of Pennsylvania Press, 1994 (sequel to the Monson collection, below).

Crowston, Clare Haru. *Fabricating Women: The Seamstresses of Old Regime France, 1675–1791.* Durham, NC: Duke University Press, 2001.

Cruz, Anne J., and Mary Elizabeth Perry, eds. *Culture and Control in Counter-Reformation Spain.* Minneapolis: University of Minnesota Press, 1992.

Datta, Satya. *Women and Men in Early Modern Venice.* Burlington, VT: Ashgate, 2003.

Davis, Natalie Zemon. *Society and Culture in Early Modern France.* Stanford, CA: Stanford University Press, 1975. Especially chapters 3 and 5.

————. *Women on the Margins: Three Seventeenth-Century Lives.* Cambridge, MA: Harvard University Press, 1995.

Daybell, James, ed. *Early Modern Women's Letter Writing, 1450–1700.* New York: Palgrave, 2001.

Dean, Trevor, and K. J. P. Lowe, eds. *Marriage in Italy, 1300–1650.* Cambridge: Cambridge University Press, 1998.

D'Elia, Anthony F. *The Renaissance of Marriage in Fifteenth-Century Italy.* Cambridge, MA: Harvard University Press, 2004.

DeJean, Joan. *Ancients against Moderns: Culture Wars and the Making of a Fin de Siècle.* Chicago: University of Chicago Press, 1997.

————. *Fictions of Sappho, 1546–1937.* Chicago: University of Chicago Press, 1989.

————. *The Reinvention of Obscenity: Sex, Lies, and Tabloids in Early Modern France.* Chicago: University of Chicago Press, 2002.

————. *Tender Geographies: Women and the Origins of the Novel in France.* New York: Columbia University Press, 1991.

Demers, Patricia. *Women's Writing in English: Early Modern England.* Toronto: University of Toronto Press, 2005.

Dictionary of Russian Women Writers. Ed. Marina Ledkovsky, Charlotte Rosenthal, and Mary Zirin. Westport, CT: Greenwood Press, 1994.

Diefendorf, Barbara. *From Penitence to Charity: Pious Women and the Catholic Reformation in Paris.* New York: Oxford University Press, 2004.

Dinan, Susan E. *Women and Poor Relief in Seventeenth-Century France: The Early History of the Daughters of Charity.* Women and Gender in the Early Modern World. Burlington, VT: Ashgate, 2006.

Dissing Elizabeth: Negative Representations of Gloriana. Ed. Julia M. Walker. Durham, NC: Duke University Press, 1998.

Dixon, Laurinda S. *Perilous Chastity: Women and Illness in Pre-Enlightenment Art and Medicine.* Ithaca, NY: Cornell University Press, 1995.

Dolan, Frances, E. *Whores of Babylon: Catholicism, Gender, and Seventeenth-Century Print Culture.* Ithaca, NY: Cornell University Press, 1999.

Donovan, Josephine. *Women and the Rise of the Novel, 1405–1726.* New York: St. Martin's Press, 1999.

Dreher, Diane Elizabeth. *Domination and Defiance: Fathers and Daughters in Shakespeare.* Lexington: University Press of Kentucky, 1986.

Dyan, Elliott. *Proving Woman: Female Spirituality and Inquisitional Culture in the Later Middle Ages.* Princeton, NJ: Princeton University Press, 2004.

Early [English] Women Writers: 1600–1720. Ed. Anita Pacheco. New York and London: Longman, 1998.

Eccles, Audrey. *Obstetrics and Gynaecology in Tudor and Stuart England.* Kent, OH: Kent State University Press, 1982.

Eigler, Friederike, and Susanne Kord, eds. *The Feminist Encyclopedia of German Literature.* Westport, CT: Greenwood Press, 1997.

Elizabeth I: Then and Now. Ed. Georgianna Ziegler. Seattle: University of Washington Press, 2003.

Emerson, Kathy Lynn. *Wives and Daughters: The Women of Sixteenth-Century England.* Troy, NY: Whitson Publishing, 1984.

An Encyclopedia of Continental Women Writers. Ed. Katharina Wilson. New York: Garland, 1991.

Encyclopedia of Women in the Renaissance: Italy, France, and England. Ed. Diana Robin, Anne R. Larsen, and Carole Levin. Santa Barbara, CA: ABC Clio, 2007.

Engendering the Early Modern Stage: Women Playwrights in the Spanish Empire. Ed. Valeria (Oakey) Hegstrom and Amy R. Williamsen. New Orleans: University Press of the South, 1999.

Erdmann, Axel. *My Gracious Silence: Women in the Mirror of Sixteenth-Century Printing in Western Europe.* Lucerne: Gilhofer and Ranschburg, 1999.

Erickson, Amy Louise. *Women and Property in Early Modern England.* London and New York: Routledge, 1993.

Evangelisti, Silvia. *Nuns: A History of Convent Life, 1450–1700.* New York: Oxford University Press, 2007.

Extraordinary Women of the Medieval and Renaissance World: A Biographical Dictionary. Ed. Carole Levin et al. Westport, CT: Greenwood Press, 2000.

Ezell, Margaret J. M. *The Patriarch's Wife: Literary Evidence and the History of the Family.* Chapel Hill: University of North Carolina Press, 1987.

———. *Social Authorship and the Advent of Print.* Baltimore: Johns Hopkins University Press, 1999.

———. *Writing Women's Literary History.* Baltimore: Johns Hopkins University Press, 1993.

Farrell, Kirby, Elizabeth H. Hageman, and Arthur F. Kinney, eds. *Women in the Renaissance: Selections from English Literary Renaissance.* Amherst: University of Massachusetts Press, 1988.

Farrell, Kirby, and Kathleen Swain, eds. *The Mysteries of Elizabeth I: Selections from English Literary Renaissance.* Amherst: University of Massachusetts Press, 2003.

Farrell, Michèle Longino. *Performing Motherhood: The Sévigné Correspondence.* Hanover, NH: University Press of New England, 1991.

Feminism and Renaissance Studies. Ed. Lorna Hutson. New York: Oxford University Press, 1999.

The Feminist Companion to Literature in English: Women Writers from the Middle Ages to the Present. Ed. Virginia Blain, Isobel Grundy, and Patricia Clements. New Haven, CT: Yale University Press, 1990.

Feminist Encyclopedia of Italian Literature. Edited by Rinaldina Russell. Westport, CT: Greenwood Press, 1997.

Feminist Thought in European History, 1400–2000. Ed. Tjitske Akkerman and Siep Sturman. London and New York: Routledge, 1997.

Feminist Perspectives on Sor Juana Inés de la Cruz. Ed. Stephanie Merrim. Detroit, MI: Wayne State University Press, 1991.

Ferguson, Margaret W. *Dido's Daughters: Literacy, Gender, and Empire in Early Modern England and France.* Chicago: University of Chicago Press, 2003.

Ferguson, Margaret W., Maureen Quilligan, and Nancy J. Vickers, eds. *Rewriting the Renaissance: The Discourses of Sexual Difference in Early Modern Europe.* Chicago: University of Chicago Press, 1987.

Feroli, Teresa. *Political Speaking Justified: Women Prophets and the English Revolution.* Newark: University of Delaware Press, 2006.

Ferraro, Joanne M. *Marriage Wars in Late Renaissance Venice.* New York: Oxford University Press, 2001.

Fisher, Sheila, and Janet E. Halley, eds. *Seeking the Woman in Late Medieval and Renaissance Writings: Essays in Feminist Contextual Criticism.* Knoxville: University of Tennessee Press, 1989.

Fisher, Will. *Materializing Gender in Early Modern English Literature and Culture.* Cambridge: Cambridge University Press, 2006.

Flandrin, Jean-Louis. *Families in Former Times: Kinship, Household, and Sexuality in Early Modern France.* Trans. Richard Southern. Cambridge: Cambridge University Press, 1979.

Fletcher, Anthony. *Gender, Sex, and Subordination in England, 1500–1800.* New Haven, CT: Yale University Press, 1995.

Franklin, Margaret. *Boccaccio's Heroines: Power and Virtue in Renaissance Society.* Women and Gender in the Early Modern World. Burlington, VT: Ashgate, 2006.

French Women Writers: A Bio-Bibliographical Source Book. Ed. Eva Martin Sartori and Dorothy Wynne Zimmerman. Westport, CT: Greenwood Press, 1991.

Froide, Amy M. *Never Married: Singlewomen in Early Modern England.* New York: Oxford University Press, 2005.

Frye, Susan, and Karen Robertson, eds. *Maids and Mistresses, Cousins and Queens: Women's Alliances in Early Modern England.* New York: Oxford University Press, 1999.

Gallagher, Catherine. *Nobody's Story: The Vanishing Acts of Women Writers in the Marketplace, 1670–1820.* Berkeley: University of California Press, 1994.

Garrard, Mary D. *Artemisia Gentileschi: The Image of the Female Hero in Italian Baroque Art.* Princeton, NJ: Princeton University Press, 1989.

George, Margaret. *Women in the First Capitalist Society: Experiences in Seventeenth-Century England.* Urbana: University of Illinois Press, 1988.

Gelbart, Nina Rattner. *The King's Midwife: A History and Mystery of Madame du Coudray.* Berkeley: University of California Press, 1998.

Gent, Lucy, and Nigel Llewellyn, eds. *Renaissance Bodies: The Human Figure in English Culture c. 1540–1660.* London: Reaktion Books, 1990.

Gibson, Wendy. *Women in Seventeenth-Century France.* New York: St. Martin's Press, 1989.

Gies, Frances. *Joan of Arc: The Legend and the Reality.* New York: Harper and Row, 1981.

Giles, Mary E., ed. *Women in the Inquisition: Spain and the New World.* Baltimore: Johns Hopkins University Press, 1999.

Gill, Catie. *Women in the Seventeenth-Century Quaker Community.* Burlington, VT: Ashgate, 2005.

Glenn, Cheryl. *Rhetoric Retold: Regendering the Tradition from Antiquity through the Renaissance.* Carbondale and Edwardsville: Southern Illinois University Press, 1997.

Gloriana's Face: Women, Public and Private, in the English Renaissance. Ed. S. P. Cerasano and Marion Wynne-Davies. Syracuse, NY: Syracuse University Press, 1992.

Goffen, Rona. *Titian's Women.* New Haven, CT: Yale University Press, 1997.

Going Public: Women and Publishing in Early Modern France. Ed. Elizabeth C. Goldsmith and Dena Goodman. Ithaca, NY: Cornell University Press, 1995.

Goldberg, Jonathan. *Desiring Women Writing: English Renaissance Examples.* Stanford, CA: Stanford University Press, 1997.

Goldsmith, Elizabeth C. *Exclusive Conversations: The Art of Interaction in Seventeenth-Century France.* Philadelphia: University of Pennsylvania Press, 1988.

Goldsmith, Elizabeth C., ed. *Writing the Female Voice.* Boston: Northeastern University Press, 1989.

Goldsmith, Elizabeth C., and Dena Goodman, eds. *Going Public: Women and Publishing in Early Modern France.* Ithaca, NY: Cornell University Press, 1995.

Grafton, Anthony, and Lisa Jardine. *From Humanism to the Humanities: Education and the Liberal Arts in Fifteenth- and Sixteenth-Century Europe.* Cambridge, MA: Harvard University Press, 1986.

Grant, Douglas. *Margaret the First: A Biography of Margaret Cavendish, Duchess of Newcastle, 1623–1673.* Toronto: University of Toronto Press, 1957.

The Graph of Sex and the German Text: Gendered Culture in Early Modern Germany, 1500–1700. Ed. Lynne Tatlock and Christiane Bohnert. Amsterdam and Atlanta: Rodolphi, 1994.

Grassby, Richard. *Kinship and Capitalism: Marriage, Family, and Business in the English-Speaking World, 1580–1740.* Cambridge: Cambridge University Press, 2001.

Gray, Catharine. *Women Writers and Public Debate in 17th-Century Britain.* New York: Palgrave Macmillan, 2007.

Greer, Margaret Rich. *Maria de Zayas Tells Baroque Tales of Love and the Cruelty of Men.* University Park: Pennsylvania State University Press, 2000.

Grossman, Avraham. *Pious and Rebellious: Jewish Women in Medieval Europe.* Trans. Jonathan Chipman. Waltham, MA: Brandeis University Press; Hanover, NH: University Press of New England, 2004.

Grossman, Marshall, ed. *Aemilia Lanyer: Gender, Genre, and the Canon.* Lexington: University Press of Kentucky, 1998.

Gutierrez, Nancy A. *"Shall She Famish Then?" Female Food Refusal in Early Modern England.* Burlington, VT: Ashgate, 2003.

Habermann, Ina. *Staging Slander and Gender in Early Modern England.* Burlington, VT: Ashgate, 2003.

Hacke, Daniela. *Women, Sex and Marriage in Early Modern Venice.* Burlington, VT: Ashgate, 2004.

Hackel, Heidi Brayman. *Reading Material in Early Modern England: Print, Gender, Literacy.* Cambridge: Cambridge University Press, 2005.

Hackett, Helen. *Women and Romance Fiction in the English Renaissance.* Cambridge: Cambridge University Press, 2000.

Hageman, Elizabeth H., and Katherine Conway, eds. *Resurrecting Elizabeth I in Seventeenth-Century England.* Madison, NJ: Fairleigh Dickinson University Press, 2007.

Hageman, Elizabeth H., and Sara Jayne Steen, eds. *Teaching Judith Shakespeare.* Special Issue of *Shakespeare Quarterly* 47 (1996).

Haigh, Christopher. *Elizabeth I.* London and New York: Longman, 1988.

Hall, Kim F. *Things of Darkness: Economies of Race and Gender in Early Modern England.* Ithaca, NY: Cornell University Press, 1995.

Hannay, Margaret P. *Philip's Phoenix: Mary Sidney, Countess of Pembroke.* New York: Oxford University Press, 1990.

Hamburger, Jeffrey. *The Visual and the Visionary: Art and Female Spirituality in Late Medieval Germany.* New York: Zone Books, 1998.

Hampton, Timothy. *Literature and the Nation in the Sixteenth Century: Inventing Renaissance France.* Ithaca, NY: Cornell University Press, 2001.

Hanawalt, Barbara A. *Women and Work in Pre-Industrial Europe*. Bloomington: Indiana University Press, 1986.

Hannay, Margaret, ed. *Silent but for the Word*. Kent, OH: Kent State University Press, 1985.

Hardwick, Julie. *The Practice of Patriarchy: Gender and the Politics of Household Authority in Early Modern France*. University Park: Pennsylvania State University Press, 1998.

Harness, Kelley Ann. *Echoes of Women's Voices: Music, Art, and Female Patronage in Early Modern Florence*. Chicago: University of Chicago Press, 2006.

Harris, Barbara J. *English Aristocratic Women, 1450–1550: Marriage and Family, Property and Careers*. New York: Oxford University Press, 2002.

Harth, Erica. *Cartesian Women: Versions and Subversions of Rational Discourse in the Old Regime*. Ithaca, NY: Cornell University Press, 1992.

———. *Ideology and Culture in Seventeenth-Century France*. Ithaca, NY: Cornell University Press, 1983.

Harvey, Elizabeth D. *Ventriloquized Voices: Feminist Theory and English Renaissance Texts*. New York: Routledge, 1992.

Haselkorn, Anne M., and Betty Travitsky, eds. *The Renaissance Englishwoman in Print: Counterbalancing the Canon*. Amherst: University of Massachusetts Press, 1990.

Hawkesworth, Celia, ed. *A History of Central European Women's Writing*. New York: Palgrave Press, 2001.

Hegstrom (Oakey), Valerie, and Amy R. Williamsen, eds. *Engendering the Early Modern Stage: Women Playwrights in the Spanish Empire*. New Orleans: University Press of the South, 1999.

Heller, Wendy. *Emblems of Eloquence: Opera and Women's Voices in Seventeenth-Century Venice*. Berkeley: University of California Press, 2004.

Hendricks, Margo, and Patricia Parker, eds. *Women, "Race," and Writing in the Early Modern Period*. New York: Routledge, 1994.

Herlihy, David. "Did Women Have a Renaissance? A Reconsideration." *Medievalia et Humanistica*, n.s. 13 (1985): 1–22.

Hibbert, Christopher. *The Virgin Queen: Elizabeth I, Genius of the Golden Age*. Reading, MA: Addison-Wesley, 1991.

Hill, Bridget. *The Republican Virago: The Life and Times of Catharine Macaulay, Historian*. New York: Oxford University Press, 1992.

Hills, Helen, ed. *Architecture and the Politics of Gender in Early Modern Europe*. Burlington, VT: Ashgate, 2003.

Hinds, Hillary. *God's Englishwoman: Seventeenth-Century Radical Sectarian Writing and Feminist Criticism*. New York: St. Martin's Press, 1996.

Hirst, Jilie. *Jane Leade: Biography of a Seventeenth-Century Mystic*. Burlington, VT: Ashgate, 2006.

A History of Central European Women's Writing. Ed. Celia Hawkesworth. New York: Palgrave Press, 2001.

A History of Women in the West. Volume 1: *From Ancient Goddesses to Christian Saints*. Ed. Pauline Schmitt Pantel. Cambridge, MA: Harvard University Press, 1992. Volume 2: *Silences of the Middle Ages*. Ed. Christiane Klapisch-Zuber. Cambridge, MA: Harvard University Press, 1992. Volume 3: *Renaissance and Enlightenment Paradoxes*. Ed. Natalie Zemon Davis and Arlette Farge. Cambridge, MA: Harvard University Press, 1993.

A History of Women Philosophers. Ed. Mary Ellen Waithe. 3 vols. Dordrecht: Martinus Nijhoff, 1987.

A History of Women's Writing in France. Ed. Sonya Stephens. Cambridge: Cambridge University Press, 2000.

A History of Women's Writing in Germany, Austria, and Switzerland. Ed. Jo Catling. Cambridge: Cambridge University Press, 2000.

A History of Women's Writing in Italy. Ed. Letizia Panizza and Sharon Wood. Cambridge: Cambridge University Press, 2000.

A History of Women's Writing in Russia. Ed. Alele Marie Barker and Jehanne M. Gheith. Cambridge: Cambridge University Press, 2002.

Hobby, Elaine. *Virtue of Necessity: English Women's Writing, 1646–1688*. London: Virago Press, 1988.

Hogrefe, Pearl. *Women of Action in Tudor England: Nine Biographical Sketches*. Ames: Iowa State University Press, 1977.

Hopkins, Lisa. *Women Who Would Be Kings: Female Rulers of the Sixteenth Century*. New York: St. Martin's Press, 1991.

Horowitz, Maryanne Cline. "Aristotle and Women." *Journal of the History of Biology* 9 (1976): 183–213.

Houlbrooke, Ralph A. *Death, Religion, and the Family in England, 1480–1760*. Oxford Studies in Social History. New York: Oxford University Press, 1998.

Howe, Elizabeth. *The First English Actresses: Women and Drama, 1660–1700*. Cambridge: Cambridge University Press, 1992.

Howell, Martha C. *The Marriage Exchange: Property, Social Place, and Gender in Cities of the Low Countries, 1300–1550*. Chicago: University of Chicago Press, 1998.

———. *Women, Production, and Patriarchy in Late Medieval Cities*. Chicago: University of Chicago Press, 1986.

Hufton, Olwen H. *The Prospect before Her: A History of Women in Western Europe, 1: 1500–1800*. New York: HarperCollins, 1996.

Hull, Suzanne W. *Chaste, Silent, and Obedient: English Books for Women, 1475–1640*. San Marino, CA: The Huntington Library, 1982.

Hulse, Clark. *Elizabeth I: Ruler and Legend*. Urbana: University of Illinois Press, 2003.

Hunt, Lynn, ed. *The Invention of Pornography: Obscenity and the Origins of Modernity, 1500–1800*. New York: Zone Books, 1996.

Hurlburt, Holly S. *The Dogaressa of Venice, 1200–1500: Wife and Icon*. The New Middle Ages. New York: Palgrave Macmillan, 2006.

Hutner, Heidi, ed. *Rereading Aphra Behn: History, Theory, and Criticism*. Charlottesville: University Press of Virginia, 1993.

Hunter, Lynette, and Sarah Hutton, eds. *Women, Science, and Medicine, 1500–1700: Mothers and Sisters of the Royal Society*. Thrupp, Stroud, Gloucester, UK: Sutton, 1997.

Hutson, Lorna, ed. *Feminism and Renaissance Studies*. New York: Oxford University Press, 1999.

———. *The Usurer's Daughter: Male Friendship and Fictions of Women in Sixteenth-Century England*. New York: Routledge, 1994.

The Impact of Feminism in English Renaissance Studies. Ed. Dympna Callaghan. New York: Palgrave Macmillan, 2007.

Ingram, Martin. *Church Courts, Sex, and Marriage in England, 1570–1640*. Cambridge: Cambridge University Press, 1987.

The Invention of Pornography: Obscenity and the Origins of Modernity, 1500–1800. Ed. Lynn Hunt. New York: Zone Books, 1996.

Italian Women Writers: A Bio-Bibliographical Sourcebook. Ed. Rinaldina Russell. Westport, CT: Greenwood Press, 1994.

Ives, E. W. *Anne Boleyn.* London: Blackwells, 1988.

Jaffe, Irma B., with Gernando Colombardo. *Shining Eyes, Cruel Fortune: The Lives and Loves of Italian Renaissance Women Poets.* New York: Fordham University Press, 2002.

James, Susan E. *Kateryn Parr: The Making of a Queen.* Burlington, VT: Ashgate, 1999.

Jankowski, Theodora A. *Women in Power in the Early Modern Drama.* Urbana: University of Illinois Press, 1992.

Jansen, Katherine Ludwig. *The Making of the Magdalen: Preaching and Popular Devotion in the Later Middle Ages.* Princeton, NJ: Princeton University Press, 2000.

Jardine, Lisa. *Still Harping on Daughters: Women and Drama in the Age of Shakespeare.* Totowa, NJ: Barnes and Noble, 1983.

Jed, Stephanie H. *Chaste Thinking: The Rape of Lucretia and the Birth of Humanism.* Bloomington: Indiana University Press, 1989.

Jones, Ann Rosalind. *The Currency of Eros: Women's Love Lyric in Europe, 1540–1620.* Bloomington: Indiana University Press, 1990.

Jones, Ann Rosalind, and Peter Stallybrass. *Renaissance Clothing and the Materials of Memory.* Cambridge: Cambridge University Press, 2000.

Jones, Michael K., and Malcolm G. Underwood. *The King's Mother: Lady Margaret Beaufort, Countess of Richymond and Derby.* Cambridge: Cambridge University Press, 1992.

Jordan, Constance. *Renaissance Feminism: Literary Texts and Political Models.* Ithaca, NY: Cornell University Press, 1990.

Kagan, Richard L. *Lucrecia's Dreams: Politics and Prophecy in Sixteenth-Century Spain.* Berkeley: University of California Press, 1990.

Kehler, Dorothea, and Laurel Amtower, eds. *The Single Woman in Medieval and Early Modern England: Her Life and Representation.* Tempe, AZ: MRTS, 2002.

Kelly, Joan. "Did Women Have a Renaissance?" In her *Women, History, and Theory.* Chicago: University of Chicago Press, 1984. Also in Renate Bridenthal, Claudia Koonz, and Susan M. Stuard, eds., *Becoming Visible: Women in European History.* 3rd ed. Boston: Houghton Mifflin, 1998.

———. "Early Feminist Theory and the *Querelle des Femmes.*" In *Women, History, and Theory.*

———. *Women, History, and Theory: The Essays of Joan Kelly.* Women in Culture and Society. Chicago: University of Chicago Press, 1984.

Kelso, Ruth. *Doctrine for the Lady of the Renaissance.* Foreword Katharine M. Rogers. Urbana: University of Illinois Press, 1956, 1978.

Kendrick, Robert L. *Celestical Sirens: Nuns and Their Music in Early Modern Milan.* New York: Oxford University Press, 1996.

Kennedy, Gwynne. *Just Anger: Representing Women's Anger in Early Modern England.* Carbondale: Southern Illinois University Press, 2000.

Kermode, Jenny, and Garthine Walker, eds. *Women, Crime, and the Courts in Early Modern England.* Chapel Hill: University of North Carolina Press, 1994.

King, Catherine E. *Renaissance Women Patrons: Wives and Widows in Italy, c. 1300–1550.* New York: St. Martin's Press, 1998.

King, Margaret L. *Women of the Renaissance.* Foreword Catharine R. Stimpson. Chicago: University of Chicago Press, 1991.

King, Thomas A. *The Gendering of Men, 1600–1700: The English Phallus.* Vol. 1. Madison: University of Wisconsin Press, 2004.

Klapisch-Zuber, Christiane. *Women, Family, and Ritual in Renaissance Italy.* Trans. Lydia G. Cochrane. Chicago: University of Chicago Press, 1985.

Kleiman, Ruth. *Anne of Austria, Queen of France.* Columbus: Ohio State University Press, 1985.

Knott, Sarah, and Barbara Taylor. *Women, Gender, and Enlightenment.* New York: Palgrave Macmillan, 2005.

Kolsky, Stephen. *The Ghost of Boccaccio: Writings on Famous Women in Renaissance Italy.* Late Medieval and Early Modern Studies 7. Turnhout: Brepols, 2005.

Krontiris, Tina. *Oppositional Voices: Women as Writers and Translators of Literature in the English Renaissance.* New York: Routledge, 1992.

Kuehn, Thomas. *Law, Family, and Women: Toward a Legal Anthropology of Renaissance Italy.* Chicago: University of Chicago Press, 1991.

Kunze, Bonnelyn Young. *Margaret Fell and the Rise of Quakerism.* Stanford, CA: Stanford University Press, 1994.

Labalme, Patricia A., ed. *Beyond Their Sex: Learned Women of the European Past.* New York: New York University Press, 1980.

Lalande, Roxanne Decker, ed. *A Labor of Love: Critical Reflections on the Writings of Marie-Catherine Desjardina (Mme de Villedieu).* Madison, NJ: Fairleigh Dickinson University Press, 2000.

Lamb, Mary Ellen. *Gender and Authorship in the Sidney Circle.* Madison: University of Wisconsin Press, 1990.

Laqueur, Thomas. *Making Sex: Body and Gender from the Greeks to Freud.* Cambridge, MA: Harvard University Press, 1990.

Larner, Christina. *Enemies of God: The Witch Hunt in Scotland.* Reprt. 1981 ed. Chester Springs, PA: Dufour Editions, 2000.

Larsen, Anne R., and Colette H. Winn, eds. *Renaissance Women Writers: French Texts/ American Contexts.* Detroit, MI: Wayne State University Press, 1994.

Laven, Mary. *Virgins of Venice: Broken Vows and Cloistered Lives in the Renaissance Convent.* New York: Viking, 2003.

Ledkovsky, Marina, Charlotte Rosenthal, and Mary Zirin, eds. *Dictionary of Russian Women Writers.* Westport, CT: Greenwood Press, 1994.

Lehfeldt, Elizabeth A. *Religious Women in Golden Age Spain: The Permeable Cloister.* Burlington, VT: Ashgate, 2005.

Leonard, Amy. *Nails in the Wall: Catholic Nuns in Reformation Germany.* Women in Culture and Society. Chicago: University of Chicago Press, 2005.

Lerner, Gerda. *The Creation of Patriarchy* and *Creation of Feminist Consciousness, 1000–1870.* Women and History, 2 vols. New York: Oxford University Press, 1986, 1994.

Levack. Brian P. *The Witch Hunt in Early Modern Europe.* London: Longman, 1987.

Levin, Carole, and Jeanie Watson, eds. *Ambiguous Realities: Women in the Middle Ages and Renaissance.* Detroit, MI: Wayne State University Press, 1987.

Levin, Carole, Jo Eldridge Carney, and Debra Barrett-Graves. *Elizabeth I: Always Her Own Free Woman.* Burlington, VT: Ashgate, 2003.

Levin, Carole, et al. *Extraordinary Women of the Medieval and Renaissance World: A Biographical Dictionary.* Westport, CT: Greenwood Press, 2000.

Levin, Carole, and Patricia A. Sullivan, eds. *Political Rhetoric, Power, and Renaissance Women.* Albany: State University of New York Press, 1995.

Levy, Allison, ed. *Widowhood and Visual Culture in Early Modern Europe.* Burlington, VT: Ashgate, 2003.

Lewalski, Barbara Kiefer. *Writing Women in Jacobean England.* Cambridge, MA: Harvard University Press, 1993.

Lewis, Gertrud Jaron. *By Women for Women about Women: The Sister-Books of Fourteenth-Century Germany.* Toronto: University of Toronto Press, 1996.

Lewis, Jayne Elizabeth. *Mary Queen of Scots: Romance and Nation.* London: Routledge, 1998.

Lindenauer, Leslie J. *Piety and Power: Gender and Religious Culture in the American Colonies, 1630–1700.* New York: Routledge, 2002.

Lindsey, Karen. *Divorced Beheaded Survived: A Feminist Reinterpretation of the Wives of Henry VIII.* Reading, MA: Addison-Wesley, 1995.

Liss, Peggy K. *Isabel the Queen: Life and Times.* Philadelphia: University of Pennsylvania Press, rev. ed. 2004.

Loades, David. *Mary Tudor: A Life.* Cambridge: Basil Blackwell, 1989.

Lochrie, Karma. *Margery Kempe and Translations of the Flesh.* Philadelphia: University of Pennsylvania Press, 1992.

Longfellow, Erica. *Women and Religious Writing in Early Modern England.* Cambridge: Cambridge University Press, 2004.

Longino Farrell, Michèle. *Performing Motherhood: The Sévigné Correspondence.* Hanover, NH: University Press of New England, 1991.

Lougee, Carolyn C. *Le paradis des femmes: Women, Salons, and Social Stratification in Seventeenth-Century France.* Princeton, NJ: Princeton University Press, 1976.

Love, Harold. *The Culture and Commerce of Texts: Scribal Publication in Seventeenth-Century England.* Amherst: University of Massachusetts Press, 1993.

Lowe, K. J. P. *Nuns' Chronicles and Convent Culture in Renaissance and Counter-Reformation Italy.* Cambridge: Cambridge University Press, 2003.

Luther on Women: A Sourcebook. Ed. Susan C. Karant-Nunn and Merry E. Wiesner-Hanks. Cambridge: Cambridge University Press, 2003.

Lux-Sterritt, Laurence. *Redefining Female Religious Life: French Ursulines and English Ladies in Seventeenth-Century Catholicism.* Burlington, VT: Ashgate, 2005.

MacCarthy, Bridget G. *The Female Pen: Women Writers and Novelists, 1621–1818.* Preface Janet Todd. New York: New York University Press, 1994. (Originally published by Cork University Press, 1946–47).

MacCurtain, Margaret, and Mary O'Dowd, eds. *Women in Early Modern Ireland.* Edinburgh: Edinburgh University Press, 1991.

Macfarlane, Alan. *Marriage and Love in England: Modes of Reproduction, 1300–1840.* New York: Basil Blackwell, 1986.

Mack, Phyllis. *Visionary Women: Ecstatic Prophecy in Seventeenth-Century England.* Berkeley: University of California Pres, 1992.

Maclean, Ian. *The Renaissance Notion of Woman: A Study of the Fortunes of Scholasticism and Medical Science in European Intellectual Life.* Cambridge: Cambridge University Press, 1980.

———. *Woman Triumphant: Feminism in French Literature, 1610–1652.* Oxford: Clarendon Press, 1977.

MacNeil, Anne. *Music and Women of the Commedia dell'Arte in the Late Sixteenth Century.* New York: Oxford University Press, 2003.

Maggi, Armando. *Uttering the Word: The Mystical Performances of Maria Maddalena de' Pazzi, a Renaissance Visionary.* Albany: State University of New York Press, 1998.

Mann, David D., Susan Garland Mann, with Camille Garnier. *Women Playwrights in England, Ireland, and Scotland, 1660–1823.* Bloomington: Indiana University Press, 1996.

Maids and Mistresses, Cousins and Queens: Women's Alliances in Early Modern England. Ed. Susan Frye and Karen Robertson. New York: Oxford University Press, 1999.

Marriage in Italy, 1300–1650. Ed. Trevor Dean and K. J. P. Lowe. Cambridge: Cambridge University Press, 1998.

Marshall, Sherrin, ed. *Women in Reformation and Counter-Reformation Europe: Public and Private Worlds.* Bloomington: Indiana University Press, 1989.

Masten, Jeffrey. *Textual Intercourse: Collaboration, Authorship, and Sexualities in Renaissance Drama.* Cambridge: Cambridge University Press, 1997.

Maternal Measures: Figuring Caregiving in the Early Modern Period. Ed. Naomi J. Miller and Naomi Yavneh. Burlington, VT: Ashgate, 2000.

Matter, E. Ann, and John Coakley, eds. *Creative Women in Medieval and Early Modern Italy.* Philadelphia: University of Pennsylvania Press, 1994 (sequel to the Monson collection, below).

McGrath, Lynette. *Subjectivity and Women's Poetry in Early Modern England.* Burlington, VT: Ashgate, 2002.

McIver, Katherine A. *Women, Art, and Architecture in Northern Italy, 1520–1580: Negotiating Power.* Women and Gender in the Early Modern World. Burlington, VT: Ashgate, 2006.

McLeod, Glenda. *Virtue and Venom: Catalogs of Women from Antiquity to the Renaissance.* Ann Arbor: University of Michigan Press, 1991.

McManus, Clare. *Women on the Renaissance Stage: Anna of Denmark and Female Masquing in the Stuart Court, 1590–1619.* Manchester: Manchester University Press, 2002.

McSheffrey, Shannon. *Gender and Heresy: Women and Men in Lollard Communities, 1420–1530.* Philadelphia: University of Pennsylvania Press, 1995.

McTavish, Lianne. *Childbirth and the Display of Authority in Early Modern France.* Burlington, VT: Ashgate, 2005.

Medieval Women's Visionary Literature. Ed. Elizabeth A. Petroff. New York: Oxford University Press, 1986.

Medwick, Cathleen. *Teresa of Avila: The Progress of a Soul.* New York: Doubleday, 1999.

Meek, Christine, ed. *Women in Renaissance and Early Modern Europe.* Dublin and Portland: Four Courts Press, 2000.

Mendelson, Sara Heller. *The Mental World of Stuart Women: Three Studies.* Amherst: University of Massachusetts Press, 1987.

Mendelson, Sara, and Patricia Crawford. *Women in Early Modern England, 1550–1720.* Oxford: Clarendon Press, 1998.

Merchant, Carolyn. *The Death of Nature: Women, Ecology, and the Scientific Revolution.* New York: HarperCollins, 1980.

Merrim, Stephanie. *Early Modern Women's Writing and Sor Juana Inés de la Cruz.* Nashville, TN: Vanderbilt University Press, 1999.

Merrim, Stephanie, ed. *Feminist Perspectives on Sor Juana Inés de la Cruz.* Detroit, MI: Wayne State University Press, 1991.

Messbarger, Rebecca. *The Century of Women: The Representations of Women in Eighteenth-Century Italian Public Discourse*. Toronto: University of Toronto Press, 2002.

Midelfort, Erik H. C. *Witchhunting in Southwestern Germany, 1562–1684: The Social and Intellectual Foundations*. Stanford, CA: Stanford University Press, 1972.

Migiel, Marilyn, and Juliana Schiesari. *Refiguring Woman: Perspectives on Gender and the Italian Renaissance*. Ithaca, NY: Cornell University Press, 1991.

Miller, Nancy K. *The Heroine's Text: Readings in the French and English Novel, 1722–1782*. New York: Columbia University Press, 1980.

Miller, Naomi J. *Changing the Subject: Mary Wroth and Figurations of Gender in Early Modern England*. Lexington: University Press of Kentucky, 1996.

Miller, Naomi J., and Gary Waller, eds. *Reading Mary Wroth: Representing Alternatives in Early Modern England*. Knoxville: University of Tennessee Press, 1991.

Miller, Naomi J., and Naomi Yavneh. *Sibling Relations and Gender in the Early Modern World: Sisters, Brothers, and Others*. Burlington, VT: Ashgate, 2006.

Miller, Naomi J., and Naomi Yavneh, eds. *Maternal Measures: Figuring Caregiving in the Early Modern Period*. Burlington, VT: Ashgate, 2000.

Monson, Craig A. *Disembodied Voices: Music and Culture in an Early Modern Italian Convent*. Berkeley: University of California Press, 1995.

Monson, Craig A., ed. *The Crannied Wall: Women, Religion, and the Arts in Early Modern Europe*. Ann Arbor: University of Michigan Press, 1992.

Monter, E. William. *Witchcraft in France and Switzerland: The Borderlands during the Reformation*. Ithaca, NY: Cornell University Press, 1976.

Montrose, Louis Adrian. *The Subject of Elizabeth: Authority, Gender, and Representation*. Chicago: University of Chicago Press, 2006.

Mooney, Catherine M. *Gendered Voices: Medieval Saints and Their Interpreters*. Philadelphia: University of Pennsylvania Press, 1999.

Moore, Cornelia Niekus. *The Maiden's Mirror: Reading Material for German Girls in the Sixteenth and Seventeenth Centuries*. Wiesbaden: Otto Harrassowitz, 1987.

Moore, Mary B. *Desiring Voices: Women Sonneteers and Petrarchism*. Carbondale: Southern Illinois University Press, 2000.

Morgan, Fidelis. *The Female Wits: Women Playwrights of the Restoration*. London: Virago Press, 1981.

Mujica, Bárbara. *Women Writers of Early Modern Spain*. New Haven, CT: Yale University Press, 2004.

Murphy, Caroline. *The Pope's Daughter: The Extraordinary Life of Felice Della Rovere*. New York: Oxford University Press, 2005.

Musacchio, Jacqueline Marie. *The Art and Ritual of Childbirth in Renaissance Italy*. New Haven, CT: Yale University Press, 1999.

The Mysteries of Elizabeth I: Selections from English Literary Renaissance. Ed. Kirby Farrell and Kathleen Swain. Amherst and Boston: University of Massachusetts Press, 2003.

Nader, Helen, ed. *Power and Gender in Renaissance Spain: Eight Women of the Mendoza Family, 1450–1650*. Urbana: University of Illinois Press, 2004.

Never Married: Singlewomen in Early Modern England. Ed. Amy M. Froide. New York: Oxford University Press, 2005.

Nevitt, Marcus. *Women and the Pamphlet Culture of Revolutionary England, 1640–1660*. Women and Gender in the Early Modern World. Burlington, VT: Ashgate, 2006.

Newman, Barbara. *God and the Goddesses: Vision, Poetry, and Belief in the Middle Ages.* Philadelphia: University of Pennsylvania Press, 2003.

Newman, Karen. *Fashioning Femininity and English Renaissance Drama.* Chicago: University of Chicago Press, 1991.

Novy, Marianne. *Love's Argument: Gender Relations in Shakespeare.* Chapel Hill: University of North Carolina Press, 1984.

O'Donnell, Mary Ann. *Aphra Behn: An Annotated Bibliography of Primary and Secondary Sources.* Burlington, VT: Ashgate, 2nd ed., 2004.

Okin, Susan Moller. *Women in Western Political Thought.* Princeton, NJ: Princeton University Press, 1979.

Ozment, Steven. *The Bürgermeister's Daughter: Scandal in a Sixteenth-Century German Town.* New York: St. Martin's Press, 1995.

————. *Flesh and Spirit: Private Life in Early Modern Germany.* New York: Penguin Putnam, 1999.

————. *When Fathers Ruled: Family Life in Reformation Europe.* Cambridge, MA: Harvard University Press, 1983.

Pacheco, Anita, ed. *Early [English] Women Writers: 1600–1720.* New York and London: Longman, 1998.

Pagels, Elaine. *Adam, Eve, and the Serpent.* New York: HarperCollins, 1988.

Panizza, Letizia, and Sharon Wood, eds. *A History of Women's Writing in Italy.* Cambridge: Cambridge University Press, 2000.

Panizza, Letizia, ed. *Women in Italian Renaissance Culture and Society.* Oxford: European Humanities Research Centre, 2000.

Pardailhé-Galabrun, Annik. *The Birth of Intimacy: Privacy and Domestic Life in Early Modern Paris.* Philadelphia: University of Pennsylvania Press, 1992.

Park, Katharine. *The Secrets of Women: Gender, Generation, and the Origins of Human Dissection.* New York: Zone Books, 2006.

Parker, Patricia. *Literary Fat Ladies: Rhetoric, Gender, and Property.* London and New York: Methuen, 1987.

Parker, Rozsika. *The Subversive Stitch: Embroidery and the Making of the Feminine.* London: The Women's Press, 1986, repr. 1989.

Paulissen, May Nelson. *The Love Sonnets of Lady Mary Wroth: A Critical Introduction.* Elizabethan and Renaissance Studies. Salzburg, Austria: Institut für Anglistik und Amerikanistik Universität Salzburg, 1982.

Perlingieri, Ilya Sandra. *Sofonisba Anguissola: The First Great Woman Artist of the Renaissance.* New York: Rizzoli, 1992.

Pernoud, Regine, and Marie-Veronique Clin. *Joan of Arc: Her Story.* Rev. and trans. Jeremy DuQuesnay Adams. New York: St. Martin's Press, 1998 (French original, 1986).

Perry, Mary Elizabeth. *Crime and Society in Early Modern Seville.* Hanover, NH: University Press of New England, 1980.

————. *Gender and Disorder in Early Modern Seville.* Princeton, NJ: Princeton University Press, 1990.

————. *The Handless Maiden: Moriscos and the Politics of Religion in Early Modern Spain.* Princeton, NJ: Princeton University Press, 2005.

Perry, Ruth. *The Celebrated Mary Astell: An Early English Feminist.* Chicago: University of Chicago Press, 1986.

Peters, Christine. *Patterns of Piety: Women, Gender, and Religion in Late Medieval and Reformation England.* Cambridge: Cambridge University Press, 2003.

Petroff, Elizabeth A., ed. *Medieval Women's Visionary Literature.* New York: Oxford University Press, 1986.

Phillippy, Patricia Berrahou. *Painting Women: Cosmetics, Canvases, and Early Modern Culture.* Baltimore: Johns Hopkins University Press, 2006.

Plowden, Alison. *Tudor Women: Queens and Commoners.* Gloucestershire, UK: Sutton Publications, rev. ed. 1998.

Political Rhetoric, Power, and Renaissance Women. Ed. Carole Levin and Patricia A. Sullivan. Albany: State University of New York Press, 1995.

Pollock, Linda. *With Faith and Physic: The Life of a Tudor Gentlewoman, Lady Grace Mildmay, 1552–1620.* London: Collins and Brown, 1993.

Poor, Sara S., and Jana K. Schulman. *Women and Medieval Epic: Gender, Genre, and the Limits of Epic Masculinity.* New York: Palgrave Macmillan, 2007.

The Practice and Representation of Reading in England. Ed. James Raven, Helen Small, and Naomi Tadmor. Cambridge: Cambridge University Press, 1996.

Price, Paola Malpezzi, and Christine Ristaino. *Lucrezia Marinella and the "Querelle des femmes" in Seventeenth-Century Italy.* Madison, NJ: Fairleigh Dickinson University Press, 2008.

Prior, Mary, ed. *Women in English Society, 1500–1800.* London and New York: Methuen, 1985.

Quilligan, Maureen. *The Allegory of Female Authority: Christine de Pizan's "Cité des Dames."* Ithaca, NY: Cornell University Press, 1991.

———. *Incest and Agency in Elizabeth's England.* Philadelphia: University of Pennsylvania Press, 2005.

Rabil, Albert. *Laura Cereta: Quattrocento Humanist.* Binghamton, NY: MRTS, 1981.

Ranft, Patricia. *Women in Western Intellectual Culture, 600–1500.* New York: Palgrave, 2002.

Rapley, Elizabeth. *The Devotés: Women and Church in Seventeenth-Century France.* Kingston, ON: McGill-Queen's University Press, 1989.

———. *A Social History of the Cloister: Daily Life in the Teaching Monasteries of the Old Regime.* Montreal: McGill-Queen's University Press, 2001.

Raven, James, Helen Small, and Naomi Tadmor, eds. *The Practice and Representation of Reading in England.* Cambridge: Cambridge University Press, 1996.

Reading Mary Wroth: Representing Alternatives in Early Modern England. Ed. Naomi Miller and Gary Waller. Knoxville: University of Tennessee Press, 1991.

Reardon, Colleen. *Holy Concord within Sacred Walls: Nuns and Music in Siena, 1575–1700.* Oxford: Oxford University Press, 2001.

Recovering Spain's Feminist Tradition. Ed. Lisa Vollendorf. New York: MLA, 2001.

Reid, Jonathan Andrew. "King's Sister—Queen of Dissent: Marguerite of Navarre (1492–1549) and Her Evangelical Network." PhD diss., University of Arizona, 2001.

Reinterpreting Christine de Pizan. Ed. Earl Jeffrey Richards, with Joan Williamson, Nadia Margolis, and Christine Reno. Athens: University of Georgia Press, 1992.

Reiss, Sheryl E., and David G. Wilkins, eds. *Beyond Isabella: Secular Women Patrons of Art in Renaissance Italy.* Kirksville, MO: Truman State University Press, 2001.

Renaissance Bodies: The Human Figure in English Culture, c. 1540–1660. Ed. Lucy Gent and Nigel Llewellyn. London: Reaktion Books, 1990.

The Renaissance Englishwoman in Print: Counterbalancing the Canon. Ed. Anne M. Haselkorn and Betty Travitsky. Amherst: University of Massachusetts Press, 1990.

Renaissance Women Writers: French Texts/American Contexts. Ed. Anne R. Larsen and Colette H. Winn. Detroit, MI: Wayne State University Press, 1994.

Rereading Aphra Behn: History, Theory, and Criticism. Ed. Heidi Hutner. Charlottesville: University Press of Virginia, 1993.

Resurrecting Elizabeth I in Seventeenth-Century England. Ed. Elizabeth H. Hageman and Katherine Conway. Madison, NJ: Fairleigh Dickinson University Press, 2007.

Rheubottom, David. *Age, Marriage, and Politics in Fifteenth-Century Ragusa.* New York: Oxford University Press, 2000.

Richards, Earl Jeffrey, ed. *Reinterpreting Christine de Pizan.* With Joan Williamson, Nadia Margolis, and Christine Reno. Athens: University of Georgia Press, 1992.

Richardson, Brian. *Printing, Writers, and Readers in Renaissance Italy.* Cambridge: Cambridge University Press, 1999.

Riddle, John M. *Contraception and Abortion from the Ancient World to the Renaissance.* Cambridge, MA: Harvard University Press, 1992.

———. *Eve's Herbs: A History of Contraception and Abortion in the West.* Cambridge, MA: Harvard University Press, 1997.

Robin, Diana. *Publishing Women: Salons, the Presses, and the Counter-Reformation in Sixteenth-Century Italy.* Chicago: University of Chicago Press, 2007.

Roelker, Nancy L. *Queen of Navarre, Jeanne d'Albret, 1528–1572.* Cambridge, MA: Harvard University Press, 1968.

Romack, Katherine, and James Fitzmaurice, eds. *Cavendish and Shakespeare: Interconnections.* Burlington, VT: Ashgate, 2006.

Roper, Lyndal. *The Holy Household: Women and Morals in Reformation Augsburg.* New York: Oxford University Press, 1989.

———. *Oedipus and the Devil: Witchcraft, Sexuality, and Religion in Early Modern Europe.* New York: Routledge, 1994.

Rose, Mary Beth. *The Expense of Spirit: Love and Sexuality in English Renaissance Drama.* Ithaca, NY: Cornell University Press, 1988.

———. *Gender and Heroism in Early Modern English Literature.* Chicago: University of Chicago Press, 2002.

Rose, Mary Beth, ed. *Women in the Middle Ages and the Renaissance: Literary and Historical Perspectives.* Syracuse, NY: Syracuse University Press, 1986.

Rosenthal, Margaret F. *The Honest Courtesan: Veronica Franco, Citizen and Writer in Sixteenth-Century Venice.* Foreword Catharine R. Stimpson. Chicago: University of Chicago Press, 1992.

Rublack, Ulinka, ed. *Gender in Early Modern German History.* Cambridge: Cambridge University Press, 2002.

Ruggiero, Guido. *Binding Passions: Tales of Magic, Marriage, and Power at the End of the Renaissance.* New York: Oxford University Press, 1993.

———. *The Boundaries of Eros: Sex Crime and Sexuality in Renaissance Venice.* New York: Oxford University Press, 1985.

Russell, Rinaldina, ed. *Feminist Encyclopedia of Italian Literature.* Westport, CT: Greenwood Press, 1997.

———. *Italian Women Writers: A Bio-Bibliographical Sourcebook.* Westport, CT: Greenwood Press, 1994.

Sackville-West, Vita. *Daughter of France: The Life of La Grande Mademoiselle.* Garden City, NY: Doubleday, 1959.

Safley, Thomas Max. *Let No Man Put Asunder: The Control of Marriage in the German Southwest: A Comparative Study, 1550–1600.* Kirksville, MO: Sixteenth Century Journal Publishers, 1984.

Sage, Lorna, ed. *Cambridge Guide to Women's Writing in English.* Cambridge: Cambridge University Press, 1999.

Sánchez, Magdalena S. *The Empress, the Queen, and the Nun: Women and Power at the Court of Philip III of Spain.* Baltimore: Johns Hopkins University Press, 1998.

Sanders, Eve Rachele. *Gender and Literacy on Stage in Early Modern England.* Cambridge: Cambridge University Press, 1998.

Sankovitch, Tilde A. *French Women Writers and the Book: Myths of Access and Desire.* Syracuse, NY: Syracuse University Press, 1988.

Sartori, Eva Martin, and Dorothy Wynne Zimmerman, eds. *French Women Writers: A Bio-Bibliographical Source Book.* Westport, CT: Greenwood Press, 1991.

Scaraffia, Lucetta, and Gabriella Zarri. *Women and Faith: Catholic Religious Life in Italy from Late Antiquity to the Present.* Cambridge, MA: Harvard University Press, 1999.

Scheepsma, Wybren. *Medieval Religious Women in the Low Countries: The "Modern Devotion," the Canonesses of Windesheim, and Their Writings.* Rochester, NY: Boydell Press, 2004.

Schiebinger, Londa. *The Mind Has No Sex? Women in the Origins of Modern Science.* Cambridge, MA: Harvard University Press, 1991.

Schleiner, Louise. *Nature's Body: Gender in the Making of Modern Science.* Boston: Beacon Press, 1993.

———. *Tudor and Stuart Women Writers.* Bloomington: Indiana University Press, 1994.

Schofield, Mary Anne, and Cecilia Macheski, eds. *Fetter'd or Free? British Women Novelists, 1670–1815.* Athens: Ohio University Press, 1986.

Schroeder, Joy A. *Dinah's Lament: The Biblical Legacy of Sexual Violence in Christian Interpretation.* Philadelphia: Fortress Press, 2007.

Schutte, Anne Jacobson, Thomas Kuehn, and Silvana Seidel Menchi, eds. *Time, Space, and Women's Lives in Early Modern Europe.* Kirksville, MO: Truman State University Press, 2001.

Schutte, Anne Jacobson. *Aspiring Saints: Pretense of Holiness, Inquisition, and Gender in the Republic of Venice, 1618–1750.* Baltimore: Johns Hopkins University Press, 2001.

Schutte, Anne Jacobson, Thomas Kuehn, and Silvana Seidel Menchi, eds. *Time, Space, and Women's Lives in Early Modern Europe.* Kirksville, MO: Truman State University Press, 2001.

Seeking the Woman in Late Medieval and Renaissance Writings: Essays in Feminist Contextual Criticism. Ed. Sheila Fisher and Janet E. Halley. Knoxville: University of Tennessee Press, 1989.

Seelig, Sharon Cadman. *Autobiography and Gender in Early Modern Literature: Reading Women's Lives, 1600–1680.* Cambridge: Cambridge University Press, 2006.

Seifert, Lewis C. *Fairy Tales, Sexuality, and Gender in France, 1690–1715: Nostalgic Utopias.* Cambridge: Cambridge University Press, 1996.

Shannon, Laurie. *Sovereign Amity: Figures of Friendship in Shakespearean Contexts.* Chicago: University of Chicago Press, 2002.

Shemek, Deanna. *Ladies Errant: Wayward Women and Social Order in Early Modern Italy.* Durham, NC: Duke University Press, 1998.

Shepherd, Simon. *Amazons and Warrior Women: Varieties of Feminism in Seventeenth-Century Drama.* New York: St. Martin's Press, 1981.

Sibling Relations and Gender in the Early Modern World: Sisters, Brothers, and Others. Ed. Naomi Miller and Naomi Yavneh. Women and Gender in the Early Modern World. Burlington, VT: Ashgate, 2006.

Silent but for the Word. Ed. Margaret Hannay. Kent, OH: Kent State University Press, 1985.

Simon, Linda. *Of Virtue Rare: Margaret Beaufort, Matriarch of the House of Tudor.* Boston: Houghton Mifflin, 1982.

The Single Woman in Medieval and Early Modern England: Her Life and Representation. Ed. Dorothea Kehler and Laurel Amtower. Tempe, AZ: MRTS, 2002.

Slater, Miriam. *Family Life in the Seventeenth Century: The Verneys of Claydon House.* London: Routledge and Kegan Paul, 1984.

Smarr, Janet L. *Joining the Conversation: Dialogues by Renaissance Women.* Ann Arbor: University of Michigan Press, 2005.

Smith, Hilda L. *Reason's Disciples: Seventeenth-Century English Feminists.* Urbana: University of Illinois Press, 1982.

———. *Women Writers and the Early Modern British Political Tradition.* Cambridge: Cambridge University Press, 1998.

Snook, Edith. *Women, Reading, and the Cultural Politics of Early Modern England.* Burlington, VT: Ashgate, 2005.

Sobel, Dava. *Galileo's Daughter: A Historical Memoir of Science, Faith, and Love.* New York: Penguin Books, 2000.

Sommerville, Margaret R. *Sex and Subjection: Attitudes to Women in Early-Modern Society.* London: Arnold, 1995.

Soufas, Teresa Scott. *Dramas of Distinction: A Study of Plays by Golden Age Women.* Lexington: University Press of Kentucky, 1997.

Spencer, Jane. *The Rise of the Woman Novelist: From Aphra Behn to Jane Austen.* Oxford: Basil Blackwell, 1986.

Spender, Dale. *Mothers of the Novel: 100 Good Women Writers before Jane Austen.* New York: Routledge, 1986.

Sperling, Jutta Gisela. *Convents and the Body Politic in Late Renaissance Venice.* Foreword Catharine R. Stimpson. Chicago: University of Chicago Press, 1999.

Staley, Lynn. *Margery Kempe's Dissenting Fictions.* University Park: Pennsylvania State University Press, 1994.

Steinbrügge, Lieselotte. *The Moral Sex: Woman's Nature in the French Enlightenment.* Trans. Pamela E. Selwyn. New York: Oxford University Press, 1995.

Stephens, Sonya, ed. *A History of Women's Writing in France.* Cambridge: Cambridge University Press, 2000.

Stephenson, Barbara. *The Power and Patronage of Marguerite de Navarre.* Burlington, VT: Ashgate, 2004.

Stevenson, Jane. *Women Latin Poets: Language, Gender, and Authority, from Antiquity to the Eighteenth Century.* New York: Oxford University Press, 2005

Stocker, Margarita. *Judith, Sexual Warrior: Women and Power in Western Culture.* New Haven, CT: Yale University Press, 1998.

Stone, Lawrence. *Family, Marriage, and Sex in England, 1500–1800.* New York: Weidenfeld and Nicolson, 1977; abr. ed. New York: Harper and Row, 1979.

Straznicky, Marta. *Privacy, Playreading, and Women's Closet Drama, 1550–1700.* Cambridge: Cambridge University Press, 2004.

Stretton, Timothy. *Women Waging Law in Elizabethan England.* Cambridge: Cambridge University Press, 1998.

Strinati, Claudio M., Carole Collier Frick, Elizabeth S. G. Nicholson, Vera Fortunati Pietrantonio, and Jordana Pomeroy. *Italian Women Artists: From Renaissance to Baroque.* Ed. National Museum of Women in the Arts, Sylvestre Verger Art Organization. New York: Skira, 2007; Distributed in North America by Rizzoli International.

Strong Voices, Weak History: Early Women Writers and Canons in England, France, and Italy. Ed. Pamela J. Benson and Victoria Kirkham. Ann Arbor: University of Michigan Press, 2005.

Stuard, Susan Mosher. *Gilding the Market: Luxury and Fashion in Fourteenth-Century Italy.* The Middle Ages Series. Philadelphia: University of Pennsylvania Press, 2006.

Summit, Jennifer. *Lost Property: The Woman Writer and English Literary History, 1380–1589.* Chicago: University of Chicago Press, 2000.

Surtz, Ronald E. *The Guitar of God: Gender, Power, and Authority in the Visionary World of Mother Juana de la Cruz (1481–1534).* Philadelphia: University of Pennsylvania Press, 1991.

———. *Writing Women in Late Medieval and Early Modern Spain: The Mothers of Saint Teresa of Avila.* Philadelphia: University of Pennsylvania Press, 1995.

Suzuki, Mihoko. *Subordinate Subjects: Gender, the Political Nation, and Literary Form in England, 1588–1688.* Burlington, VT: Ashgate, 2003.

Tatlock, Lynne, and Christiane Bohnert, eds. *The Graph of Sex and the German Text: Gendered Culture in Early Modern Germany, 1500–1700.* Amsterdam and Atlanta: Rodolphi, 1994.

Teaching Tudor and Stuart Women Writers. Ed. Susanne Woods and Margaret P. Hannay. New York: MLA, 2000.

Teague, Frances. *Bathsua Makin, Woman of Learning.* Lewisburg, PA: Bucknell University Press, 1999.

Thomas, Anabel. *Art and Piety in the Female Religious Communities of Renaissance Italy: Iconography, Space, and the Religious Woman's Perspective.* Cambridge: Cambridge University Press, 2003.

Thompson, John Lee. *John Calvin and the Daughters of Sarah: Women in Regular and Exceptional Roles in the Exegesis of Calvin, His Predecessors, and His Contemporaries.* Travaux d'Humanisme et Renaissance 259. Geneva: Librairie Droz, 1992.

Thurston, Robert W. *The Witch Hunts: A History of the Witch Persecutions in Europe and North America.* New York: Pearson Longman, 2007.

Tinagli, Paola. *Women in Italian Renaissance Art: Gender, Representation, Identity.* Manchester: Manchester University Press, 1997.

Todd, Janet. *The Secret Life of Aphra Behn.* London, New York, and Sydney: Pandora, 2000.

———. *The Sign of Angelica: Women, Writing, and Fiction, 1660–1800.* New York: Columbia University Press, 1989.

Tomas, Natalie R. *The Medici Women: Gender and Power in Renaissance Florence.* Burlington, VT: Ashgate, 2004.

Traub, Valerie. *The Renaissance of Lesbianism in Early Modern England.* Cambridge: Cambridge University Press, 2002.

Travitsky, Betty S., and Adele F. Seef, eds. *Attending to Women in Early Modern England.* Newark: University of Delaware Press, 1994.

Valenze, Deborah. *The First Industrial Woman.* New York: Oxford University Press, 1995.

Van Dijk, Susan, Lia van Gemert, and Sheila Ottway, eds. *Writing the History of Women's Writing: Toward an International Approach.* Proceedings of the Colloquium, Amsterdam, 9–11 September. Amsterdam: Royal Netherlands Academy of Arts and Sciences, 2001.

Vickery, Amanda. *The Gentleman's Daughter: Women's Lives in Georgian England.* New Haven, CT: Yale University Press, 1998.

Voicing Women: Gender and Sexuality in Early Modern Writing. Ed. Kate Chedgzoy, Melanie Hansen, and Suzanne Trill. Pittsburgh: Duquesne University Press, 1996, 1998.

Vollendorf, Lisa. *The Lives of Women: A New History of Inquisitional Spain.* Nashville: Vanderbilt University Press, 2005.

Walker, Claire. *Gender and Politics in Early Modern Europe: English Convents in France and the Low Countries.* New York: Palgrave, 2003.

Walker, Kim. *Women Writers of the English Renaissance.* Twayne's English Author Series. New York: Twayne Publishers, 1996.

Walker, Julia M. *Dissing Elizabeth: Negative Representations of Gloriana.* Durham, NC: Duke University Press, 1998.

Wall, Wendy. *The Imprint of Gender: Authorship and Publication in the English Renaissance.* Ithaca, NY: Cornell University Press, 1993.

Waller, G. F. *Mary Sidney, Countess of Pembroke: A Critical Study of Her Writings and Literary Milieu.* Elizabethan and Renaissance Studies. Salzburg, Austria: Institut für Anglistik und Amerikanistik Universität Salzburg, 1979.

Walsh, William T. *St. Teresa of Avila: A Biography.* Rockford, IL: TAN Books and Publications, 1987.

Warner, Marina. *Alone of All Her Sex: The Myth and Cult of the Virgin Mary.* New York: Knopf, 1976.

———. *Joan of Arc: The Image of Female Heroism.* Berkeley: University of California Press, 1981.

Warnicke, Retha M. *The Marrying of Anne of Cleves: Royal Protocol in Tudor England.* Cambridge: Cambridge University Press, 2000.

———. *Mary Queen of Scots.* Routledge Historical Biographies. New York: Routledge, 2006.

———. *The Rise and Fall of Anne Boleyn: Family Politics at the Court of Henry VIII.* Cambridge: Cambridge University Press, 1989.

———. *Women of the English Renaissance and Reformation.* Westport, CT: Greenwood Press, 1983.

Warren, Nancy Bradley. *Women of God and Arms: Female Spirituality and Political Conflict, 1380–1600.* Philadelphia: University of Pennsylvania Press, 2005.

Watt, Diane. *Secretaries of God: Women Prophets in Late Medieval and Early Modern England.* Cambridge: D. S. Brewer, 1997.

Weaver, Elissa B. *Convent Theatre in Early Modern Italy: Spiritual Fun and Learning for Women.* Cambridge: Cambridge University Press, 2002.

Weaver, Elissa B., ed. *Arcangela Tarabotti: A Literary Nun in Baroque Venice.* Ravenna: Longo Editore, 2006.

Weber, Alison.*Teresa of Avila and the Rhetoric of Femininity.* Princeton, NJ: Princeton University Press, 1990.

Weinstein, Donald, and Rudolph M. Bell. *Saints and Society: The Two Worlds of Western Christendom, 1000–1700.* Chicago: University of Chicago Press, 1982.

Welles, Marcia L. *Persephone's Girdle: Narratives of Rape in Seventeenth-Century Spanish Literature.* Nashville: Vanderbilt University Press, 2000.

Whitaker, Katie. *Mad Madge: The Extraordinary Life of Margaret Cavendish, Duchess of Newcastle, the First Woman to Live by Her Pen.* New York: Basic Books, 2002.

Whitehead, Barbara J., ed. *Women's Education in Early Modern Europe: A History, 1500–1800.* New York: Garland, 1999.

Widowhood and Visual Culture in Early Modern Europe. Ed. Allison Levy. Burlington, VT: Ashgate, 2003.

Widowhood in Medieval and Early Modern Europe. Ed. Sandra Cavallo and Lydan Warner. New York: Longman, 1999.

Wiesner-Hanks, Merry E. *Christianity and Sexuality in the Early Modern World: Regulating Desire, Reforming Practice.* New York: Routledge, 2000.

———. *Gender, Church, and State in Early Modern Germany: Essays.* New York: Longman, 1998.

———. *Gender in History.* Malden, MA: Blackwell, 2001.

———. *Women and Gender in Early Modern Europe.* Cambridge: Cambridge University Press, 1993.

———. *Working Women in Renaissance Germany.* New Brunswick, NJ: Rutgers University Press, 1986.

Wilcox, Helen, ed. *Women and Literature in Britain, 1500–1700.* Cambridge: Cambridge University Press, 1996.

Willard, Charity Cannon. *Christine de Pizan: Her Life and Works.* New York: Persea Books, 1984.

Williamson, Marilyn L. *Raising Their Voices: British Women Writers, 1650–1750.* Detroit, MI: Wayne State University Press, 1990.

Willis, Deborah. *Malevolent Nurture: Witch-Hunting and Maternal Power in Early Modern England.* Ithaca, NY: Cornell University Press, 1995.

Wilson, Elkin Calhoun. *England's Eliza: A Study of the Idealization of Queen Elizabeth in the Poetry of Her Age.* Harvard Studies in English. London: Frank Cass, 1966.

Wilson, Katharina, ed. *Encyclopedia of Continental Women Writers.* 2 vols. New York: Garland, 1991.

Wiltenburg, Joy. *Disorderly Women and Female Power in the Street Literature of Early Modern England and Germany.* Charlottesville: University Press of Virginia, 1992.

Winn, Colette, and Donna Kuizenga, eds. *Women Writers in Pre-Revolutionary France.* New York: Garland, 1997.

Winston-Allen, Anne. *Convent Chronicles: Women Writing about Women and Reform in the Late Middle Ages.* University Park: Pennsylvania State University Press, 2004.

Witchcraft in Early Modern Europe: Studies in Culture and Belief. Ed. Jonathan Barry, Marianne Hester and Gareth Roberts. New York: Cambridge University Press, 1996.

Women and Literature in Britain, 1500–1700. Ed. Helen Wilcox. Cambridge: Cambridge University Press, 1996.

Women and Medieval Epic: Gender, Genre, and the Limits of Epic Masculinity. Ed. Sara S. Poor and Jana K. Schulman. New York: Palgrave Macmillan, 2007.

Women and Monasticism in Medieval Europe: Sisters and Patrons of the Cistercian Reform. Ed. Constance H. Berman. Kalamazoo, MI: Western Michigan University Press, 2002.

Women, Crime, and the Courts in Early Modern England. Ed. Jenny Kermode and Garthine Walker. Chapel Hill: University of North Carolina Press, 1994.

Women in Early Modern Ireland. Ed. Margaret MacCurtain and Mary O'Dowd. Edinburgh: Edinburgh University Press, 1991.

Women in English Society, 1500–1800. Ed. Mary Prior. New York: Methuen, 1985.

Women in Italian Renaissance Culture and Society. Ed. Letizia Panizza. Oxford: European Humanities Research Centre, 2000.

Women in Reformation and Counter-Reformation Europe: Public and Private Worlds. Ed. Sherrin Marshall. Bloomington: Indiana University Press, 1989.

Women in Renaissance and Early Modern Europe. Ed. Christine Meek. Dublin and Portland: Four Courts Press, 2000.

Women in the Inquisition: Spain and the New World. Ed. Mary E. Giles. Baltimore: Johns Hopkins University Press, 1999.

Women in the Middle Ages and the Renaissance: Literary and Historical Perspectives. Ed. Mary Beth Rose. Syracuse, NY: Syracuse University Press, 1986.

Women in the Renaissance: Selections from English Literary Renaissance. Ed. Kirby Farrell, Elizabeth H. Hageman, and Arthur F. Kinney. Amherst: University of Massachusetts Press, 1988.

Women Players in England, 1500–1660: Beyond the All-Male Stage. Ed. Pamela Allen Brown and Peter Parolin. Burlington, VT: Ashgate, 2005.

Women, "Race," and Writing in the Early Modern Period. Ed. Margo Hendricks and Patricia Parker. New York: Routledge, 1994.

Women, Science, and Medicine, 1500–1700: Mothers and Sisters of the Royal Society. Ed. Lynette Hunter and Sarah Hutton. Thrupp, Stroud, Gloucestershire, UK: Sutton, 1997.

Women, Writing, and the Reproduction of Culture in Tudor and Stuart Britain. Ed. Mary E. Burke, Jane Donawerth, Linda L. Dove, and Karen Nelson. Syracuse, NY: Syracuse University Press, 2000.

Women's Letters across Europe, 1400–1700: Form and Persuasion. Ed. Jane Couchman and Ann Crabb. Women and Gender in the Early Modern World. Burlington, VT: Ashgate, 2005.

Woodbridge, Linda. *Women and the English Renaissance: Literature and the Nature of Womankind, 1540–1620.* Urbana: University of Illinois Press, 1984.

Woodford, Charlotte. *Nuns as Historians in Early Modern Germany.* Oxford: Clarendon Press, 2002.

Woods, Susanne. *Lanyer: A Renaissance Woman Poet.* New York: Oxford University Press, 1999.

Woods, Susanne, and Margaret P. Hannay, eds. *Teaching Tudor and Stuart Women Writers.* New York: MLA, 2000.

Wormald, Jenny. *Mary Queen of Scots: A Study in Failure.* London: George Philip Press, 1988.

Writing the Female Voice. Ed. Elizabeth C. Goldsmith. Boston: Northeastern University Press, 1989.

Writing the History of Women's Writing: Toward an International Approach. Ed. Susan Van Dijk, Lia van Gemert, and Sheila Ottway. Proceedings of the Colloquium, Amsterdam,

9–11 September. Amsterdam: Royal Netherlands Academy of Arts and Sciences, 2001.

Ziegler, Georgianna, ed. *Elizabeth I: Then and Now.* Seattle: University of Washington Press, 2003.

Zinsser, Judith P. *Men, Women, and the Birthing of Modern Science.* DeKalb: Northern Illinois University Press, 2005.

INDEX

certainty, principle of contradiction as foundation of, 127

Champion of Women, The (Le Franc), xix

chastity, female, xxii–xxiii

Chauvelin, Germain Louis, 62n22

Chétardie, Joachim Jacques Trotti, Marquis de la, 61

Chimborazo, Mt., 311–12

Chiselden, William, 120, 120n40

Christianity. *See* religion

Christine de Pizan, xiii, xvi–xviii, xix, xxi, xxiii, xxv

Cicero, 358

Cideville, Pierre Robert le Cornier de, 55, 56–57

Clairaut, Alexis-Claude: as advisor on commentary on Newton's *Principia*, 252, 253, 260, 261; commentary on Newton's *Principia* summarizes work on attraction's effect on Earth's shape, 252, 253; Du Châtelet's commentary on Newton's *Principia* attributed to, 18, 20; on Du Châtelet's *Foundations of Physics*, 16; Du Châtelet studies mathematics with, 9–10, 116n31; on Earth's shape, 252, 253, 319, 320, 321, 324; expedition to the pole, 319n118; on *forces vives*, 114; and Jacquier, 253, 254; König nominated as associate of Academy of Sciences by, 113n25; on reappearance of comet of 1682, 341n149

Clarke, Samuel, 8, 111, 112, 112n18, 156, 156n77

Cohen, I. Bernard, 10, 19–20, 251

Col, Gontier, xviii

Collé, Charles, 262n33

Colpresse, Samuel, 330n134

comets, 337–43; action of fire on, 94; alterations in extremities of their orbits, 342; defined, 275; Descartes on vortices and, 338; durations of periods of, 340; falling into the Sun and stars, 343; Halley's, 340; Kepler's laws obeyed by, 276; Newton on motion of, 338–39; orbits of,

339; Peripatetics take for meteors, 337–38; as planets, 275; reappearance of that of 1682, 340–41, 341n149; tails of, 341–42; Tycho on location of, 338

Commentary on Newton's Principia (Du Châtelet), 263–343; attributed to Clairaut, 18, 20; composition of, 9, 10, 252, 254; as correcting and expanding on *Principia*, 1; on Earth's shape, 315–24; on ebb and flow of the sea, 325–33; facsimile edition of, 253; first section of, 251; introduction to, 263–72; letters relating to, 253–63; manuscript sent to royal library, 10; on phenomena of principal planets, 293–314; on phenomena of secondary planets, 333–37; on precession of the equinoxes, 325; on principal phenomena of system of the world, 272–93; publication of, 252–53; title page makes no mention of Newton, 12–13

Concerning Famous Women (Boccaccio), xvii–xviii

contingent truths, principle of sufficient reason as ground of, 128–29

continuity, law of, 133–38; Descartes paid insufficient attention to, 137–38; in geometry, 133–36; in laws of motion, 136; as premise in Du Châtelet's *Foundations of Physics*, 106; proves that no perfectly hard bodies exist, 136

contradiction, principle of, 106, 126–28

cooling, 94–98

Copernicus, Nicolaus: hypothesis used by, 149, 154; Sun-centered solar system of, 149, 149n, 154, 264n37, 265; Venus's motion as objection to system of, 278

Corbaccio, Il (Boccaccio), xiv, xvii

I Corinthians, 249–50

Corneille, Pierre, 42–43

Corpus of Civil Law, x

Coste, Pierre, 8n15, 35, 46, 170n96